Aircraft Electrical and Electronic Systems
Second Edition

Introducing the principles of aircraft electrical and electronic systems, this book is written for anyone pursuing a career in aircraft maintenance engineering or a related aerospace engineering discipline, and in particular will be suitable for those studying for licensed aircraft maintenance engineer status. It systematically addresses the relevant sections of modules 11 and 13 of part-66 of the EASA syllabus, and is ideal for anyone studying as part of an EASA and FAR-147 approved course in aerospace engineering.

- Delivers the essential principles and knowledge base required by Airframe and Propulsion (A&P) Mechanics for Modules 11 and 13 of the EASA Part-66 syllabus and BTEC National awards in aerospace engineering
- Supports Mechanics, Technicians and Engineers studying for a Part-66 qualification
- Comprehensive and accessible, with self-test questions, exercises and multiple choice questions to enhance learning for both independent and tutor-assisted study

This second edition has been updated to incorporate: complex notation for the analysis of alternating current (AC) circuits; an introduction to the "all electric aircraft" utilising new battery technologies; updated sensor technology using integrated solid-state technology micro-electrical-mechanical sensors (MEMS); an expanded section on helicopter/rotary wing health usage monitoring systems (HUMS).

David Wyatt has over 45 years experience in the aviation industry. This is founded on a technician apprenticeship with BOAC, progressing onto a development engineer in technical services. He then held various positions after leaving BA, including product support engineer, key account manager and FE lecturer in avionics engineering. These roles covered a range of avionic systems including: flight management systems, auto flight control systems, flight instruments, engine sensors and fire protection systems. David's final role was Head of Airworthiness in the design office at Gama Aviation.

Mike Tooley has over 30 years of teaching electronic principles, electronics and avionics to engineers and technicians, previously as Head of Department of Engineering and Vice Principal at Brooklands College in Surrey, UK. He currently works as a consultant and freelance technical author.

Aircraft Electrical and Electronic Systems

Second Edition

David Wyatt and Mike Tooley

Routledge
Taylor & Francis Group

LONDON AND NEW YORK

Second edition published 2018
by Routledge
2 Park Square, Milton Park, Abingdon, Oxon, OX14 4RN

and by Routledge
711 Third Avenue, New York, NY 10017

Routledge is an imprint of the Taylor & Francis Group, an informa business

First edition published by Butterworth-Heinemann 2009
Published by Routledge 2011

British Library Cataloguing-in-Publication Data
A catalogue record for this book is available from the British Library

Library of Congress Cataloging-in-Publication Data
A catalog record has been requested for this book

ISBN: 978-1-138-58960-5 (hbk)
ISBN: 978-0-415-82776-8 (pbk)
ISBN: 978-0-429-50422-8 (ebk)

Typeset in Times New Roman PS MT
by Apex CoVantage, LLC

Printed and bound by CPI Group (UK) Ltd, Croydon, CR0 4YY

Contents

Preface to the second edition   **x**

Preface   **xii**

Acknowledgements   **xviii**

Chapter 1   Electrical fundamentals   **1**
1.1   Electron theory   1
1.2   Electrostatics and capacitors   3
1.3   Direct current   6
1.4   Current, voltage and resistance   7
1.5   Power and energy   12
1.6   Electromagnetism and inductors   14
1.7   Alternating current and transformers   19
1.8   Safety   36
1.9   Multiple choice questions   37

Chapter 2   Electronic fundamentals   **39**
2.1   Semiconductor theory   39
2.2   Diodes   42
2.3   Transistors   53
2.4   Integrated circuits   59
2.5   Sensors and transducers   60
2.6   Multiple choice questions   62

Chapter 3   Digital fundamentals   **65**
3.1   Logic gates   65
3.2   Combinational logic systems   66
3.3   Monostable devices   69
3.4   Bistable devices   71
3.5   Decoders   72
3.6   Encoders   75
3.7   Multiplexers   76
3.8   Bus systems   78
3.9   Computers   82
3.10   Multiple choice questions   84

Chapter 4   Generators and motors   **87**
4.1   Generator and motor principles   87
4.2   AC generators   95
4.3   Three-phase generation and distribution   99

	4.4	AC motors	101
	4.5	Practical aircraft generating systems	110
	4.6	Multiple choice questions	115

Chapter 5   Batteries   **117**

	5.1	Overview	117
5.2	Storage cells	118
5.3	Lead-acid batteries	119
5.4	Nickel-cadmium batteries	123
5.5	Lithium batteries	126
5.6	Nickel-metal hydride batteries	126
5.7	Battery locations	128
5.8	Battery venting	128
5.9	Battery connections	129
5.10	Multiple choice questions	131

Chapter 6   Power supplies   **133**

6.1	Regulators	133
6.2	External power	137
6.3	Inverters	137
6.4	Transformer rectifier units	139
6.5	Transformers	139
6.6	Auxiliary power unit (APU)	140
6.7	Emergency power	142
6.8	Multiple choice questions	142

Chapter 7   Wiring and circuit protection   **143**

7.1	Overview	143
7.2	Construction and materials	144
7.3	Specifications	145
7.4	Shielding/screening	146
7.5	Circuit protection	148
7.6	Multiple choice questions	152

Chapter 8   Distribution of power supplies   **153**

8.1	Single engine/general aviation aircraft	153
8.2	Twin engine general aviation aircraft	157
8.3	Larger aircraft systems	158
8.4	Split bus system	159
8.5	Parallel bus system	161
8.6	Split/parallel bus system	161
8.7	Standby and essential power	161
8.8	Battery charging	162
8.9	Control and protection	163
8.10	Load-shedding	165
8.11	Multiple choice questions	169

Chapter 9	**Controls and transducers**	**171**
9.1	Switches	171
9.2	Relays and contactors	174
9.3	Variable resistors	176
9.4	Linear displacement transducers	177
9.5	Fluid pressure transducers	180
9.6	Temperature transducers	181
9.7	Strain transducers	185
9.8	Rotary position transducers	186
9.9	Accelerometers	188
9.10	Solid state technology	189
9.11	Multiple choice questions	189
Chapter 10	**Engine systems**	**193**
10.1	Starting and ignition	193
10.2	Indicating systems overview	203
10.3	Primary indicating systems	203
10.4	Secondary indicating systems	213
10.5	Electronic indicating systems	214
10.6	Multiple choice questions	218
Chapter 11	**Fuel management**	**221**
11.1	Storage overview	221
11.2	Fuel quantity measurement and indication	221
11.3	Fuel feed and distribution	226
11.4	Fuel transfer	228
11.5	Refuelling and defuelling	229
11.6	Fuel jettison	229
11.7	Fuel tank venting	229
11.8	Fuel tank inerting	229
11.9	Multiple choice questions	231
Chapter 12	**Lights**	**233**
12.1	Lighting technologies	233
12.2	Flight compartment lights	234
12.3	Passenger cabin lights	238
12.4	Exterior lights	240
12.5	Multiple choice questions	246
Chapter 13	**Cabin systems**	**249**
13.1	Passenger address system	249
13.2	Galley equipment	251
13.3	In-flight entertainment (IFE)	251
13.4	Satellite communications	252
13.5	Multiplexing	255
13.6	Fibre optics	256
13.7	Air conditioning	257
13.8	Pressurization	260

13.9	Airstairs	262
13.10	Multiple choice questions	262

Chapter 14 Airframe monitoring, control and indicating systems — **265**
14.1	Landing gear	265
14.2	Trailing edge flaps	268
14.3	Control surfaces	269
14.4	Electronic indicating systems	270
14.5	Multiple choice questions	272

Chapter 15 Warning and protection systems — **273**
15.1	Stall warning and protection	273
15.2	Airframe ice and rain protection	279
15.3	Windscreen ice and rain protection	283
15.4	Anti-skid	285
15.5	Configuration warning	287
15.6	Aural warnings	289
15.7	Multiple choice questions	290

Chapter 16 Fire and overheat protection — **291**
16.1	Overview	291
16.2	Engine/APU fire detection	292
16.3	Cargo bay/baggage area	299
16.4	Fire extinguishing	302
16.5	Multiple choice questions	306

Chapter 17 Terrain awareness warning system (TAWS) — **307**
17.1	System overview	307
17.2	System warnings and protection	309
17.3	External references	311
17.4	Ground proximity modes	315
17.5	Forward-looking terrain avoidance (FLTA)	321
17.6	Rotorcraft TAWS	324
17.7	Architecture and configurations	325
17.8	Future developments	326
17.9	Multiple choice questions	326

Chapter 18 Flight data and cockpit voice recorders — **329**
18.1	Flight data recorder history	329
18.2	Mandatory equipment requirements	331
18.3	Flight data recorder (FDR) specifications	334
18.4	Cockpit voice recorders	340
18.5	Health and usage monitoring system (HUMS)	342
18.6	Multiple choice questions	344

Chapter 19   Electrical and magnetic fields   **347**
    19.1   Electromagnetic interference   347
    19.2   EMI reduction   352
    19.3   High-intensity/energy radiated fields   353
    19.4   Lightning   354
    19.5   Grounding and bonding   357
    19.6   Multiple choice questions   358

Chapter 20   Continuing airworthiness   **361**
    20.1   Wire and cable installations   361
    20.2   Bonding   367
    20.3   Static charges   368
    20.4   Earth returns   369
    20.5   Aircraft manuals   370
    20.6   Circuit testing   371
    20.7   Automatic test equipment   374
    20.8   On-board diagnostic equipment   374
    20.9   Electrostatic sensitive devices (ESSDs)   376
   20.10   Multiple choice questions   378

Appendix 1: Abbreviations and acronyms   381
Appendix 2: Revision papers   387
Appendix 3: Answers to multiple choice questions   393
Appendix 4: Electrical quantities, symbols and units   397
Appendix 5: Electrical formulae   399
Appendix 6: Decibels   401
Appendix 7: Wire and cable sizes   403
Appendix 8: ATA chapter/subsystem list   405
Appendix 9: Electrical and electronic symbols   409
Appendix 10: Wire numbering/coding   413

Index   **415**

Preface to the second edition

This second edition has been updated to incorporate requests from readers, reviewers and colleagues, together with changes in technology that have occurred since the first edition was published in 2009.

Additional theory has been added into Chapter 1 (electrical fundamentals), introducing complex notation for the analysis of alternating current (AC) circuits. Complex notation uses mathematical quantities, representing two dimensions, i.e. magnitude and direction. Electrical quantities that have both magnitude and direction are referred to as phasors (in other engineering contexts, these are called vectors.)

Technology updates have occurred since the first edition was published, including the development of the "all electric aircraft" utilising new battery technologies. At the time of writing, there are industry reports of a short-haul all-electric airliner being developed, with the goal of bringing this into service within 10 years. The concept is for a 120-seat airliner, with a T-tail and distributed electric ducted fans buried in the wings. The design features eight ducted fans per wing with underfloor battery packs based on 'advanced cell chemistry'. It remains to be seen if this concept will become viable. That said, the electric flight revolution is definitely taking place, albeit in the light aircraft sector. On larger aircraft, electrical systems are replacing tradition mechanical systems. For example, in Boeing's 787 aircraft, bleed air is only used for engine cowl ice protection and pressurization of hydraulic reservoirs. Electrical functions are utilised for: wing de-icing protection, engine starting, high-capacity hydraulic pumps, and the cabin environmental control system. Although lithium batteries were covered in the first edition, the benefits of Nanophosphate® technology have been included in the second edition. An overview of integrated solid-state technology micro-electrical-mechanical sensors (MEMS) has been included in the second edition.

Several high-profile accidents have occurred since the first edition was published, in particular flight AF447 (June 2009) and flight MH370 (March 2014).

These accidents have highlighted shortcomings in the way that flights are currently tracked. The subsequent investigations into these two accidents is beyond the scope of this book, and indeed the book series, however the subjects covered by the book series are very topical in the context of the accident investigations, e.g. flight data recorders (covered in this book), together with air traffic control (ATC); emergency locator transmitters (ELT); satellite communications/navigation (all covered in ACNS).

Lithium batteries have caused fire and overheating of during transportation in aircraft freight holds. There are now restrictions on the carrying of lithium batteries in bulk on passenger flights. At least four aircraft suffered from electrical system problems stemming from lithium-ion batteries installed on the aircraft, causing fire and overheating. The systems and products have since been modified.

Responding to feedback received from a variety of sources, this second edition has an expanded section on helicopter/rotary wing health usage monitoring systems (HUMS). The authors also considered a new section for helicopter flight data monitoring programs (HFDM). The purpose of HUMS is to detect and identify defects requiring maintenance. HFDM evaluates flight crew operations that may contribute to an incident or accident. Since HFDM is an operational tool, it is not included in this second edition. The authors are aware of integrated vehicle health monitoring (IVHM), an evolution of the diagnostic and prognostic monitoring and recording systems discussed throughout this book. In this context, IVHM can be applied to military vehicles, aircraft, or spacecraft. There are various definitions/scopes of IVHM systems; these comprise different functions, subsystems and components. The objectives of IVHM are however, generally the same, i.e. to improve safety, availability, reliability and maintenance costs through via diagnostics and prognostics. The authors decided that IVHM was beyond the scope of the book series.

As with other technical subjects covered in this series, the continuing airworthiness of aircraft and aeronautical products, parts and appliances is given for training purposes. The change of requirements, and increasing knowledge levels is covered in the applicable regulations, e.g. Commission Regulation (EU) No 1149.

Acknowledgements

- Aerossurance: aviation consultancy specialising in air safety, airworthiness, aviation regulation, expert witness & aviation contracting support services.
- True Blue Power® (a division of Mid-Continent Instrument Co., Inc.): design and manufacture of Nanophosphate® batteries.

Images and reference material remain the property of each respective organization credited in this book.

Preface

Aircraft Electrical and Electronic Systems continues the series of textbooks written for aircraft engineering students. This book addresses the electrical contents of the EASA Part 66 Modules 11 and 13; it also provides reference material for the avionic and aircraft electrical units of various BTEC National and Higher National, City and Guilds, NVQ and Foundation Degree modules.

This book is designed to cover the essential knowledge base required by certifying mechanics, technicians and engineers engaged in engineering maintenance activities on commercial aircraft and in general aviation. In addition, this book should appeal to members of the armed forces and others attending training and educational establishments engaged in aircraft maintenance and related aeronautical engineering programmes. This book will also appeal to others within the aircraft industry who need an insight into electrical and electronic systems, e.g. pilots, engineering managers, etc.

The book provides an introduction to the fundamentals of electrical, electronic and digital theory that underpins the principles of systems covered in the remainder of the book. For the reader that already has background knowledge of the fundamentals, the subsequent chapters can be read as individual subjects. For the reader that requires a deeper understanding of related fundamentals, additional material can be found in related books in the series:

- *Aircraft Engineering Principles*
- *Aircraft Digital Electronic and Computer Systems*
- *Aircraft Communications and Navigation Systems.*

The books in this series have been designed for both independent and tutor-assisted studies. They are particularly useful to the 'self-starter' and to those wishing to update or upgrade their aircraft maintenance licence. The series also provides a useful source of reference for those taking ab initio training programmes in EASA Part 147 and FAR 147 approved organizations as well as those following related programmes in further and higher education institutions.

The title of this book, *Aircraft Electrical and Electronic Systems*, has been specifically chosen to differentiate between other avionic systems such as communications, navigation, flight guidance and instruments. The term *avionics* (aviation electronics) was first used in the late 1940s to identify electrical and electronic equipment such as radar, radio navigation and communications, although the term was not in general use until the late 1960s. During the 1970s, integrated computer-based systems were being developed, e.g. ground proximity warning systems; these used a number of existing aircraft sensors that monitored parameters such as barometric altitude, vertical speed and radio altitude.

The continued development and integration of electrical and electronic systems, together with the widespread use of integrated circuits, microprocessors, data communications and electronic displays, have given new meaning to the term avionics. Aircraft engineers will be exposed to in-service aircraft using older technology, together with the new aircraft entering service based on modern technology. Using trends from the last 40 years, there will be an ever-increasing dependence on avionic systems. The eventual outcome could be the **all-electric aircraft**, a concept where traditional mechanical linkages, hydraulics and pneumatics are totally replaced by electrical and electronic systems.

This book establishes a reference point for engineering students; it does not attempt to address all system types for all aircraft types. It is also important to note that this book does not attempt to provide the level of detail found in the aircraft publications, including the maintenance and wiring diagram manuals. Although there are many examples quoted in the book that are based on specific aircraft types, this is only done to illustrate a specific point.

Throughout the book, the principles and operation of systems are summarized by numerous 'key points'. The reader will be invited at regular intervals to assess knowledge via 'test your understanding' questions. Finally, the principles and operation of systems are put into the context of aircraft maintenance engineering by numerous 'key maintenance points'. Each chapter concludes with a number of multiple choice questions; the reader will then find revision papers in the appendices for additional assessment purposes. A summary of the book is as follows.

Chapter 1 sets the scene by providing an explanation of electricity in terms of the motion of electric charge and basic electrical quantities such as current, voltage, resistance and power. The chapter provides an introduction to electrostatics and capacitors and also to electromagnetism and inductors. Here the emphasis is on the key concepts and fundamental laws that underpin the operation of the electrical systems found in aircraft. The chapter provides a detailed introduction to alternating current and transformer principles, and concludes with an essential section on safety. This chapter will be particularly useful if you have not previously studied electrical principles. It has also been designed to help fill any gaps in your knowledge and bring you quickly up to speed.

Electronic fundamentals are introduced in Chapter 2. This chapter explains the principles, construction and basic application of a variety of common semiconductor devices including diodes, thyristors, transistors and integrated circuits. The chapter includes a detailed explanation of rectifier circuits, both halfwave and full-wave types, and the use of transistors as current amplifiers.

The advent of digital techniques and integrated circuits has revolutionized the scope and applications for avionic systems. Chapter 3 provides readers with an introduction to digital techniques. The function and operation of logic gates are established before moving on to explore the use of combinational and sequential logic in several typical aircraft applications. The chapter also provides an overview of coding systems and the logic systems that are used to represent numerical data. An introduction to aircraft data bus systems is provided together with a brief overview of the architecture and principal constituents of simple computer systems.

Generators and motors are widely used in modern aircraft. Chapter 4 explains the principles on which they operate as well as the theoretical and practical aspects of aircraft power generation and distribution. Three-phase systems and methods of connection are described in some detail.

If you are in any doubt as to whether or not you should work through Chapters 1 to 4, you can always turn to the multiple choice questions at the end of each chapter to assess your knowledge.

All electrical and electronic systems require a power source. Batteries are primary sources of electrical power found on most aircraft delivering direct current (DC). Chapter 5 reviews the battery types used on aircraft, typical applications and how they are installed and maintained. There are several types of battery used on aircraft, usually defined by the types of materials used in their construction; these include lead-acid and nickel-cadmium batteries. Other types of battery are being considered for primary power on aircraft; these include lithium and nickel-metal hydride.

In Chapter 6, we review the other sources of electrical power used on aircraft and their typical applications. Electrical power can be derived from a variety of sources; these are categorized as either primary or secondary sources. Batteries and generators are primary sources of electrical power; inverters and transformer rectifier units (TRUs) are secondary sources of power. This power is either in the form of direct or alternating current depending on system requirements. Generators can either supply direct or alternating current (AC); the outputs of generators need to be regulated. Inverters are used to convert DC (usually from the battery) into alternating (AC). Transformer rectifier units (TRUs) convert AC into DC; these are often used to charge batteries from AC generators. In some installations, transformers (as described in Chapter 1) are used to convert AC into AC, typically for stepping down from 115 to 26 V AC. In addition to onboard equipment, most aircraft have the facility to be connected to an external power source during servicing or maintenance. An auxiliary power unit (APU) is normally used for starting the aircraft's main engines via the air distribution system. While the aircraft is on the ground, the APU can also provide electrical power. In the event of generator failure(s), continuous power can be provided by a ram air turbine (RAT).

The safe and economic operation of an aircraft is becoming ever more dependent on electrical and electronic systems. These systems are all interconnected with wires and cables; these take many forms. Chapter 7 describes the physical construction of wires and cables together with how they are protected from overload conditions before power is distributed to the various loads on the aircraft. Electrical wires and cables have to be treated as an integral part of the aircraft requiring careful installation; this is followed by direct ongoing inspection and maintenance requirements for continued airworthiness.

Wire and cable installations cannot be considered (or treated) as 'fit and forget'. System reliability will be seriously affected by wiring that has not been correctly installed or maintained. We need to distribute the sources of electrical power safely and efficiently and control its use on the aircraft. Once installed, the wires and cables must be protected from overload conditions that could lead to overheating, causing the release of toxic fumes, possibly leading to fire. Legislation is being proposed to introduce a new term: **electrical wire interconnection system** (EWIS); this will acknowledge the fact that wiring is just one of many components installed on the aircraft. EWIS relates to any wire, wiring device, or combination of these, including termination devices, installed in the aircraft for transmitting electrical energy between two or more termination points.

Electrical power is supplied to the various loads in the aircraft via common points called busbars. In Chapter 8, we will focus on busbar configurations and how these are arranged for the protection and management of the various power supply sources available on the aircraft. The electrical power distribution system is based on one or more busbar(s); these provide pre-determined routes to circuits and components throughout the aircraft. The nature and complexity of the distribution system depend on the size and role of the aircraft, ranging from single-engine general aviation through to multi-engine passenger transport aircraft.

The word 'bus' (as used in electrical systems) is derived from the Latin word *omnibus* meaning 'for all'. The busbar can be supplied from one or more of the power sources previously described (generator, inverter, transformer rectifier unit or battery). Protection devices, whether fuses or circuit-breakers, are connected in series with a specific system; they will remove the power from that system if an overload condition arises. There also needs to be a means of protecting the power source and feeder lines to the busbar, i.e. before the individual circuit protection devices.

There are many systems on an aircraft that need to be controlled and/or monitored, either manually by the crew, or automatically. Chapter 9 describes generic controls and transducer devices used on aircraft. A switch provides the simplest form of circuit control and monitoring. Switches can be operated manually by a person, activated by sensing movement, or controlled remotely. Many other aircraft parameters

need to be measured; this is achieved by a variety of transducers; these are devices used to convert the desired parameter, e.g. pressure, temperature, displacement, etc. into electrical energy.

The aircraft engine is installed with many systems requiring electrical power. Chapter 10 describes engine starting, ignition and indicating system for both piston and gas turbine engines. The predominant electrical requirement (in terms of current consumption) is for the engine starting system. General aviation aircraft use electrical starter motors for both piston and gas turbine engines; larger transport aircraft use an air-start system (controlled electrically) derived from ground support equipment or by air cross-fed from another engine. Electrical starting systems on piston and gas turbine engines are very different. The trend towards the all-electric aircraft will see more aircraft types using electrical starting methods. The engine also requires electrical power for the ignition system; once again, the needs of piston and gas turbine engines are quite different. Although starting and ignition systems are described in this chapter as separate systems, they are both required on a co-ordinated basis, i.e. a means to rotate the engine and ignite the air/fuel mixture.

Electrical and electronic requirements for engines also include the variety of indicating systems required to operate and manage the engine. These indicating systems include (but are not limited to) the measurement and indication of: rotational speed, thrust, torque, temperature, fuel flow and oil pressure. Indications can be provided by individual indicators or by electronic displays.

The management of fuel is essential for the safe and economic operation of the aircraft. The scope of fuel management depends on the size and type of aircraft; fuel is delivered to the engines using a variety of methods. Chapter 11 provides an overview of fuel management on a range of aircraft types. The system typically comprises fuel quantity indication, distribution, refuelling, defuelling and fuel jettison. In the first instance, we need to measure the quantity of fuel on board. Various technologies and methods are used to measure fuel quantity: this depends mainly on the type and size of aircraft. Technologies range from sight gauges through to electronic sensors. On larger aircraft, fuel is fed to the engines by electrically driven pumps. On smaller aircraft, an engine-driven pump is used with electrical pumps used as back-up devices. Solenoid or motorized valves are used to isolate the fuel supply to engines under abnormal conditions. On larger aircraft, the fuel can be transferred between tanks; this is controlled manually by the crew, or automatically by a fuel control computer.

Lighting is installed on aircraft for a number of reasons including: safety, operational needs, servicing and for the convenience of passengers. Chapter 12 reviews a number of lighting technologies and the type of equipment used in specific aircraft applications. The applications of aircraft lights can be broadly grouped into four areas: flight compartment (cockpit), passenger cabin, exterior and servicing (cargo and equipment bays). Lights are controlled by on/off switches, variable resistors or by automatic control circuits.

Passenger transport and business aircraft are fitted with a range of cabin electronic equipment for passenger safety, convenience and entertainment. Typical applications for this equipment includes lighting, audio and visual systems. Chapter 13 describes the many types of systems and equipment used for passenger safety, convenience and entertainment. Audio systems include the passenger address system used by the flight or cabin crew to give out safety announcements and other flight information. These announcements are made from hand-held microphones and are heard over loudspeakers in the cabin and passenger headsets. The same system can be used to play automatic sound tracks; this is often used for announcements in foreign languages, or to play background music during boarding and disembarkation. A range of galley equipment is installed on business and passenger aircraft. The nature of this equipment depends on the size and role of the aircraft. Air conditioning is provided in passenger aircraft for the comfort of passengers; pressurization is required for flying at high altitudes. Airstairs allow passengers, flight crew and ground personnel to board or depart the aircraft without the need for a mobile staircase or access to a terminal. All of these systems have electrical/electronic interfaces and control functions.

Chapter 14 reviews airframe systems such as landing gear control and indication, control surface position and indicating systems. Various sensors are needed for the monitoring and control of airframe systems. Broadly speaking, the sensors can be considered as detecting one of two states (or 'conditions'), or a variable position. Two-state conditions include: landing gear (up or down) or cabin doors (open or closed). Variable positions include: control surfaces and flap position. Micro-switches or prox-imity sensors detect two-state positions; variable position devices are detected by a variety of devices including synchros and variable resistors.

Chapter 15 describes a variety of systems installed on aircraft to protect them from a variety of hazards including: stalling, ice, rain, unsafe take-off configuration and skidding. Stall protection systems provide the crew with a clear and distinctive warning before an unsafe condition is reached. Flying in ice and/or rain conditions poses a number of threats to the safe operation of the aircraft; ice formation can affect the aerodynamics and/or trim of the aircraft. The configuration warning system (also known as a take-off warning system) provides a warning if the pilot attempts to take off with specific controls not selected in the correct position, i.e. an unsafe configuration. The anti-skid system (also called an anti-lock braking system: ABS) is designed to prevent the main landing gear wheels from locking up during landing, particularly on wet or icy runway surfaces.

Fire on board an aircraft is a very serious hazard; all precautions must be taken to minimize the risk of a fire starting. In the event that a fire does occur, there must be adequate fire protection on the aircraft. Chapter 16 focuses on the equipment and systems used to detect fire and smoke together with the means of delivering the fire-extinguishing agent. The subject of fire protection theory is a branch of engineering in its own right. Basic fire protection theory is covered in this chapter to provide the reader with sufficient information to understand how this theory is applied through aircraft electrical systems.

During the 1970s, studies were carried out by accident investigators and regulatory authorities to examine one of the most significant causes of aircraft accidents of the time: controlled flight into terrain (CFIT). This can be defined as an accident where a serviceable aircraft, under the control of a qualified pilot, inadvertently flies into terrain, an obstacle or water. Chapter 17 describes the generic name given to this type of protection: terrain awareness warning system (TAWS). CFIT accidents usually occur during poor visual conditions, often influenced by other factors, e.g. flight crew distraction, malfunctioning equipment or air traffic control (ATC) miscommunication. With CFIT, the pilots are generally unaware of this situation until it is too late. Ground proximity warning system (GPWS) was developed in 1967 to alert pilots that their aircraft was in immediate danger of CFIT. This system was further developed into the enhanced ground proximity warning system (EGPWS) by adding a forward-looking terrain avoidance (FLTA) feature, made possible via global positioning system technology.

One of the fundamental electrical/electronic systems associated with the aircraft industry is the crash survivable flight data recorder (FDR). This is often referred to in the press as the 'black box', even though the item is actually bright orange! Flight data recorders used for accident investigation are mandatory items of equipment in commercial transport aircraft. Efforts to introduce crash-survivable flight data recorders can be traced back to the 1940s; the FDR has now been supplemented with the cockpit voice recorder (CVR). Chapter 18 reviews the range of FDR/CVR technologies that are employed for both accident investigation and trend monitoring. Recorders are used after an incident or accident as an integral part of the investigators' efforts to establish the cause(s). Data recorders can also be used to indicate trends in aircraft and engine performance. Algorithms are established for healthy and normal conditions during the aircraft's flight-testing programme and the early period of service. These algorithms include engine parameters such as engine exhaust temperature, oil pressure and shaft vibration for given speeds and altitudes. These parameters are then monitored during the aircraft's life; any deviations from the norm are analysed to determine if the engine requires inspection, maintenance or removal.

One of the consequences of operating electrical and electronic equipment is the possibility of disturbing, or interfering with, nearby items of electronic equipment. Chapter 19 looks at some of the implications of interference from electrical and magnetic fields. The term given to this type of disturbance is electromagnetic interference (EMI). Electrical or electronic products will both radiate and be susceptible to the effects of EMI. This is a paradox since many principles of electrical engineering are based on electromagnetic waves coupling with conductors to produce electrical energy and vice versa (generators and motors). Furthermore, systems are specifically designed to transmit and receive electromagnetic energy, i.e. radio equipment. In complex avionic systems, the consequences of EMI can cause serious, if not hard-to-find–problems. The ability of an item of equipment to operate alongside other items of equipment without causing EMI is electromagnetic compatibility (EMC).

Modern digital equipment operates at very high speed and relatively low power levels. In addition to EMI, high-intensity radiated fields (HIRFs) are received from the external environment, e.g. from radio and radar transmitters, power lines and lightning. The high energy created by these radiated fields disrupts electronic components and systems in the aircraft. (This effect is also referred to as high-energy radiated fields – HERFs.) The electromagnetic energy induces large currents to flow, causing direct damage to electronic components together with the secondary effects of EMI.

Advances in electronic technology bring many new features and benefits, e.g. faster processors, higher-density memory and highly efficient displays. These advances are primarily due to the reduction in the physical size of semiconductor junctions; this leads to higher-density components in a given size of integrated circuit. One significant problem associated with certain types of semiconductor devices is that the smaller junctions are susceptible to damage from electrostatic voltages. This is a problem that can potentially affect a wide range of electronic equipment fitted in an aircraft. Effects range from weakening of semiconductor junctions through total failure of the equipment; both these effects can occur without any visible signs of damage to the naked eye! Electrostatic sensitive devices (ESSDs) are electronic components that are prone to damage from stray electrical charge produced primarily from human operators. This problem is particularly prevalent with high-density memory devices and electronic displays. Weakening and damage to static-sensitive devices can result from mishandling and inappropriate methods of storage; the practical issues for handling ESSDs are addressed in Chapter 19.

Many processes are required throughout the aircraft's operating life to ensure that it complies with the applicable airworthiness requirements and can be safely operated. The generic term for this range of processes is continuing airworthiness. Chapter 20 reviews some practical installation requirements, documentation and test equipment required by the avionics engineer to ensure the continued airworthiness of aircraft electrical and electronic systems. The term 'maintenance' is used for any combination of overhaul, repair, inspection, replacement, modification or defect rectification of an aircraft or component, with the exception of the pre-flight inspection. Particular emphasis is given to wire and cable installations since these cannot be considered (or treated) as 'fit and forget'. System reliability will be seriously affected by wiring that has not been correctly installed or maintained. Persons responsible for the release of an aircraft or a component after maintenance are the certifying staff. Maintenance of an aircraft and its associated systems requires a variety of test equipment and documentation; these are required by certifying staff to fulfil their obligations in ensuring continued airworthiness.

Supporting material for the book series (including interactive questions, media files, etc.) is available online at **www.66web.co.uk** or **www.key2study.com** and then follow the links for aircraft engineering.

Acknowledgements

The authors would like to express their thanks to the following persons for ideas, support and contributions to the book. We thank Lloyd Dingle, who had the original idea for the aircraft engineering series, and Alex Hollingsworth for commissioning this book. Thanks also to Lucy Potter and Holly Bathie, editorial assistants at Elsevier Science & Technology Books.

Thanks also to the following organizations for permission to reproduce their information:

- Advanced Technological Systems International Limited (new generation aircraft batteries)

- Aero Quality (battery maintenance details)
- Flight Display Systems (in-flight entertainment systems)
- Iridium Satellite LLC (satellite communication systems)
- Lees Avionics, Helicopter Services and Wycombe Air Centre (photo images taken at their premises)
- Specialist Electronic Services Ltd (flight data recorders).
- True Blue Power® (a division of Mid-Continent Instrument Co., Inc.)

Chapter 1

Electrical fundamentals

This chapter will provide you with an introduction to the essential electrical theory that underpins the rest of this book. It has been designed to help fill any gaps in your knowledge and bring you quickly up to speed. You will find this chapter particularly useful if you have not previously studied electrical principles. However, if you are in any doubt as to whether or not you should work through this chapter you can always turn to Section 1.9 on page 37 and see how you get on with the multiple choice questions at the end of this chapter.

1.1 Electron theory

All matter is made up of atoms or groups of atoms (**molecules**) bonded together in a particular way. In order to understand something about the nature of electrical charge we need to consider a simple model of the atom. This model, known as the **Bohr model** (see Fig. 1.1), shows a single atom consisting of a central nucleus with orbiting electrons.

Within the nucleus there are **protons** which are positively charged and **neutrons** which, as their name implies, are electrically neutral and *have no charge*. Orbiting the nucleus are electrons that have a negative charge, equal in magnitude (size) to the charge on the proton. These electrons are approximately two thousand times lighter than the protons and neutrons in the nucleus.

In a stable atom the number of protons and electrons are equal, so that overall, the atom is neutral and has no charge. However, if we rub two particular materials together, electrons may be transferred from one to another. This alters the stability of the atom, leaving it with a net positive or negative charge. When an atom within a material loses electrons it becomes positively charged and is known as a **positive ion**, when an atom gains an electron it has a surplus negative charge and so is referred to as a **negative ion**. These differences in charge can cause **electrostatic** effects. For example, combing your hair with a nylon

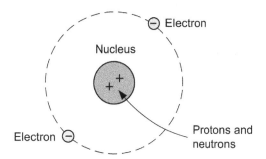

Figure 1.1 The Bohr model of the atom

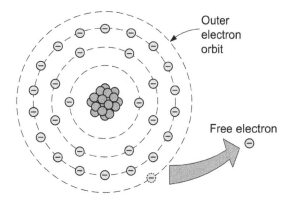

Figure 1.2 A material with a loosely bound electron in its outer shell

comb may result in a difference in charge between your hair and the rest of your body, resulting in your hair standing on end when your hand or some other differently charged body is brought close to it.

The number of electrons occupying a given orbit within an atom is predictable and is based on the position of the element within the periodic table. The electrons in all atoms sit in a particular orbit, or **shell**, dependent on their energy level. Each of these shells within the atom is filled by electrons from the nucleus outwards, as shown in Fig. 1.2. The first, innermost, of these shells can have up to two electrons, the second shell can have up to eight and the third up to 18.

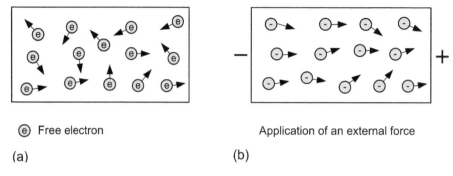

(e) Free electron   Application of an external force

(a)   (b)

Figure 1.3   Free electrons and the application of an external force

All electrons and protons carry an electrostatic **charge** but its value is so small that a more convenient unit of charge is needed for practical use which we call the **coulomb**. One coulomb (C) is the total amount of the charge carried by 6.21×10^{18} electrons. Thus a single electron has a charge of a mere 1.61×10^{-19} C!

A material which has many free electrons available to act as charge carriers, and thus allows current to flow freely, is known as a **conductor**. Examples of good conductors include aluminium, copper, gold and iron. Figure 1.2 shows a material with one outer electron that can become easily detached from the parent atom. A small amount of external energy is required to overcome the attraction of the nucleus. Sources of such energy may include heat, light or electrostatic fields. The electron once detached from the atom is able to move freely around the structure of the material and is called a **free electron**. It is these free electrons that become the **charge carriers** within a material. Materials that have large numbers of free electrons make good **conductors** of electrical energy and heat.

In a material containing free electrons their direction of motion is random, as shown in Fig. 1.3(a), but if an external force is applied that causes the free electrons to move in a uniform manner (Fig. 1.3(b)) an electric **current** is said to flow.

Metals are the best conductors, since they have a very large number of free electrons available to act as charge carriers. Materials that do not conduct charge are called **insulators**; their electrons are tightly bound to the nuclei of their atoms. Examples of insulators include plastics, glass, rubber and ceramic materials.

The effects of electric current flow can be detected by the presence of one or more of the following effects: light, heat, magnetism, chemical, pressure and friction. For example, heat is produced when an electric current is passed through a resistive heating element. Light is produced when an electric current flows through the thin filament wire in the evacuated bulb of an electric lamp.

Key point

Electrons each carry a tiny amount of negative electrical charge.

Key point

Metals such as copper and silver are good conductors of electricity and they readily support the flow of electric current. Plastics, rubber and ceramic materials on the other hand are insulators and do not support the flow of electric current.

Test your understanding 1.1

1. Explain the following terms:
    (a) electron
    (b) ion
    (c) charge
    (d) conductor
    (e) insulator.

2. State, with reasons, whether an insulator or conductor is required in each of the following applications:
    (a) the body of a fuse
    (b) the outer protective sheath of a power cable
    (c) the fuselage covering of a transport aircraft
    (d) the radiating element of an antenna.

1.2 Electrostatics and capacitors

Electric charge is all around us. Indeed, many of the everyday items that we use in the home and at work rely for their operation on the existence of electric charge and the ability to make that charge do something useful. Electric charge is also present in the natural world and anyone who has experienced an electrical storm cannot fail to have been awed by its effects. In this section we begin by explaining what electric charge is and how it can be used to produce conduction in solids, liquids and gases.

We have already found that, if a conductor has a deficit of electrons, it will exhibit a net positive charge. If, on the other hand, it has a surplus of electrons, it will exhibit a net negative charge. An imbalance in charge can be produced by friction (removing or depositing electrons using materials such as silk and fur, respectively) or induction (by attracting or repelling electrons using a second body which is respectively positively or negatively charged).

If two bodies have charges with the same polarity (i.e. either both positively or both negatively charged) the two bodies will move apart, indicating that a force of repulsion exists between them. If, on the other hand, the charges on the two bodies are unlike (i.e. one positively charged and one negatively charged) the two bodies will move together, indicating that a force of attraction exists between them. From this we can conclude that like charges repel and unlike charges attract.

Static charges can be produced by friction. In this case, electrons and protons in an insulator are separated from each other by rubbing two materials together in order to produce opposite charges. These charges will remain separated for some time until they eventually leak away due to losses in the insulating **dielectric** material or in the air surrounding the materials. Note that more charge will be lost in a given time if the air is damp.

Static electricity is something that can cause particular problems in an aircraft and special measures are taken to ensure that excessive charges do not build up on the aircraft's structure. The aim is that of equalizing the potential of all points on the aircraft's external surfaces. The static charge that builds up during normal flight can be dissipated into the atmosphere surrounding the aircraft by means of small conductive rods connected to the aircraft's trailing surfaces. These are known as **static dischargers** or **static wicks** – see Fig. 1.4.

Figure 1.4   Static discharging devices

Key point

Charged bodies with the same polarity repel one another whilst charges with opposite polarity will attract one another.

Key point

A significant amount of charge can build up between conducting surfaces when they are insulated from one another. Where this might be a problem steps are taken to dissipate the charge instead of allowing it to accumulate uncontrolled.

Key maintenance point

Stray static charges can very easily damage static-sensitive devices such as semiconductors, memory devices and other integrated circuits. Damage can be prevented by adopting the appropriate electrostatic sensitive device (ESSD) precautions (described in the aircraft maintenance manual) when handling such devices. Precautions usually involve using wrist straps and grounding leads as well as using static-dissipative packaging materials.

1.2.1 Electric fields

The force exerted on a charged particle is a manifestation of the existence of an electric field. The electric

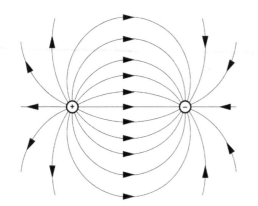

Figure 1.5 Electric field between isolated unlike charges

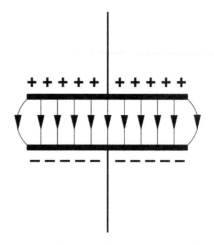

Figure 1.7 Electric field between the two charged parallel metal plates of a capacitor

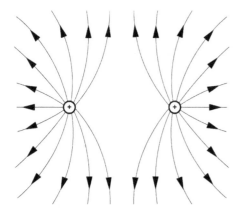

Figure 1.6 Electric field between isolated like charges

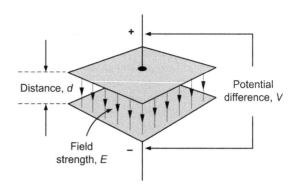

Figure 1.8 Electric field strength between two charged conducting surfaces

field defines the direction and magnitude of a force on a charged object. The field itself is invisible to the human eye but can be drawn by constructing lines which indicate the motion of a free positive charge within the field; the number of field lines in a particular region being used to indicate the relative strength of the field at the point in question.

Figures 1.5 and 1.6 show the electric fields between isolated unlike and like charges whilst Fig. 1.7 shows the field that exists between two charged parallel metal plates which forms a charge storage device known as a **capacitor**.

The strength of an electric field (E) is proportional to the applied **potential difference** and inversely proportional to the distance between the two conducting surfaces (see Fig. 1.8). The electric field strength is given by:

$$E = \frac{V}{d}$$

where E is the electric field strength (in V/m), V is the applied potential difference (in V) and d is the distance (in m).

The amount of charge that can be stored by a capacitor is given by the relationship:

$$Q = C \times V$$

where Q is the charge in coulomb, C is the capacitance in farads, F, and V is the voltage in volts, V. This relationship can be re-arranged to make C or V the subject as follows:

$$C = \frac{Q}{V} \quad \text{and} \quad V = \frac{Q}{C}$$

Example 1.1

Two parallel conductors are separated by a distance of 25 mm. Determine the electric field strength if they are fed from a 600 V DC supply.

The electric field strength will be given by:

$$E = \frac{V}{d}$$

where $V = 600$ V and $d = 25$ mm $= 0.025$ m. Thus:

$$E = \frac{600}{0.025} = 24,000 \text{ V/m} = 24 \text{ kV/m}$$

Example 1.2

The field strength between two parallel plates in a cathode ray tube is 18 kV/m. If the plates are separated by a distance of 21 mm determine the potential difference that exists between the plates.

The electric field strength will be given by:

$$E = \frac{V}{d}$$

Re-arranging this formula to make V the subject gives:

$$V = E \times d$$

Now $E = 18$ kV/m $= 18,000$ V/m and $d = 21$ mm $= 0.021$ m, thus:

$$V = 18,000 \times 0.021 = 378 \text{ V}$$

Example 1.3

A potential difference of 150 V appears across the plates of a 2 μF capacitor. What charge is present?

The charge can be calculated from:

$$Q = C \times V$$

where $C = 2\,\mu\text{F}$ and $V = 150$ V, thus:

$$Q = 2\,\mu\text{F} \times 150 \text{ V} = 300\,\mu\text{C}.$$

Example 1.4

A 6.8 μF capacitor is required to store a charge of 3.4 mC. What voltage should be applied to the capacitor?

The voltage can be calculated from:

$$V = \frac{Q}{C}$$

where $Q = 3.4$ mC and $C = 6.8\,\mu\text{F}$; thus:

$$V = \frac{3.4 \text{ mC}}{6.8\,\mu\text{F}} = \frac{3,400\,\mu\text{C}}{6.8\,\mu\text{F}} = 500 \text{ V}$$

Key maintenance point

When replacing a capacitor it is essential to ensure that the replacement component is correctly rated in terms of type, value, working voltage and temperature. Capacitors are prone to failure if their maximum working voltage is exceeded and they should be de-rated when operated at a relatively high ambient temperature according to manufacturers' specifications. It is also essential to observe the correct polarity when replacing an electrolytic (polarized) component. This is usually clearly marked on the external casing.

Key maintenance point

When working with high-voltage capacitors it is essential to ensure that the capacitor is fully discharged before attempting to replace the component. In most cases, any accumulated charge will safely drain away within a few seconds after removal of power. However, this should not be relied upon and a safe discharge path through a high-value resistor (say 1 MΩ) fitted with appropriate probes will ensure that capacitor is safe to work on.

Figure 1.9 A selection of capacitors with values ranging from 12 pF to 1000 μF and working voltages ranging from 25 V to 450 V

Figure 1.10 A typical aircraft battery

Test your understanding 1.2

1. The two plates of a parallel plate capacitor are separated by a distance of 15 mm. If the potential difference between the plates is 300 V what will the electric field strength be?
2. The electric field between two conducting surfaces is 500 V/m. If the plates are separated by a distance of 2.5 mm, determine the potential difference between the plates.

1.3 Direct current

Direct current (DC) is current that flows in one direction only. DC circuits are found in every aircraft. An understanding of how and why these circuits work is an essential prerequisite to understanding more complex circuits. Because of their negative charge, electrons will flow from a point of negative potential to a point with more positive potential (recall that like charges repel and unlike charges attract). However, when we indicate the direction of current in a circuit we show it as moving from a point that has the greatest positive potential to a point that has the most negative potential. We call this **conventional current** and, although it may seem odd, you just need to remember that it flows in the *opposite* direction to that of the motion of electrons!

The most commonly used method of generating direct current is the electrochemical cell. A **cell** is a device that produces a charge when a chemical reaction takes place. When several cells are connected together they form a **battery**.

There are two types of cell: primary and secondary. **Primary cells** produce electrical energy at the expense of the chemicals from which they are made and once these chemicals are used up, no more electricity can be obtained from the cell. In **secondary cells**, the chemical action is reversible. This means that the chemical energy is converted into electrical energy when the cell is **discharged** whereas electrical energy is converted into chemical energy when the cell is being **charged**. You will find more information on aircraft batteries in Chapter 5.

Key point

Conventional current flows from positive to negative whilst electrons travel in the opposite direction, from negative to positive.

Key point

In a primary cell the conversion of chemical energy to electrical energy is irreversible and so these cells cannot be recharged. In secondary cells, the conversion of chemical energy to electrical energy is reversible. Thus these cells can be recharged and reused many times.

Key maintenance point

When removing and replacing batteries, it is essential to observe the guidance given in the aircraft maintenance manual (AMM) when removing, charging or replacing aircraft batteries. The AMM will describe the correct procedures for isolating the battery from the aircraft's electrical system prior to its physical removal.

Test your understanding 1.3

1. Explain the difference between a primary and a secondary cell.
2. Explain the difference between electron flow and conventional current.

1.4 Current, voltage and resistance

Current, I, is defined as the rate of flow of charge and its unit is the ampere, A. One ampere is equal to one coulomb C per second, or:

$$\text{One ampere of current } I = \frac{Q}{t}$$

Where $t = $ time in seconds

So, for example: if a steady current of 3 A flows for two minutes, then the amount of charge transferred will be:

$$Q = I \times t = 3\,\text{A} \times 120\,\text{s} = 360 \text{ coulombs}$$

Example 1.5

A current of 45 mA flows from one point in a circuit to another. What charge is transferred between the two points in 10 minutes?

Here we will use $Q = It$ where $I = 45\,\text{mA} = 0.045\,\text{A}$ and $t = 10$ minutes $= 10 \times 60 = 600$ s.
Thus:

$$Q = 45\,\text{mA} \times 600\,\text{s} = 0.045\,\text{A} \times 600\,\text{s} = 27\,\text{C}$$

Example 1.6

A charge of 1.5 C is transferred to a capacitor in 30 seconds. What current is flowing in the capacitor?

Here we will use $I = Q/t$ where $Q = 1.5\,\text{C}$ and $t = 30$ s. Thus:

$$I = \frac{Q}{t} = \frac{1.5\,\text{C}}{30\,\text{s}} = 0.05\,\text{A} = 50\,\text{mA}$$

Key point

Current is the rate of flow of charge. Thus, if more charge moves in a given time, more current will be flowing. If no charge moves then no current is flowing.

1.4.1 Potential difference (voltage)

The force that creates the flow of current (or rate of flow of charge carriers) in a circuit is known as the **electromotive force** (or **e.m.f.**) and it is measured in volts (V). The **potential difference** (or **p.d.**) is the voltage difference, or voltage drop between two points.

One volt is the potential difference between two points if one Joule of energy is required to move one coulomb of charge between them. Hence:

$$V = \frac{W}{Q}$$

where $W = $ energy and $Q = $ charge, as before. Energy is defined later in Section 1.5.

Test your understanding 1.4

1. How much charge will be transferred when a current of 6 A flows for two minutes?
2. How long will it take for a charge of 0.2 C to be transferred using a current of 0.5 A?
3. If 0.4 J of energy is used to transfer 0.05 C of charge between two points what is the potential difference between the two points?

1.4.2 Resistance

All materials at normal temperatures oppose the movement of electric charge through them; this opposition to the flow of the charge carriers is known as the **resistance**, R, of the material. This resistance is due to collisions between the charge carriers (electrons) and the atoms of the material. The unit of resistance is the **ohm**, with symbol Ω.

Note that 1 V is the electromotive force (e.m.f.) required to move 6.21×10^{18} electrons (1 C) through a resistance of $1\,\Omega$ in 1 second. Hence:

$$V = \left(\frac{Q}{t}\right) \times R$$

where Q = charge, t = time, and R = resistance.

Re-arranging this equation to make R the subject gives:

$$R = \frac{V \times t}{Q}\ \Omega$$

Example 1.7

A 28 V DC aircraft supply delivers a charge of 5 C to a window heater every second. What is the resistance of the heater?

Here we will use $R = (V \times t)/Q$ where V = 28 V, Q = 5 C and t = 1 s. Thus:

$$R = \frac{V \times t}{Q} = \frac{28\,V \times 1s}{5\,C} = 5.6\,\Omega$$

___

Key point

Metals such as copper and silver are good conductors of electricity. Good conductors have low resistance whilst poor conductors have high resistance.

___

1.4.3 Ohm's law

The most basic DC circuit uses only two components; a cell (or battery) acting as a source of e.m.f., and a resistor (or *load*) through which a current is passing. These two components are connected together with

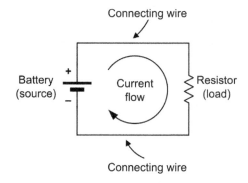

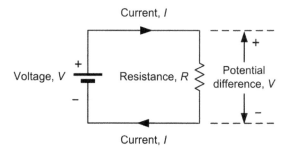

Figure 1.11   A simple DC circuit consisting of a battery (source) and resistor (load)

wire conductors in order to form a completely closed circuit as shown in Fig. 1.11.

For any conductor, the current flowing is directly proportional to the e.m.f. applied. The current flowing will also be dependent on the physical dimensions (length and cross-sectional area) and material of which the conductor is composed. The amount of current that will flow in a conductor when a given e.m.f. is applied is inversely proportional to its resistance. Resistance, therefore, may be thought of as an 'opposition to current flow'; the higher the resistance the lower the current that will flow (assuming that the applied e.m.f. remains constant).

Provided that temperature does not vary, the ratio of p.d. across the ends of a conductor to the current flowing in the conductor is a constant. This relationship is known as Ohm's law and it leads to the relationship:

$$\frac{V}{I} = \text{a constant} = R$$

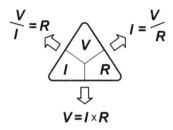

$$\frac{V}{I} = R \qquad I = \frac{V}{R}$$

$$V = I \times R$$

Figure 1.12   Relationship between V, I and R

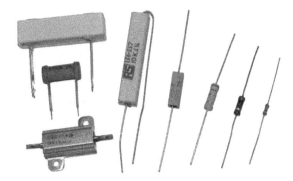

Figure 1.13   A selection of resistors with values ranging from 0.1 Ω to 10 MΩ and power ratings from 0.1 W to 15 W

where V is the potential difference (or voltage drop) in volts (V), I is the current in amps (A), and R is the resistance in ohms (Ω).

This important formula may be arranged to make V, I or R the subject, as follows:

$$V = I \times R \qquad I = \frac{V}{R} \qquad R = \frac{V}{I}$$

The triangle shown in Fig. 1.12 should help you remember these three important relationships. It is important to note that, when performing calculations of currents, voltages and resistances in practical circuits, it is seldom necessary to work with an accuracy of better than ±1% simply because component tolerances are invariably somewhat greater than this. Furthermore, in calculations involving Ohm's law, it is sometimes convenient to work in units of kΩ and mA (or MΩ and μA), in which case potential differences will be expressed directly in V.

Example 1.8

A current of 0.1 A flows in a 220 Ω resistor. What voltage drop (potential difference) will be developed across the resistor?

Here we must use $V = I \times R$ and ensure that we work in units of volts (V), amps (A), and ohms (Ω).

$$V = I \times R = 0.1\,\text{A} \times 220\,\Omega = 22\,\text{V}$$

Hence a p.d. of 22 V will be developed across the resistor.

Example 1.9

An 10 Ω resistor is connected to a 24 V battery. What current will flow in the resistor?

Here we must use $I = V/R$ (where $V = 24$ V and $R = 10\,\Omega$):

$$I = \frac{V}{R} = \frac{24\,\text{V}}{10\,\Omega} = 2.4\,\text{A}$$

Hence a current of 2.4 A will flow in the resistor.

Example 1.10

A voltage drop of 15 V appears across a resistor in which a current of 1 mA flows. What is the value of the resistance?

Here we must use $R = V/I$ (where $V = 15$ V and $I = 1$ mA $= 0.001$ A)

$$R = \frac{V}{I} = \frac{15\,\text{V}}{0.001\,\text{A}} = 15,000\,\Omega = 15\,\text{k}\Omega$$

Note that it is sometimes more convenient to work in units of mA and V, which will produce an answer directly in kΩ, i.e.

$$R = \frac{V}{I} = \frac{15\,\text{V}}{1\,\text{mA}} = 15\,\text{k}\Omega$$

Test your understanding 1.5

1. An aircraft cable has a resistance of 0.02 Ω per foot. If a 20 foot length of this cable carries a current of 0.5 A what voltage will be dropped across the ends of the cable?

2. A relay has a coil resistance of $400\,\Omega$. What current is required to operate the relay from a 24V supply?
3. A current of $125\,\mu A$ flows when an insulation tester delivers 500V to a circuit. What is the resistance of the circuit?

Key maintenance point

When replacing a resistor it is essential to ensure that the replacement component is correctly rated in terms of type, value, power and temperature. Resistors are prone to failure if their maximum power is exceeded and they should be de-rated when operated at a relatively high ambient temperature according to manufacturers' specifications.

1.4.4 Kirchhoff's laws

Used on its own, Ohm's law is insufficient to determine the magnitude of the voltages and currents present in complex circuits. For these circuits we need to make use of two further laws; Kirchhoff's current law and Kirchhoff's voltage law.

Kirchhoff's current law states that the algebraic sum of the currents present at a junction (or *node*) in a circuit is zero – see Fig. 1.14. Kirchhoff's voltage law states that the algebraic sum of the potential drops present in a closed network (or *mesh*) is zero – see Fig. 1.15.

1.4.5 Series and parallel circuits

Ohm's law and Kirchhoff's laws can be combined to solve more complex series–parallel circuits. Before we show you how this is done, however, it's important to understand what we mean by 'series' and 'parallel' circuit!

Figure 1.16 shows three circuits, each containing three resistors, R_1, R_2 and R_3:

- In Fig. 1.16(a), the three resistors are connected one after another. We refer to this as a *series circuit*. In other words the resistors are said to be connected *in series*. It's important to note that, in this arrangement, *the same current flows through each resistor*.
- In Fig. 1.16(b), the three resistors are all connected across one another. We refer to this as a *parallel circuit*. In other words the resistors are said to be connected *in parallel*. It's important to note that, in this arrangement, *the same voltage appears across each resistor*.

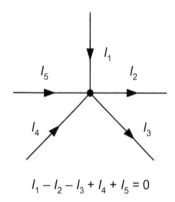

$$I_1 - I_2 - I_3 + I_4 + I_5 = 0$$

Convention:
Current flowing towards the junction is positive (+)
Current flowing away from the junction is negative (–)

Figure 1.14 Kirchhoff's current law

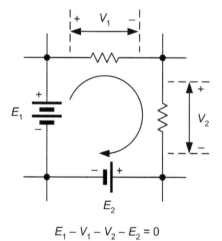

$$E_1 - V_1 - V_2 - E_2 = 0$$

Convention:
Move clockwise around the circuit starting with the positive terminal of the largest e.m.f.
Voltages acting in the same sense are positive (+)
Voltages acting in the opposite sense are negative (–)

Figure 1.15 Kirchhoff's voltage law

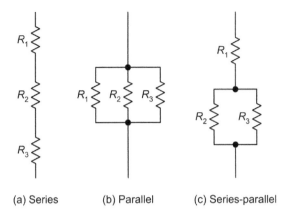

(a) Series   (b) Parallel   (c) Series-parallel

Figure 1.16 Series and parallel circuits

- In Fig. 1.16(c), we have shown a mixture of these two types of connection. Here we can say that R_1 is connected in series with the parallel combination of R_2 and R_3. In other words, R_2 and R_3 are connected *in parallel* and R_2 is connected *in series* with the parallel combination.

Example 1.11

Figure 1.17 shows a simple battery test circuit which is designed to draw a current of 2 A from a 24 V DC supply. The two test points, A and B, are designed for connecting a meter. Determine:

(a) the voltage that appears between terminals A and B (without the meter connected);
(b) the value of resistor, R.

We need to solve this problem in several small stages. Since we know that the circuit draws 2 A from the 24 V supply we know that this current must flow both through the 9 Ω resistor and through R (we hope that you have spotted that these two components are connected *in series*!).

We can determine the voltage drop across the 9 Ω resistor by applying Ohm's law:

$$V = I \times R = 2\,\text{A} \times 9\,\Omega = 18\,\text{V}$$

Next we can apply Kirchhoff's voltage law in order to determine the voltage drop, V, that appears

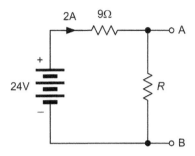

Figure 1.17 Battery test circuit – see Example 1.11

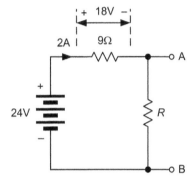

Figure 1.18 Using Ohm's law to find the voltage dropped across the 9 Ω resistor – see Example 1.11

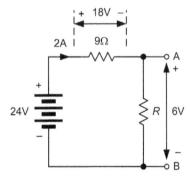

Figure 1.19 Using Kirchhoff's law to find the voltage that appears between terminals A and B – see Example 1.11

across R (i.e. the potential drop between terminals A and B):

$$+24\,\text{V} - 18\,\text{V} - V = 0$$

From which:

$$V = +6\,\text{V}$$

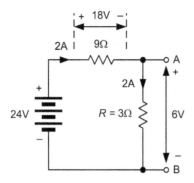

Figure 1.20 Using Ohm's law to find the value of R – see Example 1.11

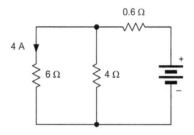

Figure 1.21 See Test your understanding 1.6

Finally, since we now know the voltage, V, and current, I, that flows in R, we can apply Ohm's law again in order to determine the value of R:

$$R = \frac{V}{I} = \frac{6\,V}{2\,A} = 3\Omega$$

Key point

Circuits with multiple branches can be solved using a combination of Kirchhoff's laws and Ohm's law.

Test your understanding 1.6

Determine the current and voltage present in each branch of the circuit shown in Fig. 1.21.

1.5 Power and energy

Power, P, is the rate at which energy is converted from one form to another and it is measured in *watts*

(W). The larger the amount of power the greater the amount of energy that is converted in a given period of time.

$$1\,\text{watt} = 1\,\text{joule per second}$$

or

$$\text{Power, } P = \frac{\text{Energy, } W}{\text{time, } t}$$

thus:

$$P = \frac{W}{t}\ \text{W}$$

Like all other forms of energy, electrical energy is the capacity to do work. Energy can be converted from one form to another. An electric fire, for example, converts electrical energy into heat. A filament lamp converts electrical energy into light, and so on. Energy can only be transferred when a difference in energy levels exists.

The unit of energy is the *joule* (J). Then, from the definition of power,

$$1\,\text{joule} = 1\,\text{watt} \times 1\,\text{second}$$

hence:

$$\text{Energy, } W = (\text{power, } P) \times (\text{time, } t)$$
$$\text{with units of (watts} \times \text{seconds)}$$

thus

$$W = Pt\ \text{J}$$

Thus joules are measured in **watt-seconds**. If the power was to be measured in kilowatts and the time in hours, then the unit of electrical energy would be the **kilowatt-hour** (kWh) (commonly knows as a **unit of electricity**). The electricity meter in your home records the amount of energy that you have used expressed in kilowatt-hours.

The power in an electrical circuit is equivalent to the product of voltage and current. Hence:

$$P = I \times V$$

where P is the power in watts (W), I is the current in amps (A), and V is the voltage in volts (V).

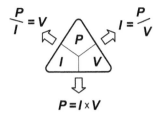

$$P = I \times V$$

Figure 1.22   Relationship between P, I and V

The formula may be arranged to make P, I or V the subject, as follows:

$$P = I \times V \qquad I = \frac{P}{V} \qquad V = \frac{P}{I}$$

The triangle shown in Fig. 1.22 should help you remember these three important relationships. It is important to note that, when performing calculations of power, current and voltages in practical circuits it is seldom necessary to work with an accuracy of better than ±1% simply because component tolerances are invariably somewhat greater than this.

Finally, we can combine the Ohm's law relationship that we met earlier with the formulae for power to arrive at two further useful relationships:

$$P = I \times V = I \times (IR) = I^2 R$$

$$P = I \times V = \left(\frac{V}{R}\right) \times V = \frac{V^2}{R}$$

Key point

Power is the rate of using energy and a power of one watt corresponds to energy being used at the rate of one joule per second.

Example 1.12

An auxiliary power unit (APU) provides an output of 1.5 kW for 20 minutes. How much energy has it supplied to the aircraft?

Here we will use $W = Pt$ where $P = 1.5\,\text{kW} = 1500\,\text{W}$ and $t = 20$ minutes $= 20 \times 60 = 1200$ s.

Thus:

$$W = 1500 \times 1200 = 1,800,000\,\text{J} = 1.8\,\text{MJ}$$

Example 1.13

The reservoir capacitor in a power supply is required to store 20 J of energy. How much power is required to store this energy in a time interval of 0.5 s?

Re-arranging $W = Pt$ to make P the subject gives:

$$P = \frac{W}{t}$$

We can now find P when $W = 20$ J and $t = 0.5$ s. Thus

$$P = \frac{W}{t} = \frac{20\,\text{J}}{0.5\,\text{s}} = 40\,\text{W}$$

Example 1.14

A main aircraft battery is used to start an engine. If the starter demands a current of 1000 A for 30 s and the battery voltage remains at 12 V during this period, determine the amount of electrical energy required to start the engine.

We need to solve this problem in two stages. First we need to find the power delivered to the starter from:

$$P = IV$$

where $I = 1000$ A and $V = 12$ V. Thus

$$P = 1000 \times 12 = 12,000\,\text{W} = 12\,\text{kW}$$

Next we need to find the energy from:

$$W = Pt$$

where $P = 12\,\text{kW}$ and $t = 30$ s. Thus

$$W = 12\,\text{kW} \times 30\,\text{s} = 360\,\text{kJ}$$

Example 1.15

A 24 V bench power unit is to be tested at its rated load current of 3 A. What value of load resistor is required and what should its minimum power rating be?

The value of load resistance required can be calculated using Ohm's law, as follows:

$$R = \frac{V}{I} = \frac{24\,\text{V}}{3\,\text{A}} = 8\,\Omega$$

The minimum power rating for the resistor will be given by:

$$P = IV = 3\,\text{A} \times 24\,\text{V} = 72\,\text{W}$$

Key point

Power is the rate at which energy is converted from one form to another. A power of one watt is equivalent to one joule of energy being converted every second.

Test your understanding 1.7

1. A window heater is rated at 150 W. How much energy is required to operate the heater for one hour?
2. A resistor is rated at 11 Ω, 2 W. What is the maximum current that should be allowed to flow in it?
3. An emergency locator transmitter is fitted with a lithium battery having a rated energy content of 18 kJ. How long can the unit be expected to operate if the transmitter consumes an input power of 2 W?

1.6 Electromagnetism and inductors

Magnetism is an effect created by moving the elementary atomic particles in certain materials such as iron, nickel and cobalt. Iron has outstanding magnetic properties, and materials that behave magnetically, in a similar manner to iron, are known as ferromagnetic materials. These materials experience forces that act on them when placed near a magnet.

A magnetic field of flux is the region in which the forces created by the magnet have influence. This field

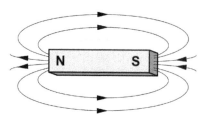

Figure 1.23   Field and flux directions for a bar magnet

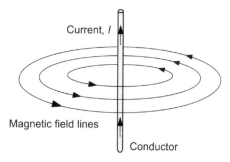

Figure 1.24   Field around a current-carrying conductor

surrounds a magnet in all directions, being strongest at the end extremities of the magnet, known as the poles. Magnetic fields are mapped by an arrangement of lines that give an indication of strength and direction of the flux as illustrated in Fig. 1.23.

Whenever an electric current flows in a conductor a magnetic field is set up around the conductor in the form of concentric circles, as shown in Fig. 1.24. The field is present along the whole length of the conductor and is strongest nearest to the conductor. Now like permanent magnets, this field also has direction. The direction of the magnetic field is dependent on the direction of the current passing through the conductor.

If we place a current-carrying conductor in a magnetic field, the conductor has a force exerted on it. Consider the arrangement shown in Fig. 1.25, in which a current-carrying conductor is placed between two magnetic poles. The direction of the current passing through it is into the page going away from us. Then by the right-hand screw rule, the direction of the magnetic field, created by the current in the conductor, is clockwise, as shown. We also know that the flux lines from the permanent magnet exit at a north

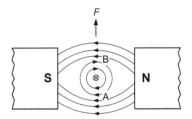

Figure 1.25 A current-carrying conductor in a magnetic field

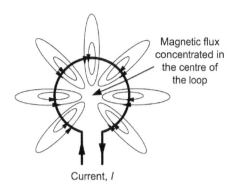

Figure 1.26 Magnetic field around a single turn loop

pole and enter at a south pole; in other words, they travel from north to south, as indicated by the direction arrows. The net effect of the coming together of these two magnetic force fields is that at position A, they both travel in the same direction and reinforce one another. While at position B, they travel in the opposite direction and tend to cancel one another. So with a stronger force field at position A and a weaker force at position B the conductor is forced upwards out of the magnetic field. If the direction of the current was reversed, i.e. if it was to travel towards us out of the page, then the direction of the magnetic field in the current-carrying conductor would be reversed and therefore so would the direction of motion of the conductor.

Key point

A magnetic field of flux is the region in which the forces created by the magnet have influence. This field surrounds a magnet in all directions and is concentrated at the north and south poles of the magnet.

Key point

Whenever an electric current flows in a conductor a magnetic field is set up in the space surrounding the conductor. The field spreads out around the conductor in concentric circles with the greatest density of magnetic flux nearest to the conductor.

The magnitude of the force acting on the conductor depends on the current flowing in the conductor, the length of the conductor in the field, and the strength of the magnetic flux (expressed in terms of its **flux density**). The size of the force will be given by the expression:

$$F = BIl$$

where F is the force in newtons (N), B is the flux density in tesla (T), I is the current (A) and l is the length (m).

Flux density is a term that merits a little more explanation. The total flux present in a magnetic field is a measure of the total magnetic intensity present in the field and it is measured in webers (Wb) and represented by the Greek symbol, Φ. The flux density, B, is simply the total flux, Φ, divided by the area over which the flux acts, A. Hence:

$$B = \frac{\Phi}{A}$$

where B is the flux density (T), Φ is the total flux present (Wb), and A is the area (m^2).

In order to increase the strength of the field, a conductor may be shaped into a loop (Fig. 1.26) or coiled to form a **solenoid** (Fig. 1.27).

Example 1.16

A flux density of 0.25 T is developed in free space over an area of 20 cm^2. Determine the total flux.

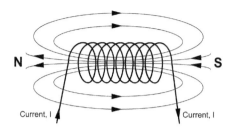

Figure 1.27 Magnetic field around a coil or solenoid

Re-arranging the formula $B = \dfrac{\Phi}{A}$ to make Φ the subject gives:

$$\Phi = BA$$

thus:

$$\Phi = 0.25\,\text{T} \times 20\,\text{cm}^2 = 0.25\,\text{T} \times 0.002\,\text{m}^2$$
$$= 0.0005\,\text{Wb or } 0.5\,\text{mWb}$$

Key point

If we place a current-carrying conductor in a magnetic field, the conductor has a force exerted on it. If the conductor is free to move this force will produce motion.

Key point

Flux density is found by dividing the total flux present by the area over which the flux acts.

1.6.1 Electromagnetic induction

The way in which electricity is generated in a conductor may be viewed as being the exact opposite to that which produces the motor force. In order to generate electricity we require movement in to get electricity out. In fact we need the same components to generate electricity as those needed for the electric motor,

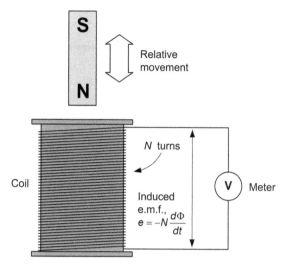

Figure 1.28 Demonstration of electromagnetic induction

namely a closed conductor, a magnetic field and movement.

Whenever relative motion occurs between a magnetic field and a conductor acting at right angles to the field, an e.m.f. is induced, or generated in the conductor. The manner in which this e.m.f. is generated is based on the principle of electromagnetic induction.

Consider Fig. 1.28, which shows relative movement between a magnet and a closed coil of wire. An e.m.f. will be induced in the coil whenever the magnet is moved in or out of the coil (or the magnet is held stationary and the coil moved). The magnitude of the induced e.m.f., e, depends on the number of turns, N, and the rate at which the flux changes in the coil, $d\Phi/dt$. Note that this last expression is simply a mathematical way of expressing the *rate of change of flux with respect to time*.

The e.m.f., e, is given by the relationship:

$$e = -N\frac{d\Phi}{dt}$$

where N is the number of turns and $d\Phi/dt$ is the rate of change of flux. The minus sign indicates that the polarity of the generated e.m.f. *opposes* the change.

Now the number of turns N is directly related to the length of the conductor, l, moving through a magnetic field with flux density, B. Also, the velocity with which

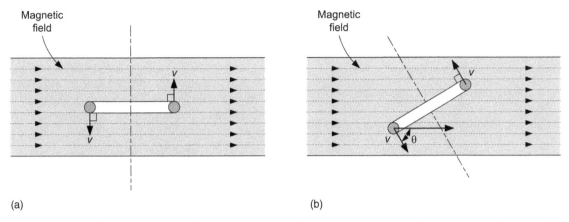

Figure 1.29 Cutting lines of flux and the e.m.f. generated: (a) cutting lines of flux at 90°, $e = Blv$; (b) cutting lines of flux at θ, $e = Blv \sin \theta$

the conductor moves through the field determines the rate at which the flux changes in the coil as it cuts the flux field. Thus the magnitude of the induced (generated) e.m.f., e, is proportional to the flux density, length of conductor and relative velocity between the field and the conductor.

The magnitude of the induced e.m.f. also depends on:

- the length of the conductor l in m
- the strength of the magnetic field, B, in tesla (T)
- the velocity of the conductor, v, in m/s.

Hence:

$$e \propto Blv$$

where B is the strength of the magnetic field (T), l is the length of the conductor in the field (m), and v is the velocity of the conductor (m/s).

Now you are probably wondering why the above relationship has the proportionality sign. In order to generate an e.m.f. the conductor must cut the lines of magnetic flux. If the conductor cuts the lines of flux at right angles (Fig. 1.29(a)) then the maximum e.m.f. is generated; cutting them at any other angle θ (Fig. 1.29(b)), reduces this value until $\theta = 0°$, at which point the lines of flux are not being cut at all and no e.m.f. is induced or generated in the conductor. So the magnitude of the induced e.m.f. is also dependent on $\sin \theta$. So we may write:

$$e = Blv \sin \theta$$

1.6.2 Faraday's and Lenz's laws

When a magnetic flux through a coil is made to vary, an e.m.f. is induced. The magnitude of this e.m.f. is proportional to the rate of change of magnetic flux. What this law is saying in effect is that relative movement between the magnetic flux and the conductor is essential to generate an e.m.f. The voltmeter shown in Fig. 1.29 indicates the induced (generated) e.m.f. and if the direction of motion changes the polarity of the induced e.m.f. in the conductor changes. Faraday's law also tells us that the magnitude of the induced e.m.f. is dependent on the relative velocity with which the conductor cuts the lines of magnetic flux.

Lenz's law states that the current induced in a conductor opposes the changing field that produces it. It is therefore important to remember that the induced current *always* acts in such a direction so as to oppose the change in flux. This is the reason for the minus sign in the formula that we met earlier:

$$e = -N \frac{d\Phi}{dt}$$

Key point

The induced e.m.f. tends to oppose any change of current and because of this we often refer to it as a *back e.m.f.*

Example 1.17

A closed conductor of length 15 cm cuts the magnetic flux field of 1.25 T with a velocity of 25 m/s. Determine the induced e.m.f. when:

(a) the angle between the conductor and field lines is 60°
(b) the angle between the conductor and field lines is 90°.

(a) The induced e.m.f. is found using $e = Blv \sin\theta$, hence:

$$e = 1.25 \times 0.15 \times 25 \times \sin 60°$$
$$= 4.688 \times 0.866 = 4.06\,\text{V}$$

(b) The maximum induced e.m.f. occurs when the lines of flux are cut at 90°. In this case $e = Blv \sin\theta = Blv$ (recall that sin 90°= 1), hence:

$$e = 1.25 \times 0.15 \times 25 = 4.69\,\text{V}$$

1.6.3 Self-inductance and mutual inductance

We have already shown how an induced e.m.f. (i.e. a **back e.m.f.**) is produced by a flux change in an inductor. The back e.m.f. is proportional to the rate of change of current (from Lenz's law), as illustrated in Fig. 1.30.

This effect is called *self-inductance* (or just *inductance*) which has the symbol L. Self-inductance is measured in henries (H) and is calculated from:

$$e = -L\frac{di}{dt}$$

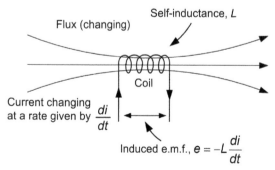

NB: Induced e.m.f. opposes current change

Figure 1.30   Self-inductance

where L is the self-inductance, di/dt is the rate of change of current and the minus sign indicates that the polarity of the generated e.m.f. *opposes* the change (you might like to compare this relationship with the one shown earlier for electromagnetic induction).

The unit of inductance is the henry (H) and a coil is said to have an inductance of 1 H if a voltage of 1 V is induced across it when a current changing at the rate of 1 A/s is flowing in it.

Example 1.18

A coil has a self-inductance of 15 mH and is subject to a current that changes at a rate of 450 A/s. What e.m.f. is produced?

Now $e = -L\,di/dt$ and hence:

$$e = -15 \times 10^{-3} \times 450 = -6.75\,\text{V}$$

Note the minus sign. This reminds us that a *back e.m.f.* of 6.75 V is induced.

Example 1.19

A current increases at a uniform rate from 2 A to 6 A in a time of 250 ms. If this current is applied to an inductor determine the value of inductance if a back e.m.f. of 15 V is produced across its terminals.

Now $e = -L\,di/dt$ and hence $L = -e\,di/dt$

Thus

$$L = -(-15)\times\frac{250\times 10^{-3}}{(6-2)} = 15 \times 62.5 \times 10^{-3}$$
$$= 937.5 \times 10^{-3}$$
$$= 0.94\,\text{H}$$

Finally, when two inductors are placed close to one another, the flux generated when a changing current flows in the first inductor will cut through the other inductor (see Fig. 1.31). This changing flux will, in turn, induce a current in the second inductor. This effect is known as **mutual inductance** and it occurs whenever two inductors are inductively coupled. This is the principle of a very useful component, the **transformer**, which we shall meet later.

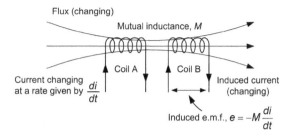

Figure 1.31   Mutual inductance

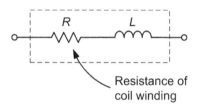

Figure 1.32   A real inductor has resistance as well as inductance

Figure 1.33   A selection of inductors with values ranging from 100 nH to 4 H and current ratings ranging from 0.1 A to 10 A

Key point

The current induced in a conductor always opposes the change that produces it.

1.6.4 Inductors

Inductors provide us with a means of storing electrical energy in the form of a magnetic field. Typical applications include chokes, filters, and frequency selective circuits. The electrical characteristics of an inductor are determined by a number of factors including the material of the core (if any), the number of turns, and the physical dimensions of the coil.

In practice every coil comprises both inductance and resistance and the circuit of Fig. 1.32 shows these as two discrete components. In reality the inductance, L, and resistance, R, are both distributed throughout the component but it is convenient to treat the inductance and resistance as separate components in the analysis of the circuit.

Key point

An e.m.f. is produced when the magnetic flux passing through an inductor changes.

Test your understanding 1.8

1. A 1.5 m length of wire moves perpendicular to a magnetic flux field of 0.75 T. Determine the e.m.f. that will be induced across the ends of the wire if it moves at 10 m/s.
2. An e.m.f. of 30 V is developed across the terminals of an inductor when the current flowing in it changes from zero to 10 A in half a second. What is the value of inductance?

1.7 Alternating current and transformers

Direct currents are currents which, even though their magnitude may vary, essentially flow only in one direction. In other words, direct currents are unidirectional. Alternating currents, on the other hand, are bi-directional and continuously reversing their direction of flow, as shown in Fig. 1.34.

A graph showing the variation of voltage or current present in a circuit is known as a **waveform**.

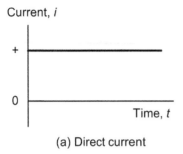

(a) Direct current

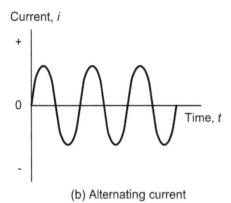

(b) Alternating current

Figure 1.34 Comparison of direct and alternating current

There are many common types of waveform encountered in electrical circuits including sine (or sinusoidal), square, triangle, ramp or sawtooth (which may be either positive or negative), and pulse. Complex waveforms like speech or music usually comprise many components at different frequencies. Pulse waveforms found in digital circuits are often categorized as either repetitive or non-repetitive (the former comprises a pattern of pulses which regularly repeats whilst the latter comprises pulses which constitute a unique event). Several of the most common waveform types are shown in Fig. 1.35.

1.7.1 Frequency and periodic time

The **frequency** of a repetitive waveform is the number of cycles of the waveform that occur in unit time. Frequency is expressed in **hertz (Hz)**. A frequency of 1 Hz is equivalent to one cycle per second. Hence, if an aircraft supply has a frequency of 400 Hz, 400 cycles of the supply will occur in every second (Fig. 1.36).

The **periodic time** (or **period**) of a waveform is the time taken for one complete cycle of the wave

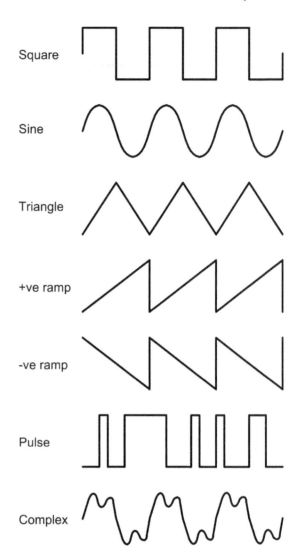

Figure 1.35 Various waveforms

(see Fig. 1.37). The relationship between periodic time and frequency is thus:

$$t = \frac{1}{f} \quad \text{or} \quad f = \frac{1}{t}$$

where t is the periodic time (in seconds) and f is the frequency (in Hz).

Example 1.20

An aircraft generator operates at a frequency of 400 Hz. What is the periodic time of the voltage generated?

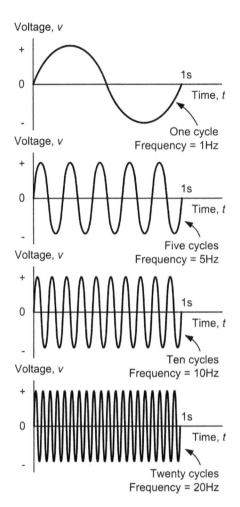

Figure 1.36 Waveforms with different frequencies

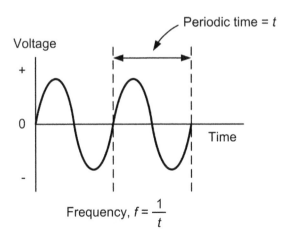

Figure 1.37 Periodic time

Now

$$t = \frac{1}{f} = \frac{1}{400\,\text{Hz}} = 0.0025\,\text{s} = 2.5\,\text{ms}$$

Hence the voltage will have a periodic time of 2.5 ms.

Example 1.21

A bench AC supply has a periodic time of 20 ms. What is the frequency of the supply?

Now

$$f = \frac{1}{t} = \frac{1}{20\,\text{ms}} = \frac{1}{0.02\,\text{s}} = 50\,\text{Hz}$$

Hence the supply has a frequency of 50 Hz.

1.7.2 Average, peak, peak–peak, and r.m.s. values

The **average value** of an alternating current which swings symmetrically above and below zero will obviously be zero when measured over a long period of time. Hence average values of currents and voltages are invariably taken over one complete half cycle (either positive or negative) rather than over one complete full cycle (which would result in an average value of zero).

The **peak value** (or **maximum value** or **amplitude**) of a waveform is a measure of the extent of its voltage or current excursion from the resting value (usually zero). The peak-to-peak value for a wave which is symmetrical about its resting value is twice its peak value.

The **root mean square (r.m.s.)** or **effective value** of an alternating voltage or current is the value which would produce the same heat energy in a resistor as a direct voltage or current of the same magnitude. Since the r.m.s. value of a waveform is very much dependent upon its shape, values are only meaningful when dealing with a waveform of known shape. Where the shape of a waveform is not specified, r.m.s. values are normally assumed to refer to sinusoidal conditions.

For a given waveform, a set of fixed relationships exist between average, peak, peak–peak, and r.m.s. values. The required multiplying factors for sinusoidal voltages and currents are summarized in the table shown below.

		Wanted quantity			
		average	peak	peak–peak	r.m.s.
Given quantity	average	1	1.57	3.14	1.11
	peak	0.636	1	2	0.707
	peak–peak	0.318	0.5	1	0.353
	r.m.s.	0.9	1.414	2.828	1

From the table we can conclude that, for example:

$$V_{av} = 0.636 \times V_{pk}, V_{pk-pk} = 2 \times V_{pk},$$
$$\text{and } V_{r.m.s.} = 0.707 \times V_{pk}$$

Similar relationships apply to the corresponding alternating currents, thus:

$$I_{av} = 0.636 \times I_{pk}, I_{pk-pk} = 2 \times I_{peak},$$
$$\text{and } I_{r.m.s.} = 0.707 \times I_{pk}$$

Example 1.22

A generator produces an r.m.s. sine wave output of 110 V. What is the peak value of the voltage?

Now

$$V_{pk} = 1.414 \times V_{r.m.s.} = 1.414 \times 110\,V = 155.5\,V$$

Hence the voltage has a peak value of 155.5 V.

Example 1.23

A sinusoidal current of 40 A peak–peak flows in a circuit. What is the r.m.s. value of the current?

Now

$$I_{r.m.s} = 0.353 \times I_{pk-pk} = 0.353 \times 40\,A = 14.12\,A$$

Hence the current has an r.m.s. value of 14.12 A.

Key point

The root mean square (r.m.s.) value of an alternating voltage will produce the same amount of heat in a resistor as a direct voltage of the same magnitude.

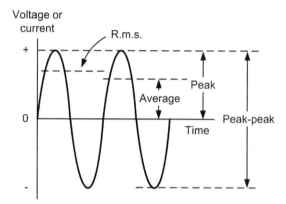

Figure 1.38   Average, r.m.s., peak and peak–peak values of a sine wave

1.7.3 Three-phase supplies

The most simple method of distributing an AC supply is a system that uses two wires. In fact, this is how AC is distributed in your home (the third wire present is simply an earth connection for any appliances that may require it for safety reasons). In many practical applications, including aircraft, it can be advantageous to use a **multiphase** supply rather than a **single-phase** supply (here the word *phase* simply refers to an AC voltage source that may not be rising and falling at the same time as other voltage sources that may be present).

The most common system uses three separate voltage sources (and three wires) and is known as **three-phase**. The voltages produced by the three sources are spaced equally in time such that the angle between them is 120° (or 360°/3). The waveforms for a three-phase supply are shown in Fig. 1.37 (note that each is a sine wave and all three sine waves have the same frequency and periodic time). We shall be returning to this topic in greater detail in Chapter 4 when we introduce three-phase power generation.

1.7.4 Reactance

In an AC circuit, **reactance**, like resistance, is simply the ratio of applied voltage to the current flowing. Thus:

$$I = \frac{V}{X}$$

where X is the reactance in ohms (Ω), V is the alternating potential difference in volts (V) and I is the alternating current in amps (A).

In the case of **capacitive reactance** (i.e. the reactance of a capacitor) we use the suffix, C, so that the reactance equation becomes:

$$X_C = \frac{V_C}{I_C}$$

Similarly, in the case of **inductive reactance** (i.e. the reactance of an inductor) we use the suffix, L, so that the reactance equation becomes:

$$X_L = \frac{V_L}{I_L}$$

The voltage and current in a circuit containing pure reactance (either capacitive or inductive) will be out of phase by 90°. In the case of a circuit containing pure capacitance the current will lead the voltage by 90° (alternatively we can say that the voltage lags the current by 90°). This relationship is illustrated by the waveforms shown in Fig. 1.40.

In the case of a circuit containing pure inductance the voltage will lead the current by 90° (alternatively we can also say that the current lags the voltage by 90°). This relationship is illustrated by the waveforms shown in Fig. 1.41.

Key point

A good way of remembering leading and lagging phase relationships is to recall the word *CIVIL*, as shown in Fig. 1.42. Note that, in the case of a circuit containing pure capacitance (*C*) the current (*I*) will lead the voltage (*V*) by 90° whilst in the case of a circuit containing pure inductance (*L*) the voltage (*V*) will lead the current (*I*) by 90°.

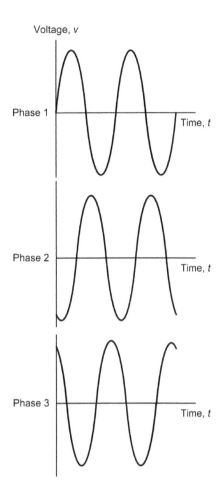

Figure 1.39 Waveforms for a three-phase AC supply

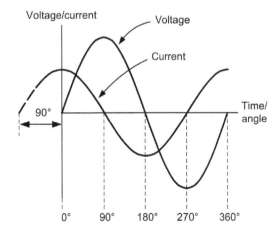

Figure 1.40 Voltage and current waveforms for a pure capacitor (the current leads the voltage by 90°)

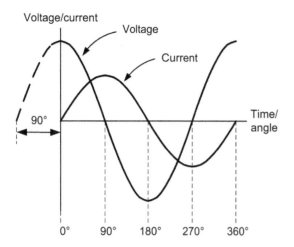

Figure 1.41 Voltage and current waveforms for a pure capacitor (the voltage leads the current by 90°)

In a capacitor, C, current, I, leads voltage, V

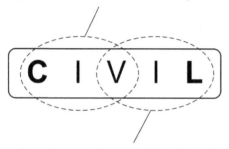

In an inductor, L, current, I, leads voltage, V

Figure 1.42 Using CIVIL to determine phase relationships in circuits containing capacitance and inductance

The reactance of an inductor (**inductive reactance**) is directly proportional to the frequency of the applied alternating current and can be determined from the following formula:

$$X_L = 2\pi fL$$

where X_L is the reactance in Ω, f is the frequency in Hz, and L is the inductance in H.

Since inductive reactance is directly proportional to frequency ($X_L \propto f$), the graph of inductive reactance plotted against frequency takes the form of a straight line (see Fig. 1.43).

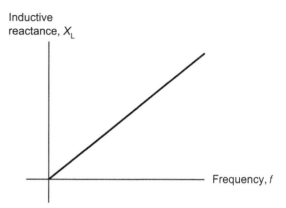

Figure 1.43 Variation of inductive reactance, X_L, with frequency, f

Example 1.24

Determine the reactance of a 0.1 H inductor at (a) 100 Hz and (b) 10 kHz.

(a) At 100 Hz, $X_L = 2\pi \times 100 \times 0.1 = 62.8\,\Omega$
(b) At 10 kHz, $X_L = 2\pi \times 10{,}000 \times 0.1 = 6280\,\Omega = 6.28\,k\Omega$.

The reactance of a capacitor (**capacitive reactance**) is inversely proportional to the frequency of the applied alternating current and can be determined from the following formula:

$$X_C = \frac{1}{2\pi fC}$$

where X_C is the reactance in Ω, f is the frequency in Hz, and C is the capacitance in F.

Since capacitive reactance is inversely proportional to frequency ($X_L \propto 1/f$), the graph of inductive reactance plotted against frequency takes the form of a rectangular hyperbola (see Fig. 1.44).

Example 1.25

Determine the reactance of a 1 μF capacitor at (a) 100 Hz and (b) 10 kHz.

(a) At 100 Hz,

$$X_C = \frac{1}{2\pi fC} = \frac{1}{2\pi \times 100 \times 1 \times 10^{-6}}$$
$$= \frac{0.159}{10^{-4}} = 1.59\,k\Omega$$

Capacitive
reactance, X_C

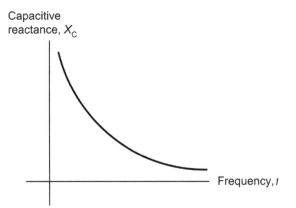

Figure   1.44 Variation   of   capacitive
reactance, X_C, with frequency, f

(b) At 10 kHz,

$$X_C = \frac{1}{2\pi fC} = \frac{1}{2\pi \times 10 \times 10^3 \times 1 \times 10^{-6}}$$

$$= 0.159 \times 10^2 = 15.9\,\Omega$$

Key point

When alternating voltages are applied to capacitors or inductors the magnitude of the current flowing will depend upon the value of capacitance or inductance and on the frequency of the voltage. In effect, capacitors and inductors oppose the flow of current in much the same way as a resistor. The important difference being that the effective resistance (or *reactance*) of the component varies with frequency (unlike the case of a conventional resistor where the magnitude of the current does not change with frequency).

1.7.5 Impedance

Circuits that contain a mixture of both resistance and reactance (either capacitive reactance or inductive reactance or both) are said to exhibit **impedance**. Impedance, like resistance and reactance, is simply the ratio of applied voltage to the current flowing. Thus:

$$Z = \frac{V}{I}$$

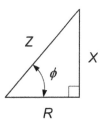

Figure 1.45 The impedance triangle

where Z is the impedance in ohms (Ω), V is the alternating potential difference in volts (V) and I is the alternating current in amps (A).

Because the voltage and current in a pure reactance are at 90° to one another (we say that they are in **quadrature**) we can't simply add up the resistance and reactance present in a circuit in order to find its impedance. Instead, we can use the **impedance triangle** shown in Fig. 1.45. The impedance triangle takes into account the 90° phase angle and from it we can infer that the impedance of a series circuit (R in series with X) is given by:

$$Z = \sqrt{R^2 + X^2}$$

where Z is the impedance (in Ω), X is the reactance, either capacitive or inductive (expressed in Ω), and R is the resistance (also in Ω).

We shall be explaining the significance of the **phase angle**, ϕ, later on. For now you simply need to be aware that ϕ is the angle between the impedance, Z, and the resistance, R. Later on we shall obtain some useful information from the fact that:

$$\sin\phi = \frac{opposite}{hypotenuse} = \frac{X}{Z} \text{ from which}$$

$$\phi = \arcsin\left(\frac{X}{Z}\right)$$

$$\cos\phi = \frac{adjacent}{hypotenuse} = \frac{R}{Z} \text{ from which}$$

$$\phi = \arccos\left(\frac{R}{Z}\right)$$

and

$$\tan\phi = \frac{opposite}{adjacent} = \frac{X}{R} \text{ from which}$$

$$\phi = \arctan\left(\frac{X}{R}\right)$$

Key point

Resistance and reactance combine together to make *impedance*. In other words, impedance is the *resultant* of combining resistance and reactance in the impedance triangle. Because of the *quadrature* relationship between voltage and current in a pure capacitor or inductor, the angle between resistance and reactance in the impedance triangle is always 90°.

Example 1.26

A resistor of 30 Ω is connected in series with a capacitive reactance of 40 Ω. Determine the impedance of the circuit and the current flowing when the circuit is connected to a 115 V supply.

First we must find the impedance of the C–R series circuit:

$$Z = \sqrt{R^2 + X^2} = \sqrt{30^2 + 40^2}$$
$$= \sqrt{2,500} = 50\,\Omega$$

The current taken from the supply can now be found:

$$I = \frac{V}{Z} = \frac{115}{50} = 2.3\,\text{A}$$

Example 1.27

A coil is connected to a 50 V AC supply at 400 Hz. If the current supplied to the coil is 200 mA and the coil has a resistance of 60 Ω, determine the value of inductance.

Like most practical forms of inductor, the coil in this example has both resistance *and* reactance (see Fig. 1.31). We can find the impedance of the coil from:

$$Z = \frac{V}{I} = \frac{50}{0.2} = 250\,\Omega$$

Since $Z = \sqrt{R^2 + X^2}$,

$$Z^2 = R^2 + X^2 \quad \text{and} \quad X^2 = Z^2 - R^2$$

from which:

$$X^2 = Z^2 - R^2 = 250^2 - 60^2$$
$$= 62,500 - 3,600$$
$$= 58,900$$

Thus

$$X = \sqrt{58,900} = 243\,\Omega$$

Now since $X_L = 2\pi f L$,

$$L = \frac{X_L}{2\pi f} = \frac{243}{6.28 \times 400}$$
$$= \frac{243}{2,512} = 0.097\,\text{H}$$

Hence

$$L = 97\,\text{mH}$$

1.7.6 Resonance

It is important to note that a special case occurs when $X_C = X_L$ in which case the two equal but opposite reactances effectively cancel each other out. The result of this is that the circuit behaves as if only resistance, R, is present (in other words, the impedance of the circuit, $Z = R$). In this condition the circuit is said to be **resonant**. The frequency at which resonance occurs is given by:

$$X_C = X_L$$

thus

$$\frac{1}{2\pi f C} = 2\pi f L$$

from which

$$f^2 = \frac{1}{4\pi^2 LC}$$

and thus

$$f = \frac{1}{2\pi \sqrt{LC}}$$

where f is the resonant frequency (in Hz), L is the inductance (in H) and C is the capacitance (in F).

Example 1.28

A series circuit comprises an inductor of 10 mH, a resistor of 50 Ω and a capacitor of 40 nF. Determine the frequency at which this circuit is resonant and the current that will flow in it when it is connected to a 20 V AC supply at the resonant frequency.

Using:

$$f = \frac{1}{2\pi\sqrt{LC}}$$

where $L = 10 \times 10^{-3}$ H and $C = 40 \times 10^{-9}$ F gives:

$$\begin{aligned} f &= \frac{1}{6.28\sqrt{10 \times 10^{-3} \times 40 \times 10^{-9}}} \\ &= \frac{0.159}{\sqrt{4 \times 10^{-10}}} = \frac{0.159}{2 \times 10^{-5}} \\ &= \frac{0.159}{2 \times 10^{-5}} = 7,950 = 7.95 \text{ kHz} \end{aligned}$$

At the resonant frequency the circuit will behave as a pure resistance (recall that the two reactances will be equal but opposite) and thus the supply current at resonance can be determined from:

$$I = \frac{V}{Z} = \frac{V}{R} = \frac{20}{50} = 0.4 \text{ A}$$

1.7.7 Power factor

The power factor in an AC circuit containing resistance and reactance is simply the ratio of true power to apparent power. Hence:

$$\text{Power factor} = \frac{\text{true power}}{\text{apparent power}}$$

The true power in an AC circuit is the power that is actually dissipated as heat in the resistive component. Thus:

$$\text{True power} = I^2 R$$

where I is r.m.s. current and R is the resistance. True power is measured in watts (W).

The apparent power in an AC circuit is the power that is apparently consumed by the circuit and is the product of the supply current and supply voltage (which may not be in phase). Note that, unless the voltage and current are in phase (i.e. $\phi = 0°$), the apparent power *will not* be the same as the power which is actually dissipated as heat. Hence:

$$\text{Apparent power} = IV$$

where I is r.m.s. current and V is the supply voltage. To distinguish apparent power from true power, apparent power is measured in volt-amperes (VA).

Now since $V = IZ$ we can re-arrange the apparent power equation as follows:

$$\text{Apparent power} = IV = I \times IZ = I^2 Z$$

Now returning to our original equation:

$$\text{Power factor} = \frac{\text{true power}}{\text{apparent power}}$$

$$= \frac{I^2 R}{IV} = \frac{I^2 R}{I \times IZ} = \frac{I^2 R}{I^2 Z} = \frac{R}{Z}$$

From the *impedance triangle* shown earlier in Fig. 1.45, we can infer that:

$$\text{Power factor} = \frac{R}{Z} = \cos\phi$$

Example 1.29

An AC load has a power factor of 0.8. Determine the true power dissipated in the load if it consumes a current of 2 A at 110 V.

Now since:

$$\text{Power factor} = \cos\phi = \frac{\text{true power}}{\text{apparent power}}$$

True power = power factor × apparent power = power factor × VI. Thus:

$$\text{True power} = 0.8 \times 2 \times 110 = 176 \text{ W}$$

Example 1.30

A coil having an inductance of 150 mH and resistance of 250 Ω is connected to a 115 V 400 Hz AC supply. Determine:

(a) the power factor of the coil
(b) the current taken from the supply
(c) the power dissipated as heat in the coil.

(a) First we must find the reactance of the inductor, X_L, and the impedance, Z, of the coil at 400 Hz.

$$X_L = 2\pi \times 400 \times 150 \times 10^{-3} = 376 \, \Omega$$

and

$$Z = \sqrt{R^2 + X_L^2} = \sqrt{250^2 + 376^2} = 452 \, \Omega$$

We can now determine the power factor from:

$$\text{Power factor} = \frac{R}{Z} = \frac{250}{452} = 0.553$$

(b) The current taken from the supply can be determined from:

$$I = \frac{V}{Z} = \frac{115}{452} = 0.254 \, \text{A}$$

(c) The power dissipated as heat can be found from:

$$\text{True power} = \text{power factor} \times VI$$
$$= 0.553 \times 115 \times 0.254$$
$$= 16.15 \, \text{W}$$

Key point

In an AC circuit the power factor is the ratio of true power to apparent power. The power factor is also the cosine of the phase angle between the supply current and supply voltage.

Test your understanding 1.9

1. Determine the reactance of a 60 mH inductor at (a) 20 Hz and (b) 4 kHz.
2. Determine the reactance of a 220 nF capacitor at (a) 400 Hz and (b) 20 kHz.

3. A 0.5 μF capacitor is connected to a 110 V 400 Hz supply. Determine the current flowing in the capacitor.
4. A resistor of 120 Ω is connected in series with a capacitive reactance of 160 Ω. Determine the impedance of the circuit and the current flowing when the circuit is connected to a 200 V AC supply.
5. A capacitor of 2 μF is connected in series with a 100 Ω resistor across a 24 V 400 Hz AC supply. Determine the current that will be supplied to the circuit and the voltage that will appear across each component.
6. An 80 mH coil has a resistance of 10 Ω. Calculate the current flowing when the coil is connected to a 250 V 50 Hz supply.
7. Determine the phase angle and power factor for Question 6.
8. An AC load has a power factor of 0.6. If the current supplied to the load is 5 A and the supply voltage is 110 V determine the true power dissipated by the load.
9. An AC load comprises a 110 Ω resistor connected in parallel with a 20 μF capacitor. If the load is connected to a 220 V 50 Hz supply, determine the apparent power supplied to the load and its power factor.
10. A filter consists of a 2 μF capacitor connected in series with a 50 mH inductor. At what frequency will the filter be resonant?

1.7.8 Complex notation

Complex notation allows us to represent electrical quantities that have both *magnitude* and *direction* (you may already know that in other contexts we call these *vectors* but in electrical circuits we refer to them as *phasors*). The *magnitude* is simply the amount of resistance, reactance, voltage or current, etc. In order to specify the *direction* of the quantity, we use an *operator* to represent the phase shift relative to the *reference* quantity (usually current for a series circuit and voltage for a parallel circuit). We call this 'operator-j'.

Every complex number consists of a *real part* and an *imaginary part*. In an electrical context, the *real part* is that part of the complex quantity that is in-phase with the reference quantity (current for series circuits and voltage for parallel circuits). The *imaginary part* (denoted by the j-operator) is that part of the complex quantity that is at 90° to the reference quantity. You can think of the j-operator as a device that allows us to indicate a rotation or *phase shift* of 90°. A phase shift of +90° is represented by +j whilst a phase shift of −90° is denoted by −j, as illustrated in Fig. 1.46.

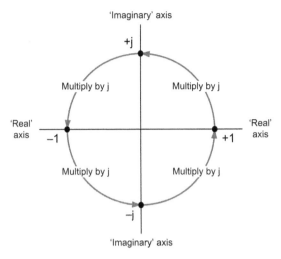

Figure 1.46 Phase rotation and the j-operator

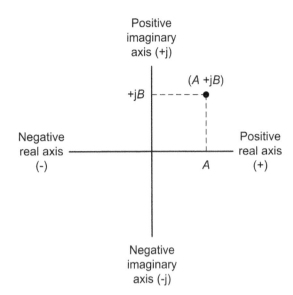

Figure 1.47 The Argand diagram

The j-operator has a value which is equal to $\sqrt{-1}$. We can therefore conclude that $j^2 = \sqrt{-1} \times \sqrt{-1} = -1$, $j^3 = \sqrt{-1} \times \sqrt{-1} \times \sqrt{-1} = -1 \times \sqrt{-1} = -j$ and $j^4 = \sqrt{-1} \times \sqrt{-1} \times \sqrt{-1} \times \sqrt{-1} = -1 \times -1 = +1$. Fig. 1.46 shows that the j-operator can be used to *rotate* a phasor where each successive multiplication by j has the effect of rotating the phasor through a further 90°. Note that because j is an operator and not a constant or variable, it should always precede the quantity to which it relates. Thus, for example, a positive 10 Ω reactance would be referred to as j10, *not* 10j. The sign that appears before the j-term (i.e. + or –) can be important because it tells us whether the phase shift is positive (leading) or negative (lagging).

Key point

Complex quantities have both real and imaginary parts. These two components are perpendicular to one another so that the imaginary part has no effect along the line of action of the real part, and vice versa.

Key point

The j-operator has the effect of rotating the direction of a voltage or current through an angle of 90°. The j-operator has a value equal to $\sqrt{-1}$.

The Argand diagram (see Fig. 1.47) provides a useful method of visualising complex quantities and allowing us to solve problems graphically. In common with any ordinary 'x-y' graph, the Argand diagram has two sets of axes at right angles, as shown in Fig. 1.47. The horizontal axis is known as the *real axis* whilst the *vertical axis* is known as the imaginary axis (don't panic – the imaginary axis isn't really imaginary we simply use the term to indicate that we are using this axis to plot values that are multiples of the j-operator).

In Fig. 1.47 we have plotted an impedance which has a real (resistive) part, A, and an imaginary (reactive) part, B. We can refer to this impedance as $(A + jB)$. The brackets help us to remember that the impedance is made up from two components; one imposing no phase shift whilst the other changes phase by 90°. Examples 1.31 and 1.32 show you how this works.

Example 1.31

A resistance of 4 Ω is connected in series with a capacitive reactance of 3 Ω.

(a) Sketch a circuit and express this impedance in complex form
(b) Plot the impedance on an Argand diagram
(c) Sketch the impedance triangle for the circuit.

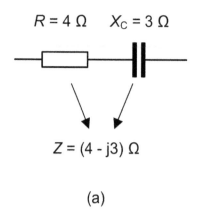

$R = 4\ \Omega \qquad X_C = 3\ \Omega$

$Z = (4 - j3)\ \Omega$

(a)

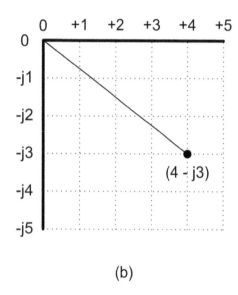

(b)

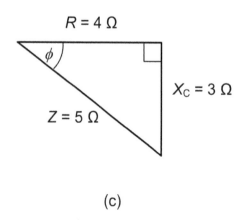

$R = 4\ \Omega$

$X_C = 3\ \Omega$

$Z = 5\ \Omega$

(c)

Figure 1.48 See Example 1.31

Example 1.32

A resistance of 4 Ω is connected in series with an inductive reactance of 3 Ω.

(a) Sketch a circuit and express the impedance in complex form
(b) Plot the impedance on an Argand diagram
(c) Sketch the impedance triangle for the circuit.

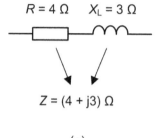

$R = 4\ \Omega \qquad X_L = 3\ \Omega$

$Z = (4 + j3)\ \Omega$

(a)

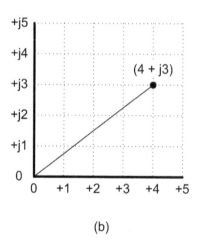

(b)

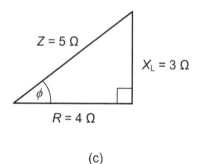

$Z = 5\ \Omega$

$X_L = 3\ \Omega$

$R = 4\ \Omega$

(c)

Figure 1.49 See Example 1.32

As you've just seen, the j-operator and the Argand diagram provide us with a useful way of representing impedances. Any complex impedance can be represented by the relationship:

$$Z = R \pm jX$$

where Z represents impedance, R represents resistance and X represents reactance.

All three quantities being measured in ohms.

The $\pm$ j term simply allows us to indicate whether the reactance is attributable to inductance, in which case the j term is positive (i.e. $+$ j) or whether it is attributable to capacitance, in which case the j term is negative (i.e. $-$ j).

Consider, for example, the following impedances:

1. $Z_1 = 20 + j\,10$ this impedance comprises a resistance of 20 Ω connected in series with an inductive reactance (note the positive sign before the j-term) of 10 Ω.

2. $Z_2 = 15 - j\,25$ this impedance comprises a resistance of 15 Ω connected in series with a capacitive reactance (note the negative sign before the j-term) of 25 Ω.

3. $Z_3 = 30 + j\,0$ this impedance comprises a pure resistance of 30 Ω (there is no reactive component).

These three impedances are shown plotted on an Argand diagram in Fig. 1.50.

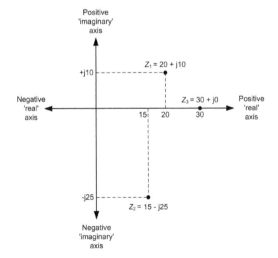

Figure 1.50 Argand diagram showing complex impedances

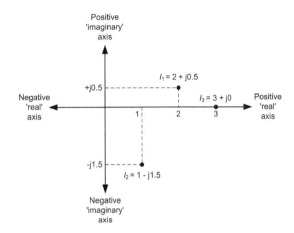

Figure 1.51 Argand diagram showing complex currents

Voltages and currents can also take complex values. Consider the following:

1. $I_1 = 2 + j\,0.5$ this current is the result of an in-phase component of 2 A and a reactive component (at $+90°$) of 0.5 A.

2. $I_2 = 1 - j\,1.5$ this current is the result of an in-phase component of 1 A and a reactive component (at $-90°$) of 1.5 A.

3. $I_3 = 3 + j\,0$ this current is in-phase and has a value of 3 A.

These three currents are shown plotted on an Argand diagram in Fig. 1.51.

To determine the voltage dropped across a complex impedance we can apply the usual relationship:

$$V = I\,Z$$

where V is the voltage (V), I is the current (A) and Z is the impedance (Ω), all expressed in complex form.

Example 1.33

A current of 2 A flows in an impedance of (100 + j 120) Ω. Derive an expression, in complex form, for the voltage that will appear across the impedance.

Since $V = I \times Z$

$$v = 2 \times (100 + j\,120) = 200 + j\,240 \text{ V}$$

Note that, in this example we have assumed that the supply current is the *reference*. In other words, it could be expressed in complex form as (2 + j 0) A.

Example 1.34

An impedance of (200 + j 100) Ω is connected to a 100 V AC supply. Determine the current flowing and express your answer in complex form.

Since $I = \dfrac{V}{Z}$

$$I = \frac{100}{(200+j100)} = \frac{100\times(200-j100)}{(200+j100)\times(200-j100)}$$

Note that we have multiplied the top and bottom by the *complex conjugate* in order to simplify the expression, reducing the denominator to a real number (i.e. no j-term).

$$I = \frac{100\times(200-j100)}{(200^2+100^2)} = \frac{(2\times10^4 - j\times10^4)}{(4\times10^4+1\times10^4)}$$

$$= \frac{(2-j)}{5} = 0.4 - j\,0.2 \text{ A}$$

Also note that we have assumed that the supply voltage is the *reference* quantity. In other words, it the voltage could be expressed in complex form as (100 + j 0) V.

Test your understanding 1.10

1. Identify the four impedances plotted on the Argand diagram shown in Fig. 1.52.

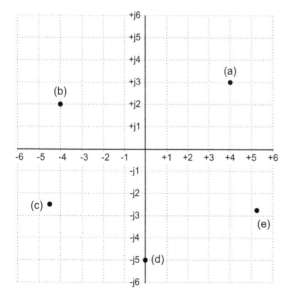

Figure 1.52   See Test your understanding 1.10

2. Plot each of the following voltages on an Argand diagram:

   (a) +j20 V
   (b) (30 + j40) V
   (c) −j10 V
   (d) (20 − j10) V.

3. A capacitor having a reactance of 30 Ω is connected in series with a resistance of 40 Ω. Express this impedance in complex form.

4. If a current of 1.5 A flows in the circuit in Question 3, determine the voltage that will appear across each component and express your answers in complex form. Also write down an expression for the voltage applied to the series circuit, once again expressed in complex form.

5. Write down the impedance, expressed in complex form, of each of the circuits shown in Fig. 1.53.

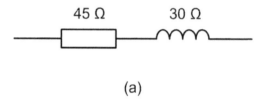

(a)

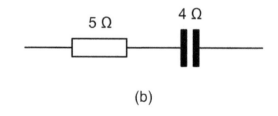

(b)

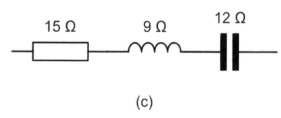

(c)

Figure 1.53   See Test your understanding 1.10

1.7.9 Transformers

Transformers provide us with a means of stepping-up or stepping down an AC voltage. For a **step-up transformer**, the output (or **secondary**) voltage will be greater than the input (or **primary**) whilst for a **step-down transformer** the secondary voltage will be less than the primary voltage. Since the primary and secondary power must be the same (no increase in power is possible), an increase in secondary voltage can only be achieved at the expense of a corresponding reduction in secondary current, and vice versa (in fact, the secondary power will be very slightly less than the primary power due to losses within the transformer).

The principle of the transformer is illustrated in Fig. 1.54. The primary and secondary windings are wound on a common low-reluctance magnetic core consisting of a number of steel laminations. All of the alternating flux generated by the primary winding is therefore coupled into the secondary winding (very little flux escapes due to leakage). A sinusoidal current flowing in the primary winding produces a sinusoidal flux within the transformer core.

At any instant the flux, Φ, in the transformer core is given by the equation:

$$\Phi = \Phi_{max} \, \sin(ft)$$

where Φ_{max} is the maximum value of flux (in Wb), f is the frequency of the applied current, and t is the time in seconds. You might like to compare this equation with the one that you met earlier for a sine wave voltage.

The r.m.s. value of the primary voltage (V_P) is given by:

$$V_P = 4.44 \, fN_P \Phi_{max}$$

Similarly, the r.m.s. value of the secondary voltage (V_S) is given by:

$$V_S = 4.44 \, fN_S \Phi_{max}$$

From these two relationships (and since the same magnetic flux appears in both the primary and secondary windings) we can infer that:

$$\frac{V_P}{V_S} = \frac{N_P}{N_S}$$

Furthermore, assuming that no power is lost in the transformer (i.e. as long as the primary and secondary powers are the same) we can conclude that:

$$\frac{I_P}{I_S} = \frac{N_S}{N_P}$$

The ratio of primary turns to secondary turns (N_P/N_S) is known as the **turns ratio**. Furthermore, since the ratio of primary voltage to primary turns is the same as the ratio of secondary turns to secondary voltage, we can conclude that, for a particular transformer:

$$\text{Turns-per-volt (t.p.v.)} = \frac{V_P}{N_P} = \frac{V_S}{N_S}$$

The turns-per-volt rating can be quite useful when it comes to designing transformers with multiple secondary windings.

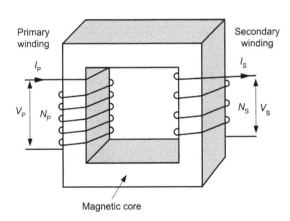

Figure 1.54   Basic arrangement of a transformer

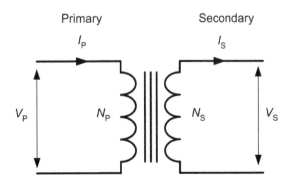

Figure 1.55   Transformer turns and voltages

Example 1.35

A transformer has 2000 primary turns and 120 secondary turns. If the primary is connected to a 220 V AC mains supply, determine the secondary voltage.

Since $\dfrac{V_P}{V_S} = \dfrac{N_P}{N_S}$ we can conclude that:

$$V_S = \frac{V_P N_S}{N_P} = \frac{220 \times 120}{2000} = 13.2\,\text{V}$$

Example 1.36

A transformer has 1200 primary turns and is designed to operate with a 110 V AC supply. If the transformer is required to produce an output of 10 V, determine the number of secondary turns required.

Since $\dfrac{V_P}{V_S} = \dfrac{N_P}{N_S}$ we can conclude that:

$$N_S = \frac{N_P V_S}{V_P} = \frac{1200 \times 10}{110} = 109$$

Example 1.37

A transformer has a turns-per-volt rating of 1.2. How many turns are required to produce secondary outputs of (a) 50 V and (b) 350 V?

Here we will use N_S = turns-per-volt × V_S.

(a) In the case of a 50 V secondary winding:

$$N_S = 1.5 \times 50 = 75\,\text{turns}$$

(b) In the case of a 350 V secondary winding:

$$N_S = 1.5 \times 350 = 525\,\text{turns}$$

Example 1.38

A transformer has 1200 primary turns and 60 secondary turns. Assuming that the transformer is loss-free, determine the primary current when a load current of 20 A is taken from the secondary.

Since $\dfrac{I_S}{I_P} = \dfrac{N_P}{N_S}$ we can conclude that:

$$I_P = \frac{I_S N_S}{N_P} = \frac{20 \times 60}{1200} = 1\,\text{A}$$

The output voltage produced at the secondary of a real transformer falls progressively as the load imposed on the transformer increases (i.e. as the secondary current increases from its no-load value). The **voltage regulation** of a transformer is a measure of its ability to keep the secondary output voltage constant over the full range of output load currents (i.e. from **no-load** to **full-load**) at the same power factor. This change, when divided by the no-load output voltage, is referred to as the **per-unit regulation** for the transformer.

Example 1.39

A transformer produces an output voltage of 110 V under no-load conditions and an output voltage of 101 V when a full load is applied. Determine the per-unit regulation.

The per-unit regulation can be determined from:

$$\begin{aligned}
\text{Per-unit regulation} &= \frac{V_{S(\text{no-load})} - V_{S(\text{full-load})}}{V_{S(\text{no-load})}} \\
&= \frac{110 - 101}{110} = 0.081 \text{ (or 8.1\%)}
\end{aligned}$$

Most transformers operate with very high values of efficiency. Despite this, in high power applications the losses in a transformer cannot be completely neglected. Transformer losses can be divided into two types of loss:

- losses in the magnetic core (often referred to as **iron loss**)
- losses due to the resistance of the coil windings (often referred to as **copper loss**).

Iron loss can be further divided into **hysteresis loss** (energy lost in repeatedly cycling the magnetic flux in the core backwards and forwards) and **eddy current loss** (energy lost due to current circulating in the steel core).

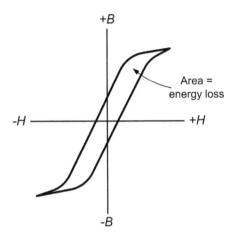

Figure 1.56 Hysteresis curves and energy loss

Hysteresis loss can be reduced by using material for the magnetic core that is easily magnetized and has a very high permeability (see Fig. 1.56 – note that energy loss is proportional to the area inside the B–H curve). Eddy current loss can be reduced by laminating the core (using E and I laminations, for example) and also ensuring that a small gap is present. These laminations and gaps in the core help to ensure that there is no closed path for current to flow. Copper loss results from the resistance of the coil windings and it can be reduced by using wire of large diameter and low resistivity.

It is important to note that, since the flux within a transformer varies only slightly between the no-load and full-load conditions, iron loss is substantially constant regardless of the load actually imposed on a transformer. Copper loss, on the other hand, is zero when a transformer is under no-load conditions and rises to a maximum at full-load.

The efficiency of a transformer is given by:

$$\text{Efficiency} = \frac{\text{output power}}{\text{input power}} \times 100\%$$

from which

$$\text{Efficiency} = \frac{\text{input power} - \text{losses}}{\text{input power}} \times 100\%$$

and

$$\text{Efficiency} = 1 - \frac{\text{losses}}{\text{input power}} \times 100\%$$

As we have said, the losses present are attributable to iron loss and copper loss but the copper loss appears in both the primary and the secondary windings. Hence:

$$\text{Efficiency} = 1 - \frac{\begin{array}{c}\text{iron loss} + \text{primary}\\ \text{copper loss} + \text{secondary}\\ \text{copper loss}\end{array}}{\text{input power}} \times 100\%$$

Example 1.40

A transformer rated at 500 VA has an iron loss of 3 W and a full-load copper loss (primary plus secondary) of 7 W. Calculate the efficiency of the transformer at 0.8 power factor.

The input power to the transformer will be given by the product of the apparent power (i.e. the transformer's *VA* rating) and the power factor. Hence:

$$\text{Input power} = 0.8 \times 5000 = 400 \text{ W}$$

Now

$$\text{Efficiency} = 1 - \frac{(7 + 3)}{400} \times 100\% = 97.5\%$$

Figure 1.57 A selection of transformers for operation at frequencies ranging from 50 Hz to 50 kHz

Key maintenance point

When replacing a transformer it is essential to ensure that the replacement component is correctly rated. The specifications for a transformer usually include the rated primary and secondary voltage and current, the power rating expressed in volt-amperes, VA (this is the maximum power that the transformer can deliver under a given set of conditions), the frequency range for the transformer (note that a transformer designed for operation at 400 Hz will not work at 50 Hz or 60 Hz), and the **per-unit regulation** of the transformer (this is the ability of the transformer to maintain its rated output when under load).

Test your understanding 1.11

1. A transformer has 480 primary turns and 120 secondary turns. If the primary is connected to a 110 V AC supply determine the secondary voltage.
2. A step-down transformer has a 220 V primary and a 24 V secondary. If the secondary winding has 60 turns, how many turns are there on the primary?
3. A transformer has 440 primary turns and 1800 secondary turns. If the secondary supplies a current of 250 mA, determine the primary current (assume that the transformer is loss-free).
4. A transformer produces an output voltage of 220 V under no-load conditions and an output voltage of 208 V when full load is applied. Determine the per-unit regulation.
5. A 1 kVA transformer has an iron loss of 15 W and a full-load copper loss (primary plus secondary) of 20 W. Determine the efficiency of the transformer at 0.9 power factor.
6. Identify the components shown in Fig. 1.58.

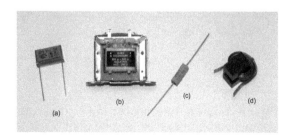

Figure 1.58   See Test your understanding 1.11

1.8 Safety

When working on aircraft electrical and electronic systems, personal safety (both yours and of those around you) should be paramount in everything that you do. Hazards can exist within many circuits – even those that, on the face of it, may appear to be totally safe. Inadvertent misconnection of a supply, incorrect earthing, reverse connection of components, and incorrect fitting can all result in serious hazards to personal safety as a consequence of fire, explosion or the generation of toxic fumes. In addition, there is a need to ensure that your work will not compromise the safety and integrity of the aircraft and not endanger the passengers and crew that will fly in it.

Potential hazards can be easily recognized and it is well worth making yourself familiar with them but perhaps the most important point to make is that electricity acts very quickly and you should always think carefully before working on circuits where mains or high voltages (i.e. those over 50 V or so) are present. Failure to observe this simple precaution can result in the very real risk of electric shock.

Voltages in many items of electronic equipment, including all items which derive their power from the aircraft's 400 Hz AC supply, are at a level which can cause sufficient current flow in the body to disrupt normal operation of the heart. The threshold will be even lower for anyone with a defective heart. Bodily contact with AC supplies and other high-voltage circuits can thus be lethal.

The most critical path for electric current within the body (i.e. the one that is most likely to stop the heart) is that which exists from one hand to the other. The hand-to-foot path is also dangerous but somewhat less dangerous than the hand-to-hand path. So, before you start to work on an item of electronic equipment, it is essential not only to switch off but to disconnect the equipment at the mains by removing the mains plug. If you have to make measurements or carry out adjustments on a piece of working (or 'live') equipment, a useful precaution is that of using one hand only to perform the adjustment or to make the measurement. Your 'spare' hand should be placed safely away from contact with anything metal (including the chassis of the equipment which may, or may not, be earthed).

The severity of electric shock depends upon several factors including the magnitude of the current, whether it is alternating or direct current, and its precise path through the body. The magnitude of the current depends upon the voltage which is applied and the resistance of the body. The electrical energy developed in the body will depend upon the time for which the current flows. The duration of contact is also crucial in determining the eventual physiological effects of the shock. As a rough guide, and assuming that the voltage applied is from the aircraft's 400 Hz AC supply, the following effects are typical:

Current	Physiological effect
less than 1 mA	Not usually noticeable
1 mA to 2 mA	Threshold of perception (a slight tingle may be felt)
2 mA to 4 mA	Mild shock (effects of current flow are felt)
4 mA to 10 mA	Serious shock (shock is felt as pain)
10 mA to 20 mA	Motor nerve paralysis may occur (unable to let go)
20 mA to 50 mA	Respiratory control inhibited (breathing may stop)
more than 50 mA	Ventricular fibrillation of heart muscle (heart failure)

Other hazards

Various other hazards, apart from electric shock, exist within the environment of an aircraft when work is being carried out on electrical and electronic systems. For example, accidental movement of the aircraft's spoilers can result in injury and/or damage to equipment. Whenever power sources are changed or when switches or circuit-breakers are opened that may cause movement of spoilers it is essential to ensure that the spoilers are deactivated or that all persons and equipment are removed from the vicinity.

Key maintenance point

The figures quoted in the table are provided as a guide – there have been cases of lethal shocks resulting from contact with much lower voltages and at relatively small values of current. The upshot of all this is simply that any potential in excess of 50 V should be considered dangerous. Lesser potentials may, under unusual circumstances, also be dangerous. As such, it is wise to get into the habit of treating all electrical and electronic systems with great care.

Key maintenance point

It is essential to remove electrical power from an aircraft before removing or installing components in the power panels. Failure to observe this precaution can result in electric shock as well as damage to components and equipment.

1.9 Multiple choice questions

1. A material in which there are no free charge carries is known as:
   (a) a conductor
   (b) an insulator
   (c) a semiconductor.

2. Conventional current flow is:
   (a) always from negative to positive
   (b) in the same direction as electron movement
   (c) in the opposite direction to electron movement.

3. Which one of the following gives the symbol and abbreviated units for electric charge?
   (a) Symbol, Q; unit, C
   (b) Symbol, C; unit, F
   (c) Symbol, C; unit, V.

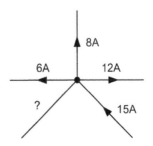

Figure 1.59 See Question 4

4. The unknown current shown in Fig. 1.59 will be:
   (a) 11 A flowing towards the junction
   (b) 17 A flowing away from the junction
   (c) 41 A flowing away from the junction.

5. The power factor in an AC circuit is the same as:
   (a) the sine of the phase angle
   (b) the cosine of the phase angle
   (c) the tangent of the phase angle.

6. An aircraft cabin has 110 passenger reading
   lamps each rated at 10 W, 28 V. What is the
   maximum load current imposed by these lamps?
   (a) 25.5 A
   (b) 39.3 A
   (c) 308 A.

7. Two isolated charges have dissimilar polarities.
   The force between them will be:
   (a) a force of attraction
   (b) a force of repulsion
   (c) zero.

8. The relationship between voltage, V, current, I,
   and resistance, R, for a resistor is:
   (a) $V = IR$
   (b) $V = R/I$
   (c) $V = I R^2$.

9. A potential difference of 7.5 V appears across a
   15 Ω resistor. Which one of the following gives
   the current flowing:
   (a) 0.25 A
   (b) 0.5 A
   (c) 2 A.

10. An aircraft supply has an RMS value of
    115 V. Which one of the following gives the
    approximate peak value of the supply voltage?
    (a) 67.5 V
    (b) 115 V
    (c) 163 V.

11. An AC waveform has a period of 4 ms. Which
    one of the following gives its frequency?
    (a) 25 Hz
    (b) 250 Hz
    (c) 4 kHz.

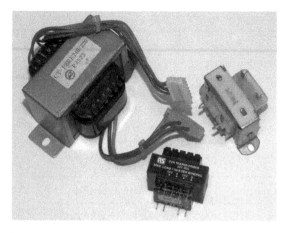

Figure 1.60   See Question 15

12. Which one of the following gives the angle
    between the successive phases of a three-phase
    supply?
    (a) 60°
    (b) 90°
    (c) 120°.

13. The peak value of current supplied to an aircraft
    TRU is 28 A. Which one of the following gives
    the approximate value of RMS current
    supplied?
    (a) 10 A
    (b) 14 A
    (c) 20 A.

14. A transformer has 2400 primary turns and 600
    secondary turns. If the primary is supplied from
    a 220 V AC supply which one of the following
    gives the resulting secondary voltage?
    (a) 55 V
    (b) 110 V
    (c) 880 V.

15. Identify the components shown in Fig. 1.60.
    (a) resistors
    (b) inductors
    (c) transformers.

Chapter 2

Electronic fundamentals

Electronics underpins all avionic systems, and semiconductor devices such as diodes, transistors and integrated circuits are the basic building blocks of these systems. This chapter provides you with an introduction to the theory and operation of semiconductors and their application in basic electronic circuits. As with Chapter 1, if you are in any doubt as to whether or not you should work through this chapter you can always turn to Section 2.6 on page 62 and see how you get on with the multiple choice questions at the end of the chapter.

2.1 Semiconductor theory

Materials that combine some of the electrical characteristics of conductors with those of insulators are known as **semiconductors**. Common types of semiconductor material are silicon, germanium, selenium and gallium. In the pure state, these materials may have relatively few free electrons to permit the flow of electric current. However, it is possible to add foreign atoms (called **impurity atoms**) to the semiconductor material in order to modify the properties of the semiconductor and allow it to conduct electricity.

You should recall that an atom contains both **negative charge carriers** (electrons) and **positive charge carriers** (protons). Electrons each carry a single unit of negative electric charge while protons each exhibit a single unit of positive charge. Since atoms normally contain an equal number of electrons and protons, the net charge present will be zero. For example, if an atom has eleven electrons, it will also contain eleven protons. The end result is that the negative charge of the electrons will be exactly balanced by the positive charge of the protons.

Electrons are in constant motion as they orbit around the nucleus of the atom. Electron orbits are organized into **shells**. The maximum number of electrons present in the first shell is two, in the second shell eight, and in the third, fourth and fifth shells it is 18, 32 and 50, respectively. In electronics, only

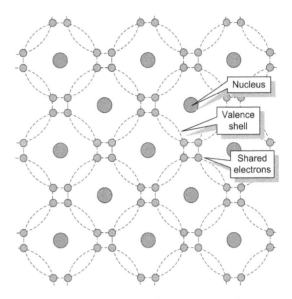

Figure 2.1 Regular lattice structure of a pure semiconductor material

the electron shell furthermost from the nucleus of an atom is important. It is important to note that the movement of electrons between atoms only involves those present in the outer **valence shell**.

If the valence shell contains the maximum number of electrons possible the electrons are rigidly bonded together and the material has the properties of an insulator. If, however, the valence shell does not have its full complement of electrons, the electrons can be easily detached from their orbital bonds, and the material has the properties associated with an electrical conductor.

In its pure state, silicon is an insulator because the covalent bonding rigidly holds all of the electrons leaving no free (easily loosened) electrons to conduct current. If, however, an atom of a different element (i.e. an **impurity**) is introduced that has five electrons in its valence shell, a surplus electron will be present (see Fig. 2.2). These free electrons become available for use as charge carriers and they can be made

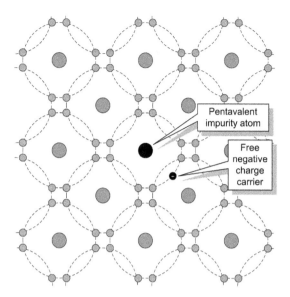

Figure 2.2 Effect of introducing a pentavalent impurity

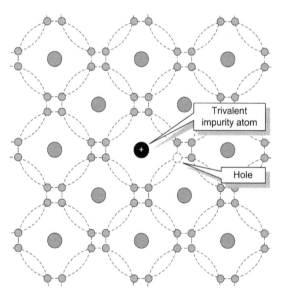

Figure 2.3 Effect of introducing a trivalent impurity

to move through the lattice by applying an external potential difference to the material.

Similarly, if the impurity element introduced into the pure silicon lattice has three electrons in its valence shell, the absence of the fourth electron needed for proper covalent bonding will produce a number of spaces into which electrons can fit (see Fig. 2.3). These gaps are referred to as **holes**. Once

again, current will flow when an external potential difference is applied to the material.

Regardless of whether the impurity element produces surplus electrons or holes, the material will no longer behave as an insulator, neither will it have the properties that we normally associate with a metallic conductor. Instead, we call the material a semiconductor – the term simply indicates that the substance is no longer a good insulator nor a good conductor but has the properties of something between the two! Examples of semiconductors include **germanium (Ge)** and **silicon (Si)**.

The process of introducing an atom of another (impurity) element into the lattice of an otherwise pure material is called **doping**. When the pure material is doped with an impurity with five electrons in its valence shell (i.e. a **pentavalent impurity**) it will become an N-type (i.e. negative type) material. If, however, the pure material is doped with an impurity having three electrons in its valence shell (i.e. a **trivalent impurity**) it will become P-type material (i.e. positive type). N-type semiconductor material contains an excess of negative charge carriers, and P-type material contains an excess of positive charge carriers.

2.1.1 Temperature effects

As explained in Chapter 1, all materials offer some resistance to current flow. In conductors the free electrons, rather than passing unobstructed through the material, collide with the relatively large and solid nuclei of the atoms. As the temperature increases, the nuclei vibrate more energetically, further obstructing the path of the free electrons, causing more frequent collisions. The result is that the resistance of a metal conductor increases with temperature.

Due to the nature of the bonding in insulators, there are no free electrons, except that when thermal energy increases as a result of a temperature increase, a few outer electrons manage to break free from their fixed positions and act as charge carriers. The result is that the resistance of an insulator decreases as temperature increases.

Semiconductors behave in a similar manner to insulators. At absolute zero ($-273°C$) both types of material act as perfect insulators. However, unlike the insulator, as temperature increases in a semiconductor large numbers of electrons break free to act as charge carriers. Therefore, as temperature

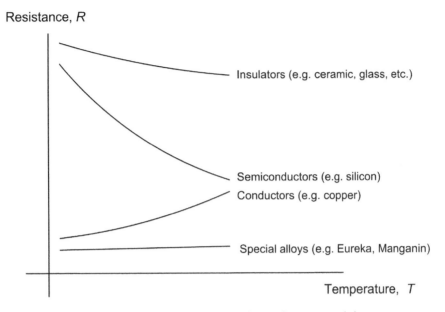

Figure 2.4 Variation of resistance with temperature for various materials

increases, the resistance of a semiconductor decreases rapidly.

By producing special alloys, such as Eureka and Manganin that combine the effects of insulators and conductors, it is possible to produce a material where the resistance remains constant with increase in temperature. Fig. 2.4 shows how the resistance of insulators, semiconductors and conductors change with temperature.

Unlike conventional resistors, the resistance of a **thermistor** is intended to change considerably with temperature. Thermistors are employed in a wide variety of temperature-sensing and temperature-compensating applications. Two basic types of thermistors are available: negative temperature coefficient (NTC) and positive temperature coefficient (PTC).

Typical NTC thermistors have resistances that vary from a few hundred (or thousand) ohms at 25°C to a few tens (or hundreds) of ohms at 100°C. PTC thermistors, on the other hand, usually have a resistance–temperature characteristic that remains substantially flat (usually at around $100\,\Omega$) over the range 0°C to around 75°C. Above this, and at a critical threshold temperature (usually in the range 80°C to 120°C) their resistances rise rapidly up to, and beyond, $10\,k\Omega$ (see Figs 2.5 and 2.6).

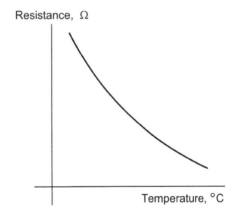

Figure 2.5 NTC thermistor characteristics

Key point

The properties of most semiconductor devices are profoundly affected by temperature. In order to maintain performance within specification, care must be taken to ensure that temperature variations are minimized. In some applications, additional thermal protection may be required to reduce the risk of **thermal runaway** (a condition in which uncontrolled temperature rise may result in failure of one or more semiconductor devices).

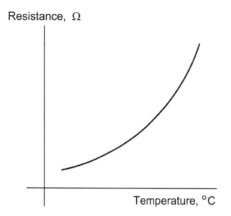

Figure 2.6   PTC thermistor characteristics

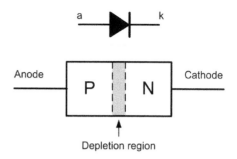

Figure 2.7   A P-N junction diode

Key point

Thermistors provide us with a means of sensing temperature changes. The resistance of an NTC thermistor falls with temperature whilst that of a PTC thermistor increases with temperature.

Key maintenance point

Many semiconductor devices are fitted with dedicated cooling systems such as heat sinks, heat pipes and fans. These components provide cooling by means of a combination of thermal conduction and convection and they are crucial to the correct operation of the devices to which they are fitted. It is therefore essential to ensure that cooling systems are correctly fitted and are fully operational whenever maintenance is carried out on the semiconductor devices themselves.

Key maintenance point

Beryllium oxide (BeO_2) is used in the construction of some power semiconductors. This material is highly toxic and special care should be taken when handling and disposing of any semiconductors that use it. It is essential to refer to semiconductor manufacturers' recommendations concerning the fitting, handling and disposal of these devices.

Test your understanding 2.1

Explain why thermal runaway is undesirable and why cooling may be necessary in conjunction with semiconductor devices.

2.2 Diodes

When a junction is formed between N-type and P-type semiconductor materials, the resulting device is called a **diode**. This component offers an extremely low resistance to current flow in one direction and an extremely high resistance to current flow in the other. This characteristic allows diodes to be used in applications that require a circuit to behave differently according to the direction of current flowing in it. An ideal diode would pass an infinite current in one direction and no current at all in the other direction.

Connections are made to each side of the diode. The connection to the P-type material is referred to as the **anode** while that to the N-type material is called the **cathode**. With no externally applied potential, electrons from the N-type material will cross into the P-type region and fill some of the vacant holes. This action will result in the production of a region either side of the junction in which there are no free charge carriers. This zone is known as the **depletion region**.

If a positive voltage is applied to the anode (see Fig. 2.8), the free positive charge carriers in the P-type material will be repelled and they will move away from the positive potential towards the junction. Likewise, the negative potential applied to the cathode will cause the free negative charge carriers in the N-type material to move away from the negative potential towards the junction.

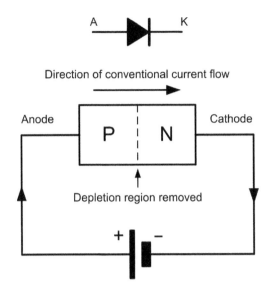

Figure 2.8 A forward-biased P-N junction diode

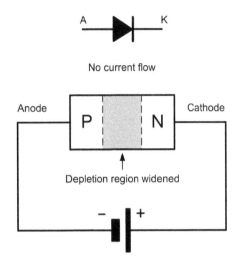

Figure 2.9 A reverse-biased P-N junction diode

When the positive and negative charge carriers arrive at the junction, they will attract one another and combine (recall that that unlike charges attract). As each negative and positive charge carrier combine at the junction, a new negative and positive charge carrier will be introduced to the semiconductor material from the voltage source. As these new charge carriers enter the semiconductor material, they will move towards the junction and combine. Thus, current flow is established and it will continue for as long as the voltage is applied. In this **forward-biased** condition, the diode freely passes current.

If a negative voltage is applied to the anode (see Fig. 2.9), the free positive charge carriers in the P-type material will be attracted and they will move away from the junction. Likewise, the positive potential applied to the cathode will cause the free negative charge carriers in the N-type material to move away from the junction. The combined effect is that the depletion region becomes wider. In this **reverse-biased** condition, the diode passes a negligible amount of current.

Key point

In the freely conducting forward-biased state, the diode acts rather like a closed switch. In the reverse-biased state, the diode acts like an open switch.

2.2.1 Diode characteristics

Typical I/V characteristics for germanium and silicon diodes are shown in Fig. 2.10. It should be noted from these characteristics that the approximate **forward conduction voltage** for a germanium diode is 0.2 V whilst that for a silicon diode is 0.6 V. This threshold voltage must be high enough to completely overcome the potential associated with the depletion region and force charge carriers to move across the junction.

Diodes are limited by the amount of forward current and reverse voltage they can withstand. This limit is based on the physical size and construction of the diode. In the case of a reverse biased diode, the P-type material is negatively biased relative to the N-type material. In this case, the negative potential at the P-type material attracts the positive carriers, drawing them away from the junction. This leaves the area depleted; virtually no charge carriers exist and therefore current flow is inhibited. If the reverse bias potential is increased above the **maximum reverse voltage** (V_{RM}) or **peak inverse voltage** (**PIV**) quoted by the manufacturer, the depletion region may suffer an irreversible breakdown. Typical values of maximum reverse voltage range from as low as 50 V to well over 500 V. Note that reverse breakdown voltage is usually very much higher than the forward threshold voltage. For example, a typical general-purpose diode may be specified as having a forward threshold voltage of 0.6 V and a reverse breakdown voltage of

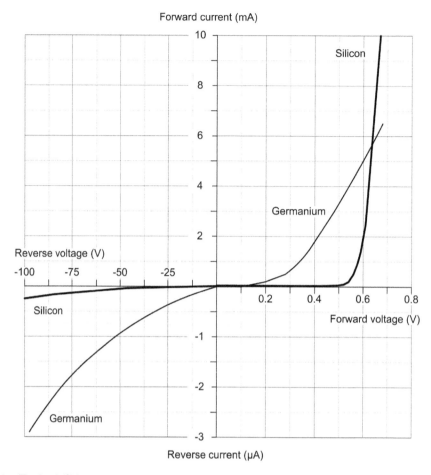

Figure 2.10   Typical *I/V* characteristics for germanium and silicon diodes

200 V. If the latter is exceeded, the diode may suffer irreversible damage.

Key point

The forward voltage for a germanium diode is approximately 0.2 V whilst that for a silicon diode is approximately 0.6 V.

2.2.2 Zener diodes

Zener diodes are heavily doped silicon diodes that, unlike normal diodes, exhibit an abrupt reverse breakdown at relatively low voltages (typically less than

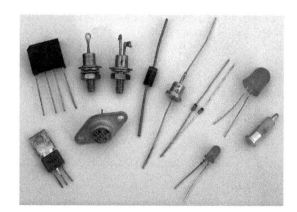

Figure 2.11 Various diodes (including rectifiers, signal diodes, Zener diodes, light-emitting diodes and silicon-controlled rectifiers)

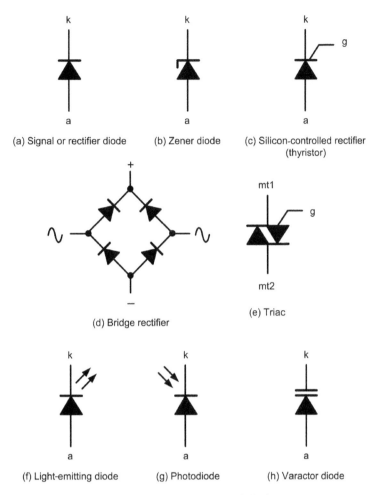

(a) Signal or rectifier diode   (b) Zener diode   (c) Silicon-controlled rectifier
(thyristor)

(d) Bridge rectifier

(e) Triac

(f) Light-emitting diode   (g) Photodiode   (h) Varactor diode

Figure 2.12   Symbols used for various common types of diode

6 V). A similar effect (avalanche) occurs in less heavily doped diodes. These avalanche diodes also exhibit a rapid breakdown with negligible current flowing below the avalanche voltage and a relatively large current flowing once the avalanche voltage has been reached. For avalanche diodes, this breakdown voltage usually occurs at voltages above 6 V. In practice, however, both types of diode are commonly referred to as Zener diodes. The symbol for a Zener diode was shown earlier in Fig. 2.12, whilst typical Zener diode characteristics are shown in Fig. 2.13.

Whereas reverse breakdown is a highly undesirable effect in circuits that use conventional diodes, it can be extremely useful in the case of Zener diodes where the breakdown voltage is precisely known. When a diode is undergoing reverse breakdown and provided its maximum ratings are not exceeded the voltage appearing across it will remain substantially constant (equal to the nominal Zener voltage) regardless of the current flowing. This property makes the Zener diode ideal for use as a **voltage regulator**.

Zener diodes are available in various families (according to their general characteristics, encapsulations and power ratings) with reverse breakdown (Zener) voltages in the range 2.4 V to 91 V.

A simple voltage regulator is shown in Fig. 2.14. The series resistor, R_S is included to limit the Zener current to a safe value when the load is disconnected. When a load (R_L) is connected, the Zener current will fall as current is diverted into the load resistance (it is usual to allow a minimum current of 2 mA to 5 mA in order to ensure that the diode regulates). The

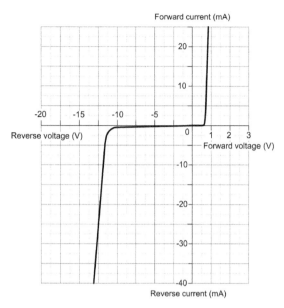

Figure 2.13 Typical Zener diode characteristic

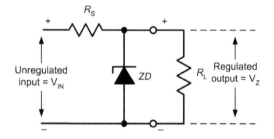

Figure 2.14 A simple Zener diode voltage regulator

output voltage will remain at the Zener voltage (V_Z) until regulation fails at the point at which the potential divider formed by R_S and R_L produces a lower output voltage that is less than V_Z. The ratio of R_S to R_L is thus important.

Key point

Zener diodes begin to conduct heavily when the applied voltage reaches a particular threshold value (known as the Zener voltage). Zener diodes can thus be used to maintain a constant voltage which is often used as a *voltage reference*.

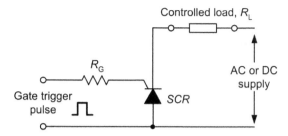

Figure 2.15 Triggering a silicon-controlled rectifier

2.2.3 Silicon-controlled rectifiers

Silicon-controlled rectifiers (or **thyristors**) are three-terminal devices which can be used for switching and AC power control. Silicon-controlled rectifiers can switch very rapidly from a conducting to a non-conducting state. In the off state, the silicon-controlled rectifier exhibits negligible leakage current, while in the on state the device exhibits very low resistance. This results in very little power loss within the silicon-controlled rectifier even when appreciable power levels are being controlled.

Once switched into the conducting state, the silicon-controlled rectifier will remain conducting (i.e. it is latched in the on state) until the forward current is removed from the device. In DC applications this necessitates the interruption (or disconnection) of the supply before the device can be reset into its non-conducting state. Where the device is used with an alternating supply, the device will automatically become reset whenever the main supply reverses. The device can then be triggered on the next half-cycle having correct polarity to permit conduction.

Like their conventional silicon diode counterparts, silicon-controlled rectifiers have anode and cathode connections; control is applied by means of a gate terminal. The symbol for a silicon-controlled rectifier was shown earlier in Fig. 2.12.

In normal use, a silicon-controlled rectifier is triggered into the conducting (on) state by means of the application of a current pulse to the gate terminal (see Fig. 2.15). The effective triggering of a silicon-controlled rectifier requires a gate trigger pulse having a fast rise time derived from a low-resistance source. Triggering can become erratic when insufficient gate current is available or when the gate current changes slowly.

Key point

Silicon-controlled rectifiers (SCRs) are diodes that can be triggered into conduction by applying a small current to their gate input. SCRs are able to control large voltages and currents from a relatively small (low-current, low-voltage) signal.

2.2.4 Light-emitting diodes

Light-emitting diodes (LEDs) can be used as general-purpose indicators and, compared with conventional filament lamps, operate from significantly smaller voltages and currents. LEDs are also very much more reliable than filament lamps. Most LEDs will provide a reasonable level of light output when a forward current of between 5 mA and 20 mA is applied.

Light-emitting diodes are available in various formats with the round types being most popular. Round LEDs are commonly available in the 3 mm and 5 mm (0.2 inch) diameter plastic packages and also in a 5 mm × 2 mm rectangular format. The viewing angle for round LEDs tends to be in the region of 20° to 40°, whereas for rectangular types this is increased to around 100°. The symbol for an LED was shown earlier in Fig. 2.12.

Key point

Light-emitting diodes produce light when a small current is applied to them. They are generally smaller and more reliable than conventional filament lamps and can be used to form larger and more complex displays.

2.2.5 Rectifiers

Semiconductor diodes are commonly used to convert alternating current (AC) to direct current (DC), in which case they are referred to as **rectifiers**. The simplest form of rectifier circuit makes use of a single diode and, since it operates on only either positive or negative half-cycles of the supply, it is known as a half-wave rectifier.

Figure 2.16 shows a simple **half-wave rectifier** circuit. The AC supply at 115 V is applied to the primary

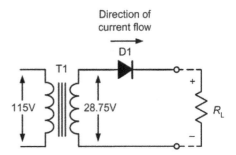

Figure 2.16 A simple half-wave rectifier circuit

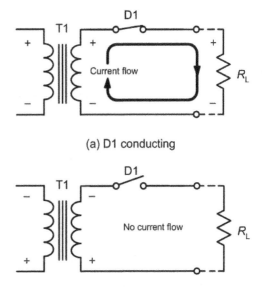

(a) D1 conducting

(b) D1 non-conducting

Figure 2.17 Switching action of the diode in the half-wave rectifier: (a) D1 forward biased, (b) D1 reverse biased

of a step-down transformer (T1). The secondary of T1 steps down the 115 V 400 Hz supply to 28.75 V RMS (the turns ratio of T1 will thus be 115/28.75 or 4:1).

Diode D1 will only allow the current to flow in the direction shown (i.e. from anode to cathode). D1 will be forward biased during each positive half-cycle and will effectively behave like a closed switch, as shown in Fig. 2.17(a). When the circuit current tries to flow in the opposite direction, the voltage bias across the diode will be reversed, causing the diode to act like an open switch, as shown in Fig. 2.17(b).

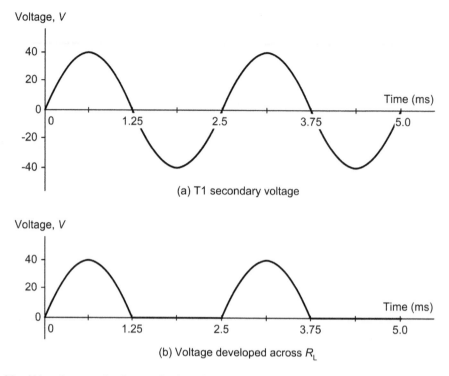

(a) T1 secondary voltage

(b) Voltage developed across R_L

Figure 2.18 Waveforms of voltages in the simple half-wave power supply

The switching action of D1 results in a pulsating output voltage, which is developed across the load resistor (R_L). Since the supply is at 400 Hz, the pulses of voltage developed across R_L will also be at 400 Hz even if only half the AC cycle is present. During the positive half-cycle, the diode will drop the 0.6 V forward threshold voltage normally associated with silicon diodes. However, during the negative half-cycle the peak AC voltage will be dropped across D1 when it is reverse biased. This is an important consideration when selecting a diode for a particular application. Assuming that the secondary of T1 provides 28.75 V RMS, the peak voltage output from the transformer's secondary winding will be given by:

$$V_{pk} = 1.414 \times V_{RMS} = 1.414 \times 28.75 \text{ V} = 40.65 \text{ V}$$

The peak voltage applied to Dl will thus be a little over 40 V. The negative half-cycles are blocked by D1 and thus only the positive half-cycles appear across R_L. Note, however, that the actual peak voltage across R_L will be the 40.65 V positive peak being supplied from the secondary on T1, minus the 0.6 V forward

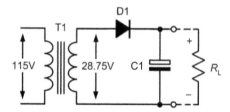

Figure 2.19 Adding a reservoir capacitor to a half-wave rectifier to make a simple DC power supply

threshold voltage dropped by D1. In other words, positive half-cycle pulses having a peak amplitude of almost exactly 40 V will appear across R_L.

Figure 2.19 shows a considerable improvement to the earlier simple rectifier. The capacitor, C1, has been added to ensure that the output voltage remains at, or near, the peak voltage even when the diode is not conducting. When the primary voltage is first applied to T1, the first positive half-cycle output from the secondary will charge C1 to the peak value seen across R_L. Hence C1 charges to 40 V at the peak of

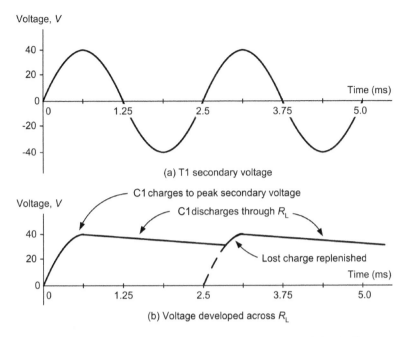

(a) T1 secondary voltage

(b) Voltage developed across R_L

Figure 2.20   Effect of adding a reservoir capacitor on the output of the half-wave rectifier

the positive half-cycle. Because C1 and R_L are in parallel, the voltage across R_L will be the same as that developed across C1 (see Fig. 2.19).

The time required for C1 to charge to the maximum (peak) level is determined by the charging circuit time constant (the series resistance multiplied by the capacitance value). In this circuit, the series resistance comprises the secondary winding resistance together with the forward resistance of the diode and the (minimal) resistance of the wiring and connections. Hence C1 charges to 40 V at the peak of the positive half-cycle. Because C1 and R_L are in parallel, the voltage across R_L will be the same as that across C1.

The time required for C1 to discharge is, in contrast, very much greater. The discharge time constant is determined by the capacitance value and the load resistance, R_L. In practice, R_L is very much larger than the resistance of the secondary circuit and hence C1 takes an appreciable time to discharge. During this time, D1 will be reverse biased and will thus be held in its non-conducting state. As a consequence, the only discharge path for C1 is through R_L.

C1 is referred to as a **reservoir capacitor**. It stores charge during the positive half-cycles of secondary voltage and releases it during the negative half-cycles. The circuit shown earlier is thus able to maintain a reasonably constant output voltage across R_L. Even

so, C1 will discharge by a small amount during the negative half-cycle periods from the transformer secondary. The figure above shows the secondary voltage waveform together with the voltage developed across R_L with and without C1 present. This gives rise to a small variation in the DC output voltage (known as **ripple**).

Since ripple is undesirable we must take additional precautions to reduce it. One obvious method of reducing the amplitude of the ripple is that of simply increasing the discharge time constant. This can be achieved either by increasing the value of C1 or by increasing the resistance value of R_L. In practice, however, the latter is not really an option because R_L is the effective resistance of the circuit being supplied and we don't usually have the ability to change it! Increasing the value of C1 is a more practical alternative and very large capacitor values (often in excess of 1000 μF) are typical.

The half-wave rectifier circuit is relatively inefficient as conduction takes place only on alternate half-cycles. A better rectifier arrangement would make use of both positive *and* negative half-cycles. These **full-wave rectifier** circuits offer a considerable improvement over their half-wave counterparts. They are not only more efficient but are significantly less demanding in terms of the reservoir and smoothing components. There are

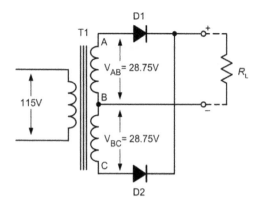

Figure 2.21 A simple bi-phase rectifier circuit

two basic forms of full-wave rectifier: the bi-phase type and the bridge rectifier type.

Figure 2.21 shows a simple **bi-phase rectifier** circuit. The AC supply at 115 V is applied to the primary of a step-down transformer (T1). This has two identical secondary windings, each providing 28.75 V RMS (the turns ratio of T1 will still be 115/28.75 or 4:1 for each secondary winding).

On positive half-cycles, point A will be positive with respect to point B. Similarly, point B will be positive with respect to point C. In this condition D1 will allow conduction (its anode will be positive with respect to its cathode) while D2 will not allow conduction (its anode will be negative with respect to its cathode). Thus D1 alone conducts on positive half-cycles.

On negative half-cycles, point C will be positive with respect to point B. Similarly, point B will be positive with respect to point A. In this condition D2 will allow conduction (its anode will be positive with respect to its cathode) while D1 will not allow conduction (its anode will be negative with respect to its cathode). Thus D2 alone conducts on negative half-cycles.

Figure 2.22 shows the bi-phase rectifier circuit with the diodes replaced by switches. In (a) D1 is shown conducting on a positive half-cycle whilst in (b) D2 is shown conducting on a negative half-cycle of the input. The result is that current is routed through the load *in the same direction* on successive half-cycles. Furthermore, this current is derived alternately from the two secondary windings.

As with the half-wave rectifier, the switching action of the two diodes results in a pulsating output voltage being developed across the load resistor (R_L).

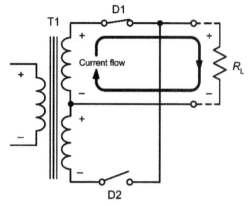

(a) D1 conducting and D2 non-conducting

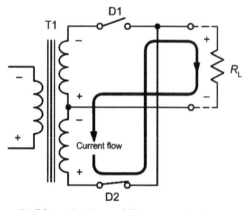

(b) D2 conducting and D1 non-conducting

Figure 2.22 Switching action of the diodes in the bi-phase rectifier: (a) D1 forward biased and D2 reverse biased, (b) D1 reverse biased and D2 forward biased

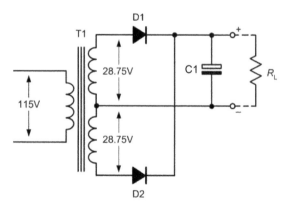

Figure 2.23 Bi-phase power supply with reservoir capacitor

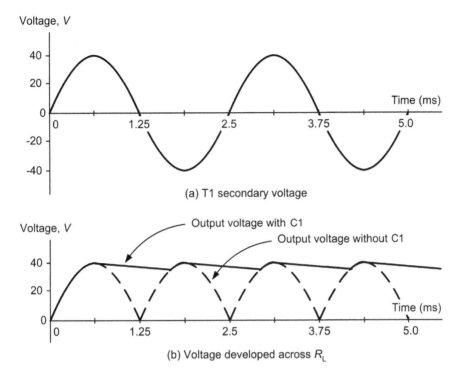

(a) T1 secondary voltage

(b) Voltage developed across R_L

Figure 2.24 Waveforms in the bi-phase power supply with and without a reservoir capacitor

However, unlike the half-wave circuit the pulses of voltage developed across R_L will occur at a frequency of 800 Hz (not 400 Hz). This doubling of the ripple frequency allows us to use smaller values of reservoir and smoothing capacitor to obtain the same degree of ripple reduction (recall that the reactance of a capacitor is reduced as frequency increases). As before, the peak voltage produced by each of the secondary windings will be approximately 17 V and the peak voltage across R_L will be about 40 V (i.e. 40.65 V less the 0.6 V forward).

An alternative to the use of the bi-phase circuit is that of using a four-diode **bridge rectifier** in which opposite pairs of diodes conduct on alternate half-cycles. This arrangement avoids the need to have two separate secondary windings. A full-wave bridge rectifier arrangement is shown in Fig. 2.25. The 115 V AC supply at 400 Hz is applied to the primary of a step-down transformer (T1). As before, the secondary winding provides 28.75 V RMS (approximately 40 V peak) and has a turns ratio of 4:1. On positive half-cycles, point A will be positive with respect to point B. In this condition D1 and D2 will allow

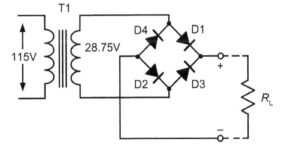

Figure 2.25 Full-wave power supply using a bridge rectifier

conduction while D3 and D4 will not allow conduction. Conversely, on negative half-cycles, point B will be positive with respect to point A. In this condition D3 and D4 will allow conduction while D1 and D2 will not allow conduction.

As with the bi-phase rectifier, the switching action of the two diodes results in a pulsating output voltage being developed across the load resistor (R_L).

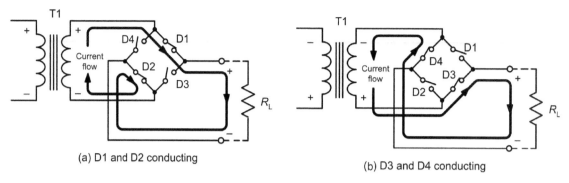

(a) D1 and D2 conducting

(b) D3 and D4 conducting

Figure 2.26 Switching action of the diodes in the full-wave bridge: (a) D1 and D2 forward biased whilst D3 and D4 are reverse biased, (b) D1 and D2 reverse biased whilst D3 and D4 are forward biased

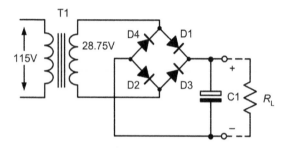

Figure 2.27 Full-wave bridge power supply with reservoir capacitor

Once again, the peak output voltage is approximately 40 V (i.e. 40.65 V less the 2×0.6 V forward threshold voltage of the two diodes).

Figure 2.27 shows how a reservoir capacitor (C1) can be added to the basic bridge rectifier circuit in order to ensure that the output voltage remains at, or near, the peak voltage even when the diodes are not conducting. This component operates in exactly the same way as for the bi-phase circuit, i.e. it charges to approximately 40 V at the peak of the positive half-cycle and holds the voltage at this level when the diodes are in their non-conducting states. These voltage waveforms are identical to those that we met earlier for the bi-phase rectifier.

Key point

The forward voltage for a germanium diode is approximately 0.2 V whilst that for a silicon diode is approximately 0.6 V.

Key point

The frequency of the ripple voltage present on the DC output of a half-wave power supply will be the same as the frequency of the AC input. The frequency of the ripple voltage present on the DC output of a full-wave power supply will be double that of the AC input.

Test your understanding 2.2

Explain the following terms when used in relation to a semiconductor diode:

(a) forward bias
(b) reverse bias
(c) forward conduction voltage.

Test your understanding 2.3

Match the following types of diode to each of the applications listed below:

Types of diode

(a) Silicon bridge rectifier
(b) Germanium signal diode
(c) Zener diode
(d) Red LED
(e) Thyristor.

Application

1. A voltage reference for use in a generator control unit

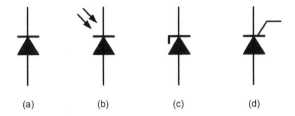

Figure 2.28 See Test your understanding 2.4

2. A power-on indicator for use in an emergency locator transmitter
3. An aircraft battery charger
4. An AC power controller
5. The detector circuit in an HF radio.

Test your understanding 2.4

Identify each of the diode symbols shown in Fig. 2.28.

2.3 Transistors

Typical applications for transistors in aircraft electrical and electronic systems are controlling generator field current, driving lights and warning displays, amplifying signals from sensors and transducers and for use as amplifying devices in cabin interphone and aircraft radio and navigation aids.

Conventional **bipolar junction transistors** (**BJTs**) generally comprise **NPN** or **PNP** junctions of either silicon (Si) or germanium (Ge) material. The junctions are produced in a single slice of silicon by diffusing impurities through a photographically reduced mask. Silicon transistors are superior when compared with germanium transistors in the vast majority of applications (particularly at high temperatures) and thus germanium devices are very rarely encountered in modern electronic equipment.

The construction of typical NPN and PNP BJTs are shown in Figs 2.29 and 2.30. In order to conduct the heat away from the junction (important in medium- and high-power applications) the collector is often connected to the metal case of the transistor.

The symbols and simplified junction models for NPN and PNP transistors are shown in Fig. 2.31. It is

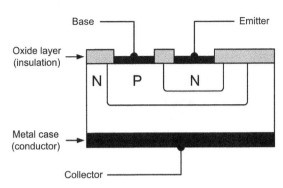

Figure 2.29 Construction of a typical NPN transistor

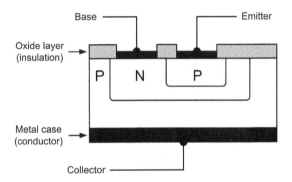

Figure 2.30 Construction of a typical PNP transistor

important to note that the base region (P-type material in the case of an NPN transistor or N-type material in the case of a PNP transistor) is extremely narrow.

2.3.1 Bias and current flow

In normal operation (i.e. for operation as a linear amplifier) the base-emitter junction of a transistor is forward biased and the collector base junction is reverse biased. The base region is, however, made very narrow so that carriers are swept across it from emitter to collector so that only a relatively small current flows in the base. To put this into context, the current flowing in the emitter circuit is typically 100 times greater than that flowing in the base. The direction of conventional current flow is from emitter to collector in the case of a PNP transistor, and collector to emitter in the case of an NPN device, as shown in Fig. 2.32.

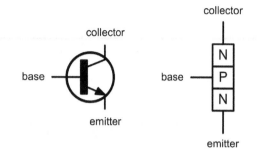

(a) NPN bipolar junction transistor (BJT)

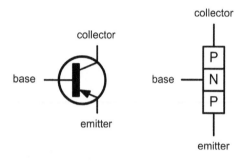

(b) PNP bipolar junction transistor (BJT)

Figure 2.31 NPN and PNP BJT symbols and simplified junction models

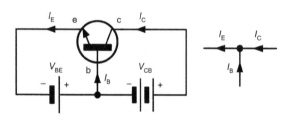

(a) NPN bipolar junction transistor (BJT)

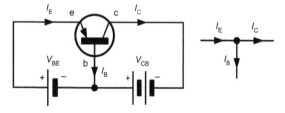

(b) PNP bipolar junction transistor (BJT)

Figure 2.32 Bias voltages and current flow in NPN and PNP bipolar junction transistors

The equation (based on Kirchhoff's current law) that relates current flow in the collector, base, and emitter of a transistor (see Fig. 2.32) is:

$$I_E = I_B + I_C$$

where I_E is the emitter current, I_B is the base current, and I_C is the collector current (all expressed in the same units)

Example 2.1

A transistor operates with a collector current of 2 A and an emitter current of 2.1 A. Determine the value of base current.

Now:

$$I_E = I_B + I_C$$

thus:

$$I_B = I_E - I_C$$

In this case, $I_C = 2$ A and $I_E = 2.1$ A, thus:

$$I_B = 2.1\,\text{A} - 2\,\text{A} = 0.1\,\text{A} = 100\,\text{mA}$$

Test your understanding 2.5

A BJT operates with an emitter current of 1.25 A and a base current of 50 mA. What will the collector current be?

Key point

The emitter current of a transistor is the sum of its base and collector currents.

2.3.2 Transistor characteristics

The characteristics of a bipolar junction transistor are usually presented in the form of a set of graphs relating voltage and current present at the transistors terminals. Figure 2.33 shows a typical **input characteristic** (I_B plotted against V_{BE}) for an NPN bipolar junction transistor operating in **common emitter mode**. In this mode, the input current is applied to the

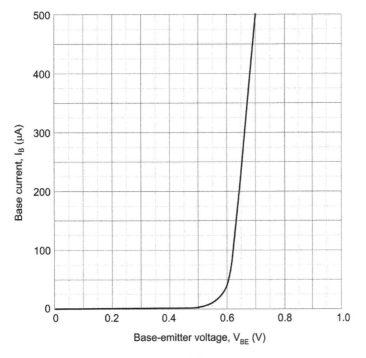

Figure 2.33 Input characteristic (I_B/V_{BE}) for an NPN bipolar junction transistor

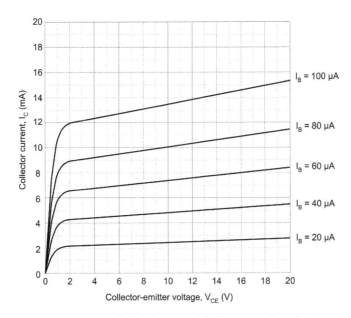

Figure 2.34 Output characteristic (I_C/V_{CE}) for an NPN bipolar junction transistor

base and the output current appears in the collector (the emitter is effectively common to both the input and output circuits).

The input characteristic shows that very little base current flows until the base emitter voltage V_{BE} exceeds 0.6 V. Thereafter, the base current increases rapidly (this characteristic bears a close resemblance to the forward part of the characteristic for a silicon diode).

Figure 2.34 shows a typical set of **output (collector) characteristics** (I_C plotted against V_{CE}) for an

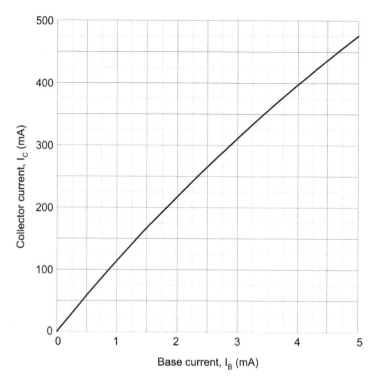

Figure 2.35   Transfer characteristic (I_C/I_B) for an NPN bipolar junction transistor

NPN bipolar transistor. Each curve corresponds to a different value of base current. Note the 'knee' in the characteristic below $V_{CE} = 2\,\text{V}$. Also note that the curves are quite flat. For this reason (i.e. since the collector current does not change very much as the collector-emitter voltage changes) we often refer to this as a *constant current characteristic*.

Figure 2.35 shows a typical **transfer characteristic** for an NPN bipolar junction transistor. Here I_C is plotted against I_B for a small-signal general purpose transistor. The slope of this curve (i.e. the ratio of I_C to I_B) is the **common emitter current gain** of the transistor. We shall explore this further in Section 2.3.4.

2.3.3 Transistor operating configurations

Three basic circuit arrangements are used for transistor **amplifiers** and these are based on the three circuit configurations that we met earlier (i.e. they depend upon which one of the three transistor connections is made common to both the input and the output). In

Figure 2.36 Various transistors (including low-frequency, high-frequency, high-voltage, small-signal and power types)

the case of bipolar transistors, the configurations are known as **common emitter**, **common collector** (or emitter follower) and **common base**. Where field effect transistors are used, the corresponding configurations are common source, common drain (or source follower) and common gate.

These basic circuit configurations depicted in Fig. 2.38 exhibit quite different performance characteristics, as shown in the table below.

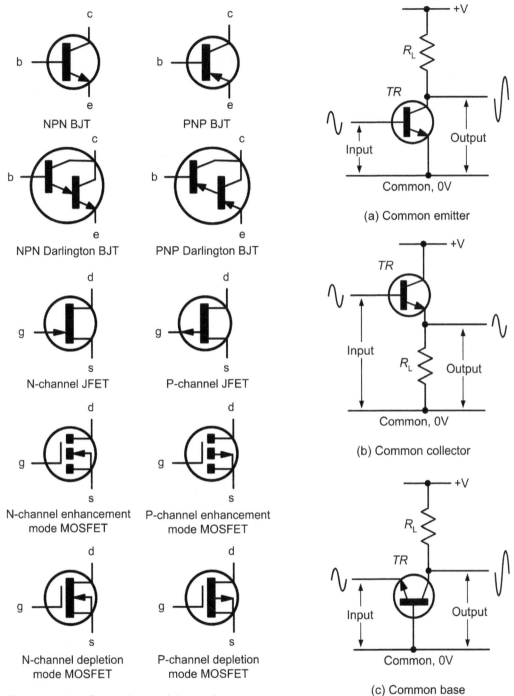

Figure 2.37 Symbols used for various types of transistor

(a) Common emitter

(b) Common collector

(c) Common base

Figure 2.38 BJT circuit configurations

Parameter	Common emitter	Common collector	Common base
Voltage gain	medium/high (40)	unity (1)	high (200)
Current gain	high (200)	high (200)	unity (1)
Power gain	very high (8000)	high (200)	high (200)
Input resistance	medium ($2.5\,k\Omega$)	high ($100\,k\Omega$)	low ($200\,\Omega$)
Output resistance	medium/high ($20\,k\Omega$)	low ($100\,\Omega$)	high ($100\,k\Omega$)
Phase shift	$180°$	$0°$	$0°$

A requirement of most amplifiers is that the output signal should be a faithful copy of the input signal or be somewhat larger in amplitude. Other types of amplifier are 'non-linear', in which case their input and output waveforms will not necessarily be similar. In practice, the degree of linearity provided by an amplifier can be affected by a number of factors including the amount of bias applied and the amplitude of the input signal. It is also worth noting that a linear amplifier will become non-linear when the applied input signal exceeds a threshold value. Beyond this value the amplifier is said to be over-driven and the output will become increasingly distorted if the input signal is further increased.

The optimum value of bias for linear (**Class A**) amplifiers is that value which ensures that the active devices are operated at the mid-point of their characteristics. In practice, this means that a static value of collector current will flow even when there is no signal present. Furthermore, the collector current will flow throughout the complete cycle of an input signal (i.e. conduction will take place over an angle of $360°$). At no stage should the transistor be **saturated** ($V_{CE} \approx 0\,V$ or $V_{DS} \approx 0\,V$) nor should it be **cut-off** ($V_{CE} \approx V_{CC}$ or $V_{DS} \approx V_{DD}$).

In order to ensure that a static value of collector current flows in a transistor, a small **bias current** must be applied to the base of the transistor. This current is usually derived from the same voltage rail that supplies the collector circuit via one or more resistors of appropriate value.

2.3.4 Current gain

In general terms, current gain is the ratio of output current to input current. When a transistor is operating in common emitter mode the input current appears at the base and the output current at the collector. Thus, for this mode of operation the current gain will simply be the ratio of collector current, I_C, to base current, I_B. We use the symbol h_{FE} to represent the static value of current gain when a transistor is connected in common emitter mode. Thus:

$$h_{FE} = \frac{I_C}{I_B}$$

Typical values of common emitter current gain vary from about 40 to 200.

Example 2.2

A current of 4A is to be supplied to a generator field winding from a BJT. If the device is operated in common emitter mode with a base current of 20 mA, what is the minimum value of current gain required?

Using $h_{FE} = I_C/I_B$ where $I_C = 4A$ and $I_B = 20\,mA$ gives:

$$h_{FE} = \frac{I_C}{I_B} = \frac{4\,A}{80\,mA} = \frac{4\,A}{0.08\,A} = 50$$

Example 2.3

A small-signal BJT has a common emitter current gain of 125. If the transistor operates with a collector current of 50 mA, determine the value of base current.

Rearranging the formula $h_{FE} = I_C/I_B$ to make I_B the subject gives $I_B = I_C/h_{FE}$ from which:

$$I_B = \frac{50 \times 10^{-3}}{125} = 400\mu A$$

Test your understanding 2.6

A BJT has a common emitter current gain of 75. If a base current of 40 mA is applied to the device what will the collector current be?

Test your understanding 2.7

Estimate the common emitter current gain of the BJT whose transfer characteristic is shown in Fig. 2.35.

Key point

The common emitter current gain of a transistor is the ratio of collector current to base current and is typically in the range 40 to 200.

2.4 Integrated circuits

Considerable cost savings can be made by manufacturing all of the components required for a particular circuit function on one small slice of semiconductor material (usually silicon). The resulting **integrated circuit** (IC) may contain as few as 10 or more than 100,000 active devices (transistors and diodes). With the exception of a few specialized applications (such as amplification at high power levels) integrated circuits have largely rendered conventional circuits (i.e. those based on **discrete components**) obsolete.

Integrated circuits can be divided into two general classes, digital (logic) and linear (analogue).

A number of devices bridge the gap between the analogue and digital world. Such devices include **analogue to digital converters (ADCs)**, **digital to analogue converters (DACs)**, and **timers**.

Digital integrated circuits have numerous applications quite apart from their obvious use in computing. Digital signals exist only in discrete steps or levels; intermediate states are disallowed. Conventional electronic logic is based on two binary states, commonly referred to as logic 0 (low) and logic 1 (high). A comparison between digital and analogue signals is shown in Fig. 2.39. We shall look at digital logic in much greater detail in Chapter 3.

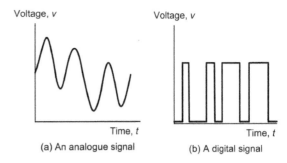

(a) An analogue signal   (b) A digital signal

Figure 2.39   Digital and analogue signals

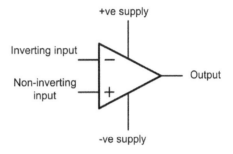

Figure 2.40   Symbol for an operational amplifier

Operational amplifiers are analogue integrated circuits designed for linear amplification that offer near-ideal characteristics (virtually infinite voltage gain and input resistance coupled with low output resistance and wide bandwidth).

Operational amplifiers can be thought of as universal 'gain blocks' to which external components are added in order to define their function within a circuit. By adding two resistors, we can produce an amplifier having a precisely defined gain. Alternatively, with three resistors and two capacitors we can realize a low-pass filter. From this you might begin to suspect that operational amplifiers are really easy to use. The good news is that they are!

The symbol for an operational amplifier is shown in Fig. 2.40. There are a few things to note about this. The device has two inputs and one output and no common connection. Furthermore, we often don't show the supply connections – it is often clearer to leave them out of the circuit altogether!

In Fig. 2.40, one of the inputs is marked '−' and the other is marked '+'. These polarity markings

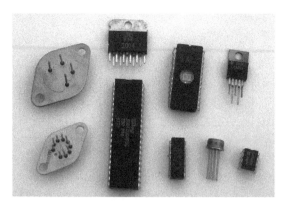

Figure 2.41 Various integrated circuits (including logic gates, operational amplifiers and memories)

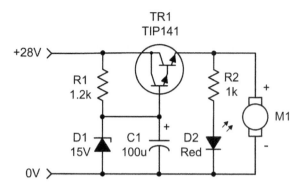

Figure 2.42 See Test your understanding 2.8

have nothing to do with the supply connections – they indicate the overall phase shift between each input and the output. The '+' sign indicates zero phase shift whilst the '−' sign indicates 180° phase shift. Since 180° phase shift produces an inverted (i.e., turned upside down) waveform, the '−' input is often referred to as the 'inverting input'. Similarly, the '+' input is known as the 'non-inverting' input.

Most (but not all) operational amplifiers require a symmetrical power supply (of typically ±6 V to ±15 V). This allows the output voltage to swing both positive (above 0 V) and negative (below 0 V). Other types of operational amplifier operate from a single supply voltage of usually between 5 V and 15 V.

Key point

Integrated circuits contain large numbers of individual components fabricated on a single slice of silicon. Integrated circuits are often classified as either digital (logic) or linear (analogue).

Key point

Operational amplifiers are linear integrated circuits that can be used as versatile 'gain blocks' within a wide variety of linear circuits.

Test your understanding 2.8

The circuit of a motor controller is shown in Fig. 2.42.

1. What type of transistor is TR1?
2. What type of diode is D1?
3. What type of diode is D2?
4. Which resistor sets the brightness of D2?
5. What voltage appears across C1?
6. What voltage appears across R1?
7. What voltage is delivered to M1?
8. What operating configuration is used for TR1?
9. What would happen if R1 is open circuit?
10. What would happen if C1 is short circuit?

2.5 Sensors and transducers

Sensors provide a means of generating signals that can be used as inputs to a wide variety of avionic systems. The physical parameters that we might need to sense include strain, temperature, and pressure. Being able to generate an electrical signal that accurately represents these physical quantities allows us not only to measure and record these physical values but also to control them.

2.5.1 Instrumentation and control systems

Fig. 2.43(a) shows the basic arrangement of an instrumentation system. The physical quantity to be measured (e.g. temperature) acts upon a sensor that produces an electrical output signal. This signal is an electrical analogue of the physical input but note that there may not be a straightforward linear relationship between the physical quantity and its electrical equivalent. Because of this, and since the output produced by the sensor may be small or may suffer from the presence of noise (i.e. unwanted signals), further signal conditioning will be required before the signal will be at an acceptable level and in an acceptable form for signal processing, display and recording. Furthermore, because the signal processing may use digital rather than analogue signals an additional stage of analogue-to-digital conversion may be necessary.

Fig. 2.43(b) shows the basic arrangement of a control system. This uses negative feedback in order to stabilise and regulate the output. It thus becomes possible to set the input or demand (i.e. what we desire the output to be) and leave the system to regulate itself by comparing the input with a signal derived from the output (via a sensor and appropriate signal conditioning).

A comparator is used to sense the difference in these two signals and where any discrepancy is detected the input to the power amplifier is adjusted accordingly. This signal is referred to as an error signal (it should be zero when the output exactly matches the demand). The input (demand) is often derived from a simple

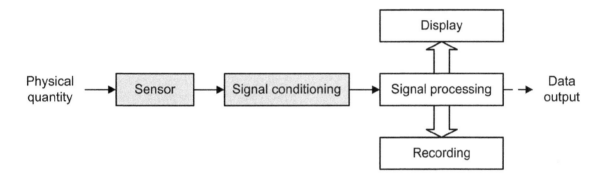

(a) An instrumentation system

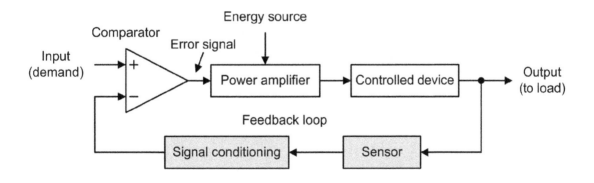

(b) A control system

Figure 2.43 Instrumentation and control systems

potentiometer connected across a stable d.c. voltage source while the controlled device can take many forms (e.g. a d.c. motor, linear actuator, heater, etc.).

2.5.2 Transducers

Transducers are devices that convert energy in the form of sound, light, heat, etc., into an equivalent electrical signal, or vice versa. Before we go further, let's consider a couple of examples that you will already be familiar with. A loudspeaker is a transducer that converts low frequency electric current into audible sounds. A microphone, on the other hand, is a transducer that performs the reverse function, i.e. that of converting sound pressure variations into voltage or current.

Transducers may be used both as inputs to electronic circuits and outputs from them. From the two previous examples, it will be apparent that a loudspeaker is an output transducer for use in conjunction with an audio system. A microphone, on the other hand, is an input transducer for use with a communication system. A variety of transducers used in aircraft systems are described and explained in detail in Chapters 9, 10 and 11 of this book.

2.5.3 Sensors

A sensor is a special kind of transducer that is used to generate an input signal to a measurement, instrumentation or control system. The signal produced by a sensor is an electrical analogy of a physical quantity, such as distance, velocity, acceleration, temperature, pressure, light level, etc. The signals returned from a sensor, together with control inputs from the user or controller (as appropriate) will subsequently be used to determine the output from the system. The choice of sensor is governed by several factors including accuracy, resolution, cost, temperature constraints and physical dimensions.

Sensors can be classed as either active or passive. An active sensor generates a current or voltage output. A passive transducer requires a source of current or voltage and it modifies this in some way (e.g. by virtue of a change in the sensor's resistance). The result may still be a voltage or current but it is not generated by the sensor on its own.

Sensors can also be classed as either digital or analogue. The output of a digital sensor can exist in only two discrete states, either *on* or *off*, *high* or *low*, *logic 1* or *logic 0* (see Chapter 3). By contrast the output of an analogue sensor can take any one of an infinite number of voltage or current levels. It is thus said to be continuously variable.

2.5.4 Thermocouples

Thermocouples (see Fig. 2.44) comprise a junction of dissimilar metals that generate a small e.m.f. proportional to the temperature differential which exists between the measuring junction and a reference junction. Since the measuring junction is usually at a greater temperature than that of the reference junction, it is sometimes referred to as the hot junction. Furthermore, the reference junction (i.e. the cold junction) is often omitted, in which case the sensing junction is simply terminated at the signal conditioning circuit. This circuit is usually maintained at, or near, normal ambient temperature.

Thermocouples are suitable for use over a very wide range of temperatures (from −100 °C to +1,100 °C). Industry standard 'type K' thermocouples comprise a positive arm (conventionally coloured brown) manufactured from nickel/chromium alloy while the negative arm (conventionally coloured blue) is manufactured from nickel/aluminium. The characteristic of a type-K thermocouple is defined in BS 4937 Part 4 of 1973 (International Thermocouple Reference Tables) and this standard gives tables of e.m.f. versus temperature over the range 0 °C to +1,100 °C. In order to minimise errors, it is usually necessary to connect thermocouples to appropriate signal conditioning using compensated cables and matching connectors.

2.6 Multiple choice questions

1. The forward voltage drop of a conducting silicon diode is approximately:
   (a) 0.2 V
   (b) 0.6 V
   (c) 2 V.

2. The region in a P-N junction diode where no free charge carriers exist is known as the:
   (a) collector
   (b) depletion region
   (c) enhancement region.

3. The device shown in Fig. 2.45 is:
   (a) a thyristor
   (b) a PNP transistor
   (c) an NPN transistor.

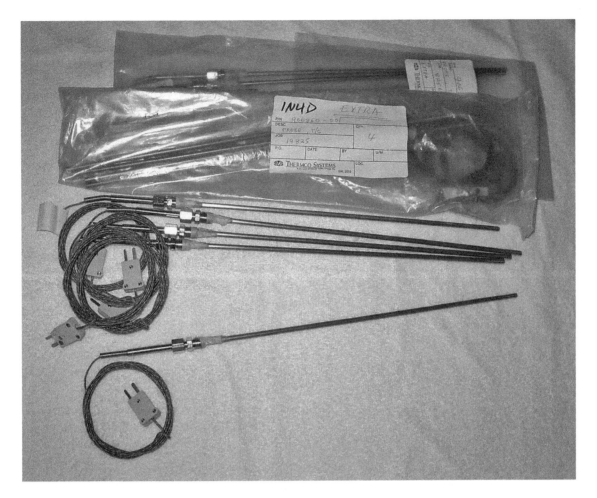

Figure 2.44   Temperature sensing thermocouple probes

Figure 2.45   See Question 3   **Figure 2.46**   See Question 6

4. The typical common emitter current gain for a bipolar transistor is:
   (a) less than 10
   (b) between 10 and 40
   (c) more than 40.

5. When a transistor is used in common emitter mode:
   (a) the input is taken to the base and the output is taken from the collector
   (b) the input is taken to the collector and the output is taken from the emitter
   (c) the input is taken to the emitter and the output is taken from the collector.

6. The device shown in Fig. 2.46 is used for:
   (a) power control
   (b) acting as a voltage reference
   (c) producing a visual indication.

7. In the bridge rectifier arrangement shown in Fig. 2.47 the DC output is taken from:
   (a) A and B
   (b) A and C
   (c) B and D.

8. The connections on a thyristor (silicon-controlled rectifier) are labelled:
   (a) anode, cathode, gate
   (b) collector, base, emitter
   (c) collector, gate, emitter.

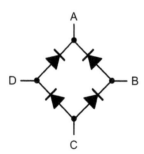

Figure 2.47 See Question 7

9. The anode of a silicon diode is connected to a +5 V DC supply and the cathode is connected to a +4 V DC supply. The diode is:
   (a) forward biased and not conducting
   (b) reverse biased and not conducting
   (c) forward biased and conducting.

10. A thyristor has which of the following?
    (a) high resistance when switched on
    (b) high resistance when switched off
    (c) a positive temperature coefficient.

11. A practical operational amplifier has:
    (a) very high input impedance and very high voltage gain
    (b) very high input impedance and very low voltage gain
    (c) very low input impedance and very high voltage gain.

12. A typical application for a Zener diode is:
    (a) voltage stabilization
    (b) power rectification
    (c) signal detection.

13. A transistor operates with a base current of 45 mA and a collector current of 1.8 A. Which of the following gives the current gain of the device when used in common emitter configuration?
    (a) 40
    (b) 90
    (c) 225.

14. Which one of the following relationships is correct for a BJT?
    (a) $I_B = I_C - I_E$
    (b) $I_C = I_E - I_B$
    (c) $I_E = I_C - I_B$.

15. The frequency of the ripple voltage present on the output of a bridge rectifier fed from a 400 Hz AC supply will be:
    (a) 200 Hz
    (b) 400 Hz
    (c) 800 Hz.

Digital fundamentals

In this chapter we introduce the basic building blocks of the digital logic systems found in modern aircraft. We begin with a review of the different types of logic gates before moving on to explore their use in several typical aircraft applications. The chapter also provides a brief overview of coding systems and the logic arrangements that are used to generate and convert the codes that are used to represent numerical data. The chapter provides an introduction to aircraft bus systems and concludes with a brief overview of the architecture and principal constituents of simple computer systems.

3.1 Logic gates

Aircraft logic systems follow the same conventions and standards as those used in other electronic applications. In particular, the MIL/ANSI standard logic symbols are invariably used and the logic elements that they represent operate in exactly the same way as those used in non-aircraft applications. MIL/ANSI standard symbols for the most common logic gates are shown together with their truth tables in Fig. 3.1.

3.1.1 Buffers

Buffers do not affect the logical state of a digital signal (i.e. a logic 1 input results in a logic 1 output whereas a logic 0 input results in a logic 0 output). Buffers are normally used to provide extra current drive at the output but can also be used to regularize the logic levels present at an interface.

Inverters are used to complement the logical state (i.e. a logic 1 input results in a logic 0 output and vice versa). Inverters also provide extra current drive and, like buffers, are used in interfacing applications where they provide a means of regularizing

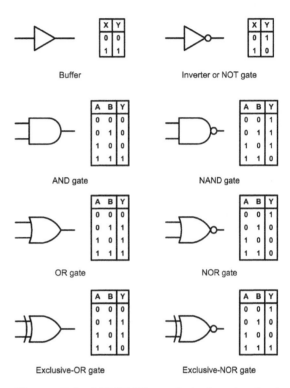

Figure 3.1 MIL/ANSI symbols for standard logic gates together with truth tables

logic levels present at the input or output of a digital system.

3.1.2 AND logic

AND gates will only produce a logic 1 output when all inputs are simultaneously at logic 1. Any other input combination results in a logic 0 output.

3.1.3 OR logic

OR gates will produce a logic 1 output whenever any one, or more, inputs are at logic 1. Putting this another way, an OR gate will only produce a logic 0 output whenever all of its inputs are simultaneously at logic 0.

3.1.4 NAND logic

NAND (i.e. NOT-AND) gates will only produce a logic 0 output when all inputs are simultaneously at logic 1. Any other input combination will produce a logic 1 output. A NAND gate, therefore, is nothing more than an AND gate with its output inverted. The circle shown at the output of the gate denotes this inversion.

3.1.5 NOR logic

NOR (i.e. NOT-OR) gates will only produce a logic 1 output when all inputs are simultaneously at logic 0. Any other input combination will produce a logic 0 output. A NOR gate, therefore, is simply an OR gate with its output inverted. A circle is again used to indicate inversion.

3.1.6 Exclusive-OR logic

Exclusive-OR gates will produce a logic 1 output whenever either one of the two inputs is at logic 1 and the other is at logic 0. Exclusive-OR gates produce a logic 0 output whenever both inputs have the same logical state (i.e. when both are at logic 0 or both are at logic 1).

3.1.7 Exclusive-NOR logic

Exclusive-NOR gates will produce a logic 0 output whenever either one of the two inputs is at logic 1 and the other is at logic 0. Exclusive-NOR gates produce a logic 1 output whenever both inputs have the same logical state (i.e. when both are at logic 0 or both are at logic 1).

3.1.8 Inverted inputs and outputs

The NAND and NOR gates that we have just met are said to have inverted outputs. In other words, they are respectively equivalent to AND and OR gates with their outputs passed through an inverter (or NOT gate) as shown in Fig. 3.2(a) and (b).

As well as inverted outputs, aircraft logic systems also tend to show logic gates in which one or more of the inputs is inverted. In Fig. 3.2(c) an AND gate is shown with one input inverted. This is equivalent to an inverter (NOT gate) connected to one input of the AND gate, as shown. In Fig. 3.2(d) an OR gate is shown with one input inverted. This is equivalent to an inverter (NOT gate) connected to one input of the OR gate, as shown.

Two further circuits with inverted inputs are shown in Fig. 3.2. In Fig. 3.2(e) both inputs of an AND gate are shown inverted. This arrangement is equivalent to the two-input NOR gate shown. In Fig. 3.2(f), both inputs of an OR gate are shown inverted. This arrangement is equivalent to the two-input NAND gate shown.

Key point

Logic circuits involve signals that can only exist in one of two mutually exclusive states. These two states are usually denoted by 1 and 0, 'on' or 'off', 'high' and 'low', 'closed' and 'open', etc.

3.2 Combinational logic systems

By using a standard range of logic levels (i.e. voltage levels used to represent the logic 1 and logic 0 states) logic circuits can be combined together in order to solve more complex logic functions. As an example, assume that a logic circuit is to be constructed that will produce a logic 1 output whenever two or more of its three inputs are at logic 1. This circuit is referred to as a majority vote circuit and its truth table is shown in Fig. 3.3. Figure 3.4 shows the logic circuitry required to satisfy the truth table.

3.2.1 Landing gear warning logic

Now let's look at a more practical example of the use of logic in the typical aircraft system shown in Fig. 3.5. The inputs to this logic system consist of five switches that detect whether or not the respective landing gear door is open. The output from the logic

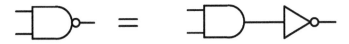

(a) AND gate with output inverted

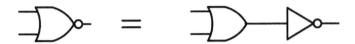

(b) OR gate with output inverted

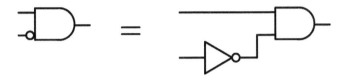

(c) AND gate with one input inverted

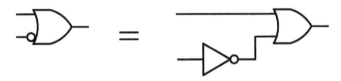

(d) OR gate with one input inverted

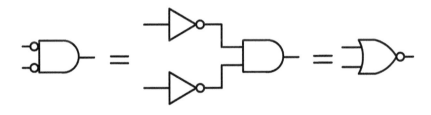

(e) AND gate with both inputs inverted

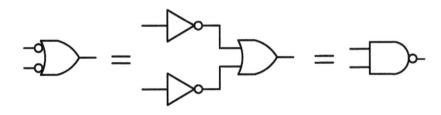

(f) OR gate with both inputs inverted

Figure 3.2 Logic gates with inverted outputs and inputs

system is used to drive six warning indicators. Four of these are located on the overhead display panel and show which door (or doors) are left open whilst an indicator located on the pilot's instrument panel provides a master landing gear door warning. A switch is also provided in order to enable or disable the five door warning indicators.

The landing gear warning logic primary module consists of the following integrated circuit devices:

A1	Regulated power supply for A5
A2	Regulated power supply for A7 and A11
A5	Ten inverting (NOT) gates
A7	Five-input NAND gate
A11	Six inverting (NOT) gates

Note that the power supply for A1 and A2 is derived from the essential services DC bus. This is a 28 V DC bus which is maintained in the event of an

A	B	C	Y
0	0	0	0
0	0	1	0
0	1	0	0
0	1	1	1
1	0	0	0
1	0	1	1
1	1	0	1
1	1	1	1

Figure 3.3   The majority vote truth table

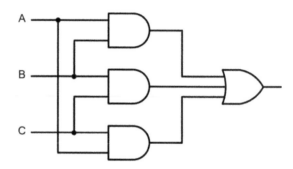

Figure 3.4   The majority vote logic

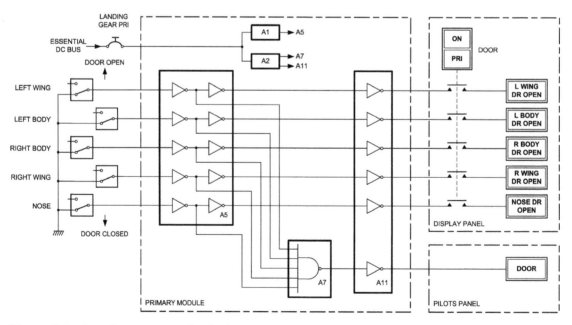

Figure 3.5   Landing gear warning logic

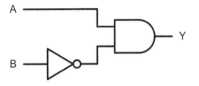

Figure 3.6 See Test your understanding 3.2

A	B	C	Y
0	0	0	0
0	0	1	0
0	1	0	1
0	1	1	1
1	0	0	0
1	0	1	1
1	1	0	1
1	1	1	1

Figure 3.7 See Test your understanding 3.2

aircraft power failure. Note also that the indicators are **active-low** devices (in other words, they require a logic 0 input in order to become illuminated).

Test your understanding 3.1

1. Show how a four-input AND gate can be made using three two-input AND gates.
2. Show how a four-input OR gate can be made using three two-input OR gates.

Test your understanding 3.2

1. Draw the truth table for the logic gate arrangement shown in Fig. 3.6.
2. Devise a logic gate arrangement that provides the output described by the truth table shown in Fig. 3.7.

3.3 Monostable devices

Monostable (or one-shot) devices provide us with a means of generating precise time delays. Such delays become important in many logic applications where logic states are not static but change with time.

Figure 3.8 Logic circuit based on transistor-transistor logic (TTL) integrated circuits

The action of a monostable is quite simple – its output is initially logic 0 until a change of state occurs at its trigger input. The level change can be from 0 to 1 (positive edge trigger) or 1 to 0 (negative edge trigger). Immediately the trigger pulse arrives, the output of the monostable changes state to logic 1. The output then remains at logic 1 for a pre-determined period before reverting back to logic 0.

3.3.1 APU starter logic

An example of the use of a monostable is shown in the auxiliary power unit (APU) starter logic shown in Fig. 3.9. This arrangement has three inputs (APU START, APU SHUTDOWN, and APU RUNNING) and one output (APU STARTER MOTOR). The inputs are all active-high (in other words, a logic 1 is generated when the pilot operates the APU START switch, and so on). The output of the APU starter motor control logic goes to logic 1 in order to apply power to the starter motor via a large relay.

There are a few things to note about the logic arrangement shown in Fig. 3.9:

1. When the APU runs on its own we need to disengage the starter motor. In this condition the APU MOTOR signal needs to become inactive (i.e. it needs to revert to logic 0).
2. We need to avoid the situation that might occur if the APU does not start but the starter motor runs continuously (as this will drain the aircraft batteries). Instead, we should run the starter motor for a reasonable time (say, 60 seconds) before disengaging the starter motor. The 60 second timing is provided by means of a positive edge triggered monostable device. This device is triggered from the APU START signal.
3. Since the pilot is only required to momentarily press the APU START switch, we need to hold the condition until such time as the engine starts or times out (i.e. at the end of the 60 second period).

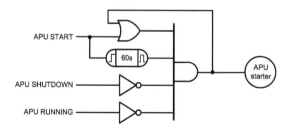

Figure 3.9   APU starter logic

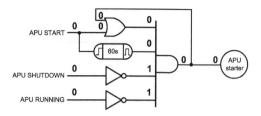

(a) Normal flight; engine power generation; APU not running

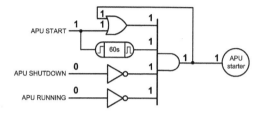

(b) APU starter switch operated; APU starter motor begins to run

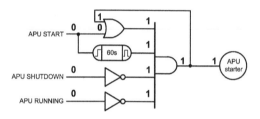

(c) APU starter motor continues to run for up to 60s

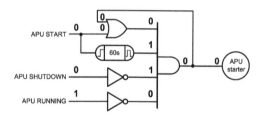

(d) APU runs before 60s timeout; starter motor stops when APU runs

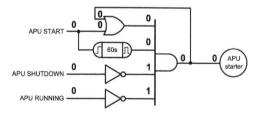

(e) APU fails to run during 60s period; further APU START signal awaited

Figure 3.10   APU starter operation

We can achieve this by OR'ing the momentary APU START signal with the APU STARTER MOTOR signal.

4. We need to provide a signal that the pilot can use to shut down the APU (for example, when the aircraft's main engines are running or perhaps in the event of a fault condition).

In order to understand the operation of the APU starter motor logic system we can once again trace through the logic system using 1's and 0's to represent the logical condition at each point (just as we did for the landing gear door warning logic).

In Fig. 3.10(a) the APU is in normal flight and the APU is not running. In this condition the main engines are providing the aircraft's electrical power.

In Fig. 3.10(b) the pilot is operating the APU START switch. The monostable is triggered and output of the OR and AND gates both go to logic 1 in order to assert the APU STARTER MOTOR signal.

In Fig. 3.10(c) the APU START signal is removed but the output of the AND gate is held at logic 1 by feeding back its logical state via the OR gate. The monostable remains triggered and continues to produce a logic 1 output for its 60 second period.

In Fig. 3.10(d) the APU is now running and the APU RUNNING signal has gone to logic 1 in order to signal this condition. This results in the output of the AND gate going to logic 0 and the APU STARTER MOTOR signal is no longer made active. The starter motor is therefore disengaged.

In Fig. 3.10(e) the APU has failed to run during the 60 second monostable period. In this timed out condition the output of the AND gate goes to logic 0 and the APU STARTER MOTOR signal becomes inactive. The system then waits for the pilot to operate the APU START button for a further attempt at starting!

Test your understanding 3.3

Part of an aircraft's generator control logic is shown in Fig. 3.11. Construct a truth table for the two outputs (GCB trip and GCR trip) and use this to explain the operation of the circuit. Under what conditions are the trip signals generated?

3.4 Bistable devices

A bistable device is a logic arrangement that is capable of 'remembering' a transient logic state such as a key press or a momentary overload condition. The output of a bistable circuit has two stable states (logic 0 or logic 1). Once **set** in one or other of these states, the output of a bistable will remain at a particular logic level for an indefinite period until **reset**. A bistable thus forms a simple form of memory as it remains in its latched state (either **set** or **reset**) until a signal is applied to it in order to change its state (or until the supply is disconnected).

The simplest form of bistable is the **R-S bistable**. This device has two inputs, SET and RESET, and complementary outputs, Q and /Q. A logic 1 applied to the SET input will cause the Q output to become (or remain at) logic 1 while a logic 1 applied to the RESET input will cause the Q output to become (or remain at) logic 0. In either case, the bistable will remain in its SET or RESET state until an input is applied in such a sense as to change the state.

Two simple forms of R-S bistable based on cross-coupled logic gates are shown in Fig. 3.12. Figure 3.12(a) is based on cross-coupled two-input NAND gates while Fig. 3.12(b) is based on cross-coupled two-input NOR gates.

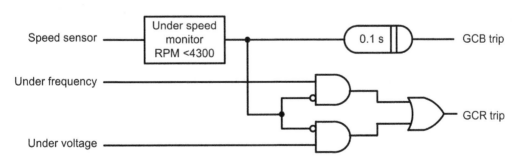

Figure 3.11 Generator control logic

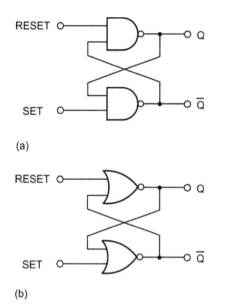

(a)

(b)

Figure 3.12 Simple R-S bistables based on: (a) NAND gates and (b) NOR gates

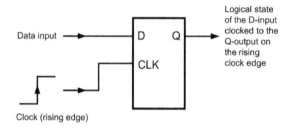

Clock (rising edge)

Figure 3.13 D-type bistable

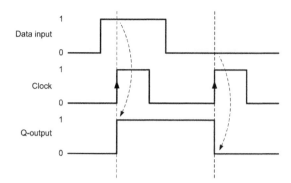

Figure 3.14 Timing diagram for the D-type bistable

memory) and as binary dividers. The simple circuit arrangement in Fig. 3.13 together with the **timing diagram** shown in Fig. 3.14 illustrate the operation of D-type bistables.

J-K bistables (see Fig. 3.15) have two clocked inputs (J and K), two direct inputs (PRESET and CLEAR), a CLOCK (CLK) input, and complementary outputs (Q and /Q). As with R-S bistables, the two outputs are complementary (i.e. when one is 0 the other is 1, and vice versa). Similarly, the PRESET and CLEAR inputs are invariably both active low (i.e. a 0 on the PRESET input will set the Q output to 1 whereas a 0 on the CLEAR input will set the Q output to 0). Figure 3.16 summarizes the input and corresponding output states of a J-K bistable for various input states. J-K bistables are the most sophisticated and flexible of the bistable types and they can be configured in various ways for use in binary dividers, shift registers, and latches.

3.5 Decoders

A variety of different coding schemes are used to represent numerical data in avionic systems. They include binary (or, more correctly, **natural binary**), **binary coded decimal (BCD)**, **Gray code**, octal (base 8), and **hexadecimal** (base 16) as shown in Fig. 3.17.

BCD uses four digits to represent each numerical character. Thus decimal 11 is represented by 0001 0001 (or just 10001 omitting the three leading zeros). Gray code is an important code because only one digit changes at a time. This property assists with error correction. Notice also how the least significant three bits of each Gray coded number become reflected after a decimal count of seven. Because of this, Gray code is often referred to as a **reflected code**.

Unfortunately, the simple cross-coupled logic gate bistable has a number of serious shortcomings (consider what would happen if a logic 1 was simultaneously present on both the SET and RESET inputs!) and practical forms of bistable make use of much improved purpose-designed logic circuits such as D-type and J-K bistables.

The **D-type bistable** has two inputs: D (standing variously for 'data' or 'delay') and CLOCK (CLK). The data input (logic 0 or logic 1) is clocked into the bistable such that the output state only changes when the clock changes state. Operation is thus said to be synchronous. Additional subsidiary inputs (which are invariably active low) are provided which can be used to directly set or reset the bistable. These are usually called PRESET (PR) and CLEAR (CLR). D-type bistables are used both as latches (a simple form of

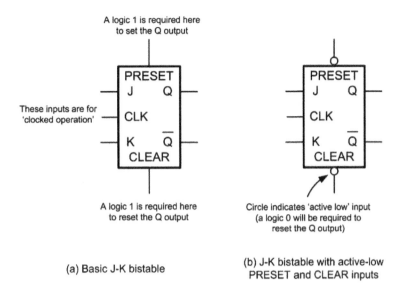

A logic 1 is required here
to set the Q output

These inputs are for
'clocked operation'

A logic 1 is required here
to reset the Q output

Circle indicates 'active low' input
(a logic 0 will be required to
reset the Q output)

(a) Basic J-K bistable

(b) J-K bistable with active-low
PRESET and CLEAR inputs

Figure 3.15   J-K bistable symbols

(a) PRESET and CLEAR inputs

Inputs		Output	Comment
PRESET	CLEAR	Q	
0	0	?	Indeterminate
0	1	0	Q output changes to 0 (i.e. Q is reset) regardless of the clock
1	0	1	Q output changes to 1 (i.e. Q is reset) on the next clock transition
1	1	-	Enables clocked operation - refer to the next truth table

Note that the PRESET and CLEAR inputs are unaffected by the state of the clock

(b) Clocked operation using the J and K inputs

Inputs		Output	Comment
J	K	Q_{N+1}	
0	0	Q_N	No change in state of the Q output on the next clock transition
0	1	0	Q output changes to 0 (i.e. Q is reset) on the next clock transition
1	0	1	Q output changes to 1 (i.e. Q is reset) on the next clock transition
1	1	Q_N	Q output changes to the opposite state on the next clock transition

Note that Q_N means 'Q in whatever state it was before' whilst Q_{N+1} means 'Q after the next clock transition'

Figure 3.16   Truth tables for the J-K bistable

Decoders are used to convert information from one number system to another, such as binary to octal or binary to decimal. A simple two to four line decoder is shown in Fig. 3.18. In this arrangement there are two inputs, A and B, and four outputs, Y_0, Y_1, Y_2 and Y_3.

The binary code appearing on A and B is decoded into one of the four possible states and corresponding output appears on the four output lines with Y_3 being the most significant. Because two to four and three to eight line decoders are frequently used as address

decoders in computer systems (where memory and I/O devices are invariably enabled by a low rather than a high state), the outputs are active-low, as indicated by the circles on the logic diagrams.

The internal logic arrangement of a two to four line decoder is shown in Fig. 3.19. This arrangement uses two inverters and three two-input NAND gates. All four outputs are active-low; Y_0 will go low when A and B are both at logic 0, Y_1 will go low when A is at logic 0 and B is at logic 1, Y_2 will go low when A is

Dec.	Binary	BCD	Gray	Octal	Hex.
0	0000	0000 0000	0000	0	0
1	0001	0000 0001	0001	1	1
2	0010	0000 0010	0011	2	2
3	0011	0000 0011	0010	3	3
4	0100	0000 0100	0110	4	4
5	0101	0000 0101	0111	5	5
6	0110	0000 0110	0101	6	6
7	0111	0000 0111	0100	7	7
8	1000	0000 1000	1100	10	8
9	1001	0000 1001	1101	11	9
10	1010	0001 0000	1111	12	A
11	1011	0001 0001	1110	13	B
12	1100	0001 0010	1010	14	C
13	1101	0001 0011	1011	15	D
14	1110	0001 0100	1001	16	E
15	1111	0001 0101	1000	17	F

Figure 3.17 Various coding schemes for numerical data

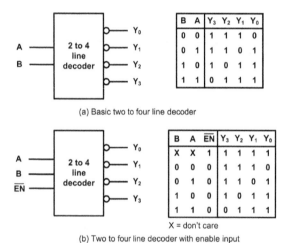

B	A	Y_3	Y_2	Y_1	Y_0
0	0	1	1	1	0
0	1	1	1	0	1
1	0	1	0	1	1
1	1	0	1	1	1

(a) Basic two to four line decoder

B	A	$\overline{EN}$	Y_3	Y_2	Y_1	Y_0
X	X	1	1	1	1	1
0	0	0	1	1	1	0
0	1	0	1	1	0	1
1	0	0	1	0	1	1
1	1	0	0	1	1	1

X = don't care

(b) Two to four line decoder with enable input

Figure 3.18 Two to four line decoders and their corresponding truth tables. The arrangement shown in (b) has a separate active-low enable input (when this input is high all of the outputs remain in the high state regardless of the A and B inputs)

at logic 1 and B is at logic 0, and Y_3 will go low when both A and B are at logic 1.

Test your understanding 3.4

1. Convert decimal 20 to (a) natural binary, (b) BCD, (c) octal, and (d) hexadecimal.
2. Convert natural binary 100001 to (a) decimal, (b) BCD, (c) octal, and (d) hexadecimal.

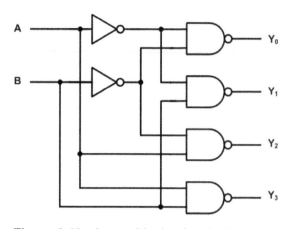

Figure 3.19 Internal logic of a simple two to four line decoder

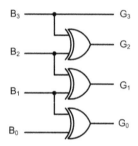

(a) Binary to Gray code conversion

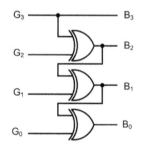

(b) Gray code to binary conversion

Figure 3.20 Natural binary to Gray code conversion

Test your understanding 3.5

Figure 3.20 shows simple logic arrangements that will perform natural binary to Gray code conversion and vice versa. Construct a truth table for each arrangement and use them to verify the operation of each circuit.

3.5.1 Gillham interface and Gillham code

Early altitude encoders were mechanical devices integrated into the barometric altimeter. The electrical output was a 10-bit parallel bus although more modern implementations use an expanded 11-bit bus. To ensure that only one bit of the bus changes on each transition (so that no transient false values are generated as the output changes between successive values)

the altitude is represented in a form of Gray code, not natural binary code (see table below). The interface required to perform this operation is commonly referred to as a **Gillham interface** and, as a result, the altitude code that it produces (which operates in increments of 100') is often referred to as **Gillham code**, see below:

Altitude (feet)	Data line									
	D4	A1	A2	A4	B1	B2	B4	C1	C2	C4
0	0	0	0	0	0	1	1	0	1	0
100	0	0	0	0	0	1	1	1	1	0
200	0	0	0	0	0	1	1	1	0	0
300	0	0	0	0	0	1	0	1	0	0
400	0	0	0	0	0	1	0	1	1	0
500	0	0	0	0	0	1	0	0	1	0
600	0	0	0	0	0	1	0	0	1	1
700	0	0	0	0	0	1	0	0	0	1
800	0	0	0	0	1	1	0	0	0	1
900	0	0	0	0	1	1	0	0	1	1
1000	0	0	0	0	1	1	0	0	1	0
1100	0	0	0	0	1	1	0	1	1	0
30500	0	1	0	0	0	0	0	0	1	0
30600	0	1	0	0	0	0	0	0	1	1
30700	0	1	0	0	0	0	0	0	0	1
30800	1	1	0	0	0	0	0	0	0	1
30900	1	1	0	0	0	0	0	0	1	1
31000	1	1	0	0	0	0	0	0	1	0
31100	1	1	0	0	0	0	0	1	1	0

Key maintenance point

Unfortunately, there has been at least one case when the failure of a single data line in a Gillham interface has generated erroneous altitude data which has caused an aircraft's **Traffic Alert and Collision Avoidance System (TCAS)** to guide an aircraft into a **near miss** situation. Hence there is now a requirement to test and verify the operation of Gillham-based systems. Furthermore, the use of Gillham code interfaces has actively been discouraged in favour of more modern digital systems.

3.6 Encoders

Encoders provide the reverse function to that of a decoder. In other words, they accept a number of inputs and then generate a binary code corresponding to the state of those inputs. Typical applications for encoders include generating a binary code corresponding to the state of a keyboard/keypad or generating BCD from a decade (ten-position) rotary switch.

A particularly useful form of encoder is one that can determine the priority of its inputs. This device is known as a **priority encoder** and its inputs are arranged in priority order, from lowest to highest priority. If more

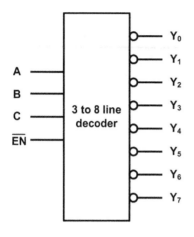

C	B	A	$\overline{EN}$	Y_7	Y_6	Y_5	Y_4	Y_3	Y_2	Y_1	Y_0
X	X	X	1	1	1	1	1	1	1	1	1
0	0	0	0	1	1	1	1	1	1	1	0
0	0	1	0	1	1	1	1	1	1	0	1
0	1	0	0	1	1	1	1	1	0	1	1
0	1	1	0	1	1	1	1	0	1	1	1
1	0	0	0	1	1	1	0	1	1	1	1
1	0	1	0	1	1	0	1	1	1	1	1
1	1	0	0	1	0	1	1	1	1	1	1
1	1	1	0	0	1	1	1	1	1	1	1

X = don't care

Figure 3.21 An eight to three line priority encoder and its corresponding truth table. Note that the enable (EN) input is active-low

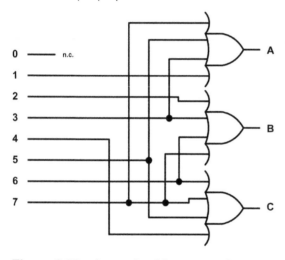

Figure 3.22 An octal to binary encoder

than one input becomes active, the input with the highest priority will be encoded and its binary coded value will appear on the outputs. The state of the other (lower priority) inputs will be ignored.

Figure 3.21 shows an eight to three line priority encoder together with its corresponding truth table. Note that this device has active-low inputs as well as active-low outputs.

Test your understanding 3.6

Figure 3.22 shows the circuit of an octal to binary encoder. Verify the operation of this circuit by constructing its truth table.

3.7 Multiplexers

Like encoders, multiplexers have several inputs. However, unlike encoders, they have only one output. Multiplexers provide a means of selecting data from one of several sources. Because of this, they are often referred to as **data selectors**. Switch equivalent circuits of some common types of multiplexer are shown in Fig. 3.23.

The single two-way multiplexer in Fig. 3.23(a) is equivalent to a simple SPDT (changeover) switch. The dual two-way multiplexer shown in Fig. 3.23(b) performs the same function but two independent circuits are controlled from the same select signal. A single four-way multiplexer is shown in Fig. 3.23(c). Note that two digital select inputs are required, A and B, in order to place the switch in its four different states.

Block schematic symbols, truth tables and simplified logic circuits for two to one and four to one multiplexers are shown in Figs 3.24 to 3.27.

Test your understanding 3.7

1. Verify the operation of the two to one multiplexer circuit shown in Fig. 3.25 by constructing its truth table.
2. Verify the operation of the four to one multiplexer circuit shown in Fig. 3.27 by constructing its truth table.

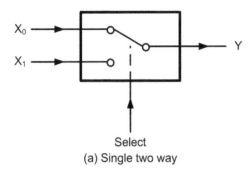

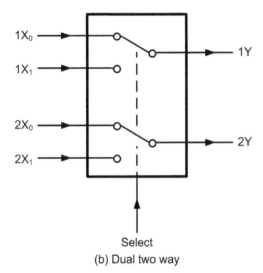

(a) Single two way

(b) Dual two way

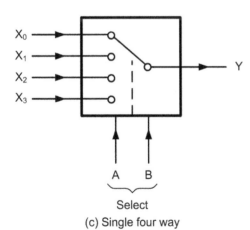

(c) Single four way

Figure 3.23 Switch equivalent circuits for some common types of multiplexer or data selector

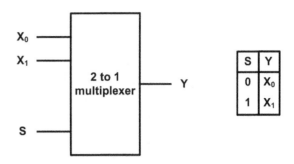

S	Y
0	X_0
1	X_1

Figure 3.24 A basic two to one multiplexer arrangement with its corresponding truth table

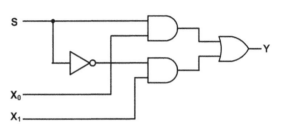

Figure 3.25 Logic circuit arrangement for the basic two to one multiplexer

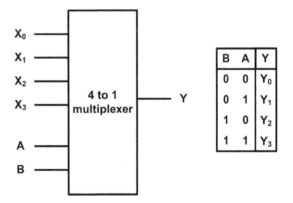

B	A	Y
0	0	Y_0
0	1	Y_1
1	0	Y_2
1	1	Y_3

Figure 3.26 A four to one multiplexer. The logic state of the A and B inputs determines which of the four logic inputs (X_0 to X_3) appears at the output

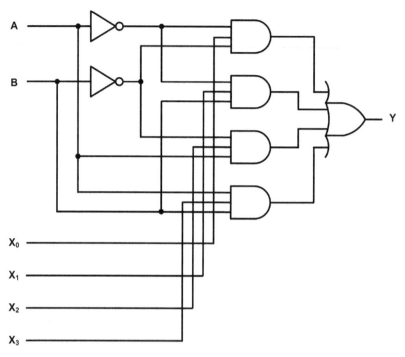

Figure 3.27   Logic gate arrangement for the four to one multiplexer

3.8 Bus systems

Aircraft data bus systems allow a wide variety of avionics equipment to communicate with one another and exchange data. Bus systems can be either **bidirectional** (two way) or **unidirectional** (one way), as shown in Fig. 3.28. They can also be serial (one bit of data transmitted at a time) or **parallel** (where often 8, 16 or 32 bits of data appear as a group on a number of data lines at the same time). Because of the constraints imposed by conductor length and weight, all practical aircraft bus systems are based on serial (rather than parallel) data transfer.

Bus systems provide an efficient means of exchanging data between the diverse avionic systems found in a modern aircraft (see Fig. 3.29). Individual **line replaceable units** (**LRUs**), such as the engine data interface or flap/slat electronics units shown in Fig. 3.29, are each connected to the bus by means of a dedicated **bus coupler** and **serial interface module** (not shown in Fig. 3.29). Within the LRU, the dedicated digital logic and microprocessor systems that process data locally each make use of their own **local bus** system. These local bus systems invariably use

parallel data transfer which is ideal for moving large amounts of data very quickly but only over short distances.

Key point

Modern aircraft use multiple redundant bus systems for exchanging data between the various avionic systems and sub-systems. These bus systems use serial data transfer because it minimizes the size and weight of aircraft cabling.

3.8.1 Serial bus principles

A simple system for serial data transfer between two line replaceable units, each of which comprises an avionic system in its own right, is shown in Fig. 3.30. Within the LRU data is transferred using an internal parallel data bus (either 8, 16, 32 or 64 bits wide). The link between the two LRUs is made using a simple serial cable (often with only two, four or six conductors). The required parallel-to-serial and serial-to-parallel data

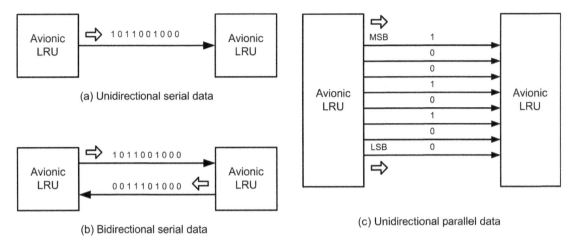

Figure 3.28 Unidirectional and bidirectional serial and parallel data

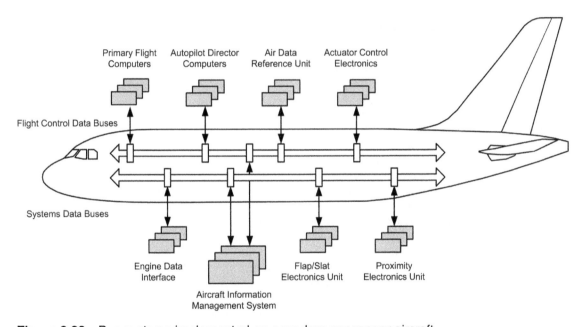

Figure 3.29 Bus systems implemented on a modern passenger aircraft

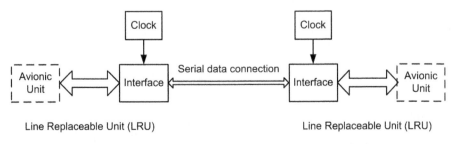

Figure 3.30 A simple system for serial data transfer between two avionic systems

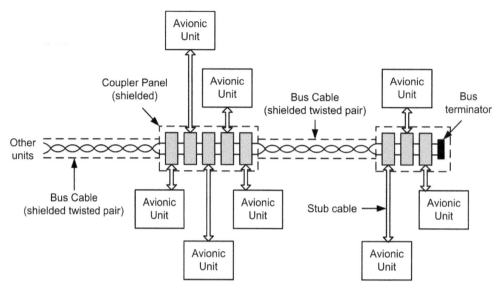

Figure 3.31   A practical aircraft data bus

conversion is carried out by a bus interface (often this is a single card or module within the LRU). The data to be transferred can be **synchronous** (using clock signals generated locally within each LRU) or it may be **asynchronous** (i.e. self-clocking).

The system shown in Fig. 3.30 has the obvious limitation that data can only be exchanged between two devices. In practice we need to share the data between many LRU/avionic units. This can be achieved by the bus system illustrated in Fig. 3.31. In this system, data is transferred using a **shielded twisted pair** (STP) **bus cable** with a number of **coupler panels** that are located at appropriate points in the aircraft (e.g. the flight deck, avionics bay, etc.). Each coupler panel allows a number of avionic units to be connected to the bus using a **stub cable**. In order to optimize the speed of data transfer and minimize problems associated with reflection and mismatch, the bus cable must be terminated at each end using a matched **bus terminator**.

Bus couplers are produced as either **voltage mode** or **current mode** units depending upon whether they use voltage or current sensing devices. Within each LRU/avionics unit, an interface is provided that performs the required serial-to-parallel or parallel-to-serial data conversion, as shown in Fig. 3.32.

3.8.2 ARINC 429

The ARINC 429 data bus has proved to be one of the most popular bus standards used in commercial

aircraft. The ARINC 429 specification defines the electrical and data characteristics and protocols that are used.

ARINC 429 employs a unidirectional data bus standard known as Mark 33 Digital Information Transfer System (DITS). Messages are transmitted in packets of 32 bits at a bit rate of either 12.5 or 100 kilobits per second (referred to as low and high bit rate respectively). Because the bus is unidirectional, separate ports, couplers and cables will be required when an LRU wishes to be able to both transmit and receive data. Note that a large number of bus connections may be required on an aircraft that uses sophisticated avionic systems.

ARINC 429 has been installed on a wide variety of commercial transport aircraft including; Airbus A310/A320 and A330/A340; Boeing 727, 737, 747, 757, and 767; and McDonnell Douglas MD-11. More modern aircraft (e.g. Boeing 777 and Airbus A380) use significantly enhanced bus specifications in order to reduce the weight and size of cabling and to facilitate higher data rates than are possible with ARINC 429. Despite these moves to faster, bidirectional bus standards, the ARINC 429 standard has proved to be highly reliable and so is likely to remain in service for many years to come.

ARINC 429 is a two wire **differential** bus which can connect a single transmitter or source to one or more receivers. The term 'differential' simply means that neither of the two twisted wires is grounded and

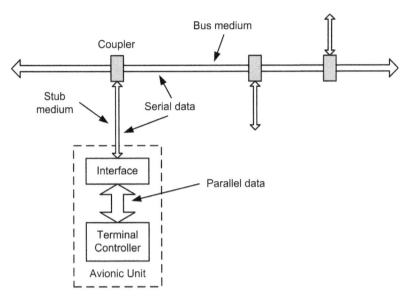

Figure 3.32 A basic bus interface

both convey signal voltages but of opposite polarity. This arrangement improves noise immunity which appears as a common mode signal induced on both of the conductors.

Two **bus speeds** are available, 12.5 kbps and 100 kbps. The data bus uses two signal wires to transmit 32-bit words. Transmission of sequential words is separated by at least four bit times of NULL (zero voltage). This eliminates the need for a separate clock signal and it makes the system **self-clocking**.

The ARINC 429 electrical characteristics are summarized below:

Voltage levels:	+5 V, 0 V, −5 V (each conductor with respect to ground) +10 V, 0 V, −10 V (conductor A with respect to conductor B)
Data encoding:	Bi-polar return to zero
Word size:	32 bits
Bit rate:	100 K bits per second (high), 12.5 K bits per second (low)

It is important to note that the received voltage on a serial bus depends on line length and the number of receivers connected to the bus. With ARINC 429, no more than 20 receivers should be connected to a single bus. Since each bus is unidirectional, a system needs to have its own transmit bus if it is required to respond or to send messages. Hence, to achieve bidirectional data transfer it is necessary to have two separate bus connections.

Since there can be only one transmitter on a twisted wire pair, ARINC 429 uses a very simple, point-to-point **protocol**. The transmitter is continuously sending 32-bit data words or is placed in the NULL state. Note that although there may only be one receiver on a particular bus cable the ARINC specification supports up to 20.

Other, faster and more sophisticated, bus systems are found on modern aircraft. They include ARINC 629 which supports a 20 Mbps data rate (20 times faster than ARINC 429), FDDI (Boeing's Fiber Distributed Data Interface), and 10 Mbps Ethernet.

Test your understanding 3.8

In relation to an aircraft bus system explain the meaning of each of the following terms:

(a) bus coupler
(b) bus terminator
(c) differential bus
(d) self-clocking
(e) NULL state.

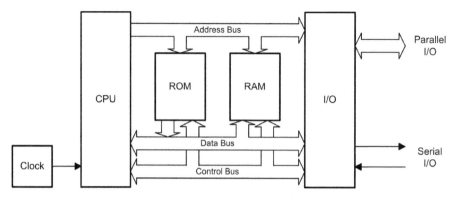

Figure 3.33 Basic components of a computer system

3.9 Computers

Modern aircraft use increasingly sophisticated avionic systems which involve the use of microprocessor-based computer systems. These systems combine hardware and software and are capable of processing large amounts of data in a very small time.

The basic components of a computer system are shown in Fig. 3.33. The main components are:

(a) a central processing unit (CPU)
(b) a memory, comprising both 'read/write' and 'read only' devices (commonly called RAM and ROM respectively)
(c) a means of providing input and output (I/O). For example, a keypad for input and a display for output.

In a microprocessor system the functions of the CPU are provided by a single very large scale integrated (VLSI) microprocessor chip. This chip is equivalent to many thousands of individual transistors.

Semiconductor devices are also used to provide the read/write and read-only memory. Strictly speaking, both types of memory permit 'random access' since any item of data can be retrieved with equal ease regardless of its actual location within the memory. Despite this, the term 'RAM' has become synonymous with semiconductor read/write memory.

The basic components of the system (CPU, RAM, ROM and I/O) are linked together using a multiple-wire connecting system know as a **bus** (see Fig. 3.33). Three different buses are present, these are:

(a) the **address bus** used to specify memory locations;

(b) the **data bus** on which data is transferred between devices; and

(c) the **control bus** which provides timing and control signals throughout the system.

The number of individual lines present within the address bus and data bus depends upon the particular microprocessor employed. Signals on all lines, no matter whether they are used for address, data, or control, can exist in only two basic states: logic 0 (**low**) or logic 1 (**high**). Data and addresses are represented by **binary numbers** (a sequence of 1 s and 0 s) that appear respectively on the data and address bus.

Some basic microprocessors designed for control and instrumentation applications have an 8-bit data bus and a 16-bit address bus. More sophisticated processors can operate with as many as 64 or 128 bits at a time.

The largest binary number that can appear on an 8-bit data bus corresponds to the condition when all eight lines are at logic 1. Therefore the largest value of data that can be present on the bus at any instant of time is equivalent to the binary number 11111111 (or 255). Similarly, the highest address that can appear on a 16-bit address bus is 1111111111111111 (or 65,535). The full range of data values and addresses for a simple microprocessor of this type is thus:

Data	From	00000000
	To	11111111
Addresses	From	0000000000000000
	To	1111111111111111

Finally, a locally generated clock signal provides a time reference for controlling the transfer of synchronous data within the system. The clock signal usually

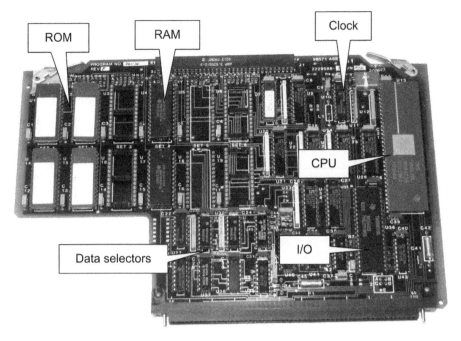

Figure 3.34   A typical aircraft computer system

consists of a high-frequency square wave pulse train derived from an accurate quartz crystal controlled oscillator.

Key point

A computer system consists of a central processing unit (CPU), a read-only memory (ROM), a read/write (random access) memory (RAM), and one or more input/output (I/O) devices. These elements are linked together using a local bus system that comprises an address bus, a data bus, and a control bus.

Key point

Depending on the size of their internal data bus, avionic computer systems usually manipulate data in groups of 8, 16 or 32 bits. When this data is obtained from an external serial bus this data must first be assembled into parallel form in order to facilitate internal processing.

3.9.1 Memories and data storage

The semiconductor read-only memory (ROM) within a microprocessor system provides storage for the program code as well as any permanent data that requires storage. All of this data is referred to as **non-volatile** because it remains intact when the power supply is disconnected.

The semiconductor RAM within a microprocessor system provides storage for the transient data and variables that are used by programs. Part of the RAM is also used by the microprocessor as a temporary store for data whilst carrying out its normal processing tasks.

It is important to note that any program or data stored in RAM will be lost when the power supply is switched off or disconnected. The only exception to this is low-power CMOS RAM that is kept alive by means of a small battery. This **battery-backed memory** is used to retain important data, such as the time and date.

When expressing the amount of storage provided by a memory device we usually use kilobytes (Kbytes). It is important to note that a kilobyte of memory is actually 1024 bytes (not 1000 bytes). The reason for choosing the Kbyte rather than the kbyte (1000 bytes) is that 1024 happens to be the nearest power of 2 (note that $2^{10} = 1024$).

The capacity of a semiconductor ROM is usually specified in terms of an address range and the number of bits stored at each address. For example, $2\,K \times 8$ bits (capacity $2\,K$bytes), $4\,K \times 8$ bits (capacity $4\,K$bytes), and so on. Note that it is not always necessary (or desirable) for the entire memory space of a computer to be populated by memory devices.

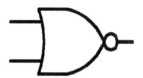

Figure 3.35 See Question 1

Key point

The term 'random access' simply refers to a memory device in which data may be retrieved from all locations with equal ease (i.e. access time is independent of actual memory address). This is important since our programs often involve moving sizeable blocks of data into and out of memory.

Key maintenance point

Stray static charges can very easily damage static-sensitive devices such as microprocessors and memory chips. Damage can be prevented by adopting the appropriate ESD procedures which usually involve using grounded wrist straps when handling chips and boards and using specially treated conductive packaging for transport and storage of component parts.

Test your understanding 3.9

In relation to a basic computer system explain the function of each of the following components:

(a) CPU
(b) clock
(c) RAM
(d) ROM
(e) I/O.

3.10 Multiple choice questions

1. The logic symbol shown in Fig. 3.35 is for:
   (a) a NAND gate
   (b) a NOR gate
   (c) an OR gate.

2. A logic 1 is present at the output of a two-input NOR gate. Which one of the following is true?
   (a) both of its inputs must be at logic 1
   (b) both of its inputs must be at logic 0
   (c) one or more of its inputs must be at logic 1.

3. The decimal number 10 is equivalent to the binary number:
   (a) 1000
   (b) 1010
   (c) 1100.

4. The BCD number 10001001 is equivalent to the decimal number:
   (a) 89
   (b) 141
   (c) 137.

5. The output of an R-S bistable is at logic 1. This means that the bistable is:
   (a) set
   (b) reset
   (c) in a high impedance state.

6. In a volatile memory:
   (a) the memory is lost as soon as power is removed
   (b) the memory is retained indefinitely
   (c) the memory needs to be refreshed constantly, even when power is on.

7. The main elements of a computer system are:
   (a) CPU, memory and I/O ports
   (b) ALU, registers and a ROM
   (c) RAM, ROM and I/O.

8. A NOR gate with both inputs inverted becomes a
   (a) NAND gate
   (b) AND gate
   (c) OR gate.

9. Another name for a data selector is:
   (a) a monostable
   (b) a multiplexer
   (c) an encoder.

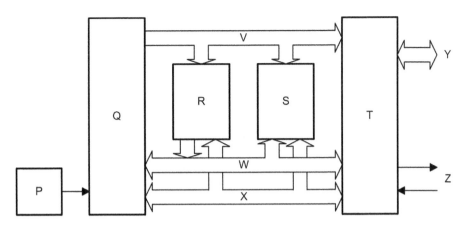

Figure 3.36 See Question 15

10. A group of bits transmitted one after another is:
    (a) a clock signal
    (b) parallel data
    (c) serial data.

11. A bus that supports the transfer of data in both directions is referred to as:
    (a) universal
    (b) bidirectional
    (c) asynchronous.

12. The main advantage of using a serial bus in an aircraft is:
    (a) there is no need for data conversion
    (b) it supports the highest possible data rates
    (c) reduction in the size and weight of cabling.

13. The maximum speed of an ARINC 429 bus is:
    (a) 12.5 Kbps
    (b) 100 Kbps
    (c) 10 Mbps.

14. What is the maximum number of receivers that may be connected to an ARINC 429 bus cable?
    (a) 1
    (b) 16
    (c) 20.

15. The feature marked Q in Fig. 3.36 is:
    (a) CPU
    (b) RAM
    (c) I/O.

Chapter 4

Generators and motors

Engine-driven generators are a primary source of electrical power in an aircraft. Generators can supply either direct or alternating current (DC or AC), as appropriate to the needs of an individual aircraft type. Motors (which can also be either DC or AC types) are fitted in aircraft in order to satisfy a wide range of needs. Generators and motors share many similarities and this chapter provides an introduction to their operating principles by looking in more detail at the construction and operation of these indispensable aircraft electrical components. The chapter also includes a brief introduction to three-phase AC supplies, including the theoretical and practical aspects of their generation and distribution.

4.1 Generator and motor principles

In Chapter 1 we introduced the concept of electromagnetic induction. Put simply, this is the generation of an e.m.f. across the ends of a conductor when it passes through a change in magnetic flux. In a similar fashion, an e.m.f. will appear across the ends of a conductor if it remains stationary whilst the field moves. In either case, the action of cutting through the lines of magnetic flux results in a generated e.m.f. – see Fig. 4.1. The amount of e.m.f., e, *induced* in the conductor will be directly proportional to:

- the density of the magnetic flux, B, measured in tesla (T)
- the effective length of the conductor, l, within the magnetic flux
- the speed, v, at which the lines of flux cut through the conductor measured in metres per second (m/s)
- the sine of the angle, θ, between the conductor and the lines of flux.

The induced e.m.f. is given by the formula:

$$e = Blv \sin \theta$$

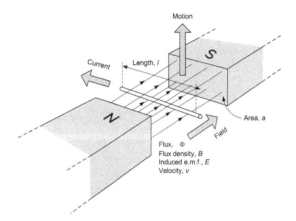

Figure 4.1 A conductor moving inside a magnetic field

Note that if the conductor moves at right angles to the field (as shown in Fig. 4.1) maximum e.m.f. will be induced. Conversely, if the conductor moves along the lines of flux (instead of at right angles) the induced e.m.f. will be zero.

Electricity and magnetism often work together to produce motion. In an electric **motor**, current flowing in a conductor placed inside a magnetic field produces motion. A **generator**, on the other hand, produces a voltage when a conductor is moved inside a magnetic field. These two effects are, as you might suspect, closely related to one another and they are vitally important in the context of aircraft electrical systems!

Key point

An e.m.f. will be induced across the ends of a conductor when there is relative motion between it and a magnetic field. The induced voltage will take its greatest value when moving at right angles to the magnetic field lines and its least value (i.e. zero) when moving along the direction of the field lines.

4.1.1 A simple AC generator

Being able to generate a voltage by moving a conductor through a magnetic field is extremely useful as it provides us with an easy way of generating electricity. Unfortunately, moving a wire at a constant linear velocity through a uniform magnetic field presents us with a practical problem simply because the mechanical power that can be derived from an aircraft engine is available in rotary (rather than linear) form!

The solution to this problem is that of using the rotary power available from the engine (via a suitable gearbox and transmission) to rotate a conductor shaped into the form of loop as shown in Fig. 4.2. The loop is made to rotate inside a permanent magnetic field with opposite poles (N and S) on either side of the loop.

There now remains the problem of making contact with the loop as it rotates inside the magnetic field but this can be overcome by means of a pair of carbon **brushes** and copper **slip-rings**. The brushes are spring loaded and held against the rotating slip-rings so that, at any time, there is a path for current to flow from the loop to the load to which it is connected.

The opposite sides of the loop consist of conductors that move through the field. At 0° (with the loop vertical as shown at A in Fig. 4.4) the opposite sides of the loop will be moving in the same direction as the lines of flux. At that instant, the angle, θ, at which the field is cut is 0° and since the sine of 0° is 0 the generated voltage (from $E = Blv\sin\theta$) will consequently also be zero.

If the loop has rotated to a position which is 90° (position B in Fig. 4.4) the two conductors will effectively be moving at right angles to the field. At that instant, the generated e.m.f. will take a maximum value (since the sine of 90° is 1).

At 180° from the starting position the generated e.m.f. will have fallen back to zero since, once again, the conductors are moving along the flux lines (but in the direction opposite to that at 0°, as shown in C of Fig. 4.4).

At 270° the conductors will once again be moving in a direction which is perpendicular to the flux lines (but in the direction opposite to that at 90°). At this point (D of Fig. 4.4), a maximum generated e.m.f. will once again be produced. It is, however, important to note that the e.m.f. generated at this instant will be of opposite polarity to that which was generated at 90°.The reason for this is simply that the relative direction of motion (between the conductors and flux lines) has effectively been reversed.

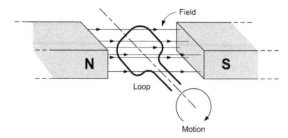

Figure 4.2   A loop rotating within a magnetic field

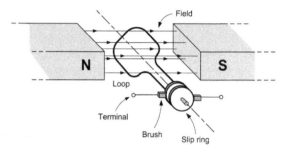

Figure 4.3   Brush arrangement

Since $E = Blv\sin\theta$, the e.m.f. generated by the arrangement shown in Fig. 4.4 will take a sinusoidal form, as shown in Fig. 4.5. Note that the maximum values of e.m.f. occur at 90° and 270° and that the generated voltage is zero at 0°, 180° and 360°.

In practice, the single loop shown in Fig. 4.2 would comprise a coil of wire wound on a suitable non-magnetic former. This coil of wire effectively increases the length of the conductor within the magnetic field and the generated e.m.f. will then be directly proportional to the number of turns on the coil.

Key point

In a simple AC generator a loop of wire rotates inside the magnetic field produced by two opposite magnetic poles. Contact is made to the loop as it rotates by means of slip-rings and brushes.

4.1.2 DC generators

When connected to a load, the simple generator shown in Fig. 4.3 produces a sinusoidal alternating current (AC) output. In many applications a steady direct current (DC) output may be preferred.

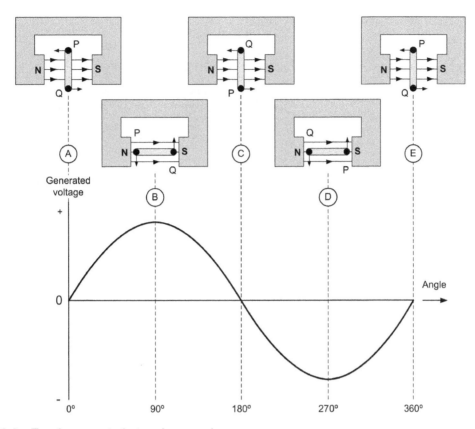

Figure 4.4   E.m.f. generated at various angles

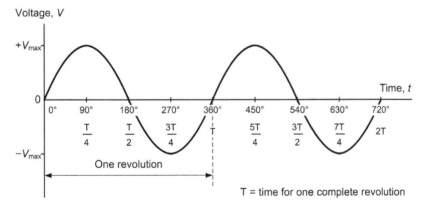

Figure 4.5   Sinusoidal voltage produced by the rotating loop

This can be achieved by modifying the arrangement shown in Fig. 4.3, replacing the brushes and slip-rings with a **commutator** arrangement, as shown in Fig. 4.6. The commutator arrangement functions as a rotating reversing switch which ensures that the e.m.f. generated by the loop is reversed after rotating through 180°. The generated e.m.f. for this arrangement is shown in Fig. 4.7. It's worth comparing this waveform with that shown in Fig. 4.5 – you should be able to spot the difference immediately!

The generated e.m.f. shown in Fig. 4.7, whilst **unipolar** (i.e. having only one polarity, either all positive or all negative), is clearly far from ideal since a DC power source should provide a constant voltage output rather than a series of pulses. One way of overcoming this problem is with the use of a second loop (or coil) at right angles to the first, as shown in Fig. 4.8. The commutator is then divided into four (rather than two) **segments** and the generated e.m.f. produced by this arrangement is shown in Fig. 4.9.

In real generators, a coil comprising a large number of turns of copper wire replaces the single-turn rotating loop. This arrangement effectively increases the total length of the conductor within the magnetic field and, as a result, also increases the generated output voltage. The output voltage also depends on the density of the magnetic flux through which the current-carrying conductor passes. The denser the field the greater the output voltage will be.

Key point

A simple DC generator uses an arrangement similar to that used for an AC generator but with the slip-rings and brushes replaced by a commutator that reverses the current produced by the generator every 180°.

4.1.3 DC motors

A simple DC motor consists of a very similar arrangement to that of the DC generator that we met earlier.

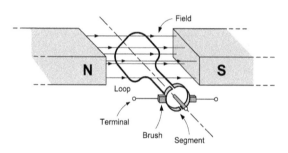

Figure 4.6 Commutator arrangement

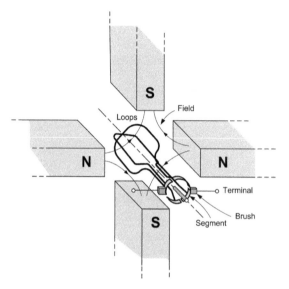

Figure 4.8 An improved DC generator

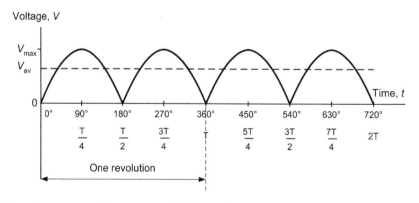

Figure 4.7 E.m.f. generated (compare with Fig. 4.5)

A loop of wire that is free to rotate is placed inside a permanent magnetic field (see Fig. 4.10). When a DC current is applied to the loop of wire, two equal and opposite forces are set up which act on the conductor in the directions indicated in Fig. 4.10.

The direction of the forces acting on each arm of the conductor can be established by again using the right-hand grip rule and Fleming's left-hand rule. Now because the conductors are equidistant from their pivot point and the forces acting on them are *equal and opposite*, then they form a **couple**. The **moment** of this couple is equal to the magnitude of a single force multiplied by the distance between them and this moment is known as **torque**, *T*. Now,

$$T = Fd$$

where *T* is the torque (in newton-metres, Nm), *F* is the force (N) and *d* is the distance (m).

We already know that the magnitude of the force *F* is given by *F = BIl*, therefore the torque produced by the current-carrying thus the torque expression can be written:

$$T = BIld$$

where *T* is the torque (Nm), *B* is the flux density (T), *I* is the current (A), *l* is the length of conductor in the magnetic field (m), and *d* is the distance (m).

The torque produces a turning moment such that the coil or loop rotates within the magnetic field. This rotation continues for as long as a current is applied. A more practical form of DC motor consists of a rectangular coil of wire (instead of a single turn loop of wire) mounted on a former and free to rotate about a shaft in a permanent magnetic field, as shown in Fig. 4.11.

In real motors, this rotating coil is know as the **armature** and consists of many hundreds of turns of conducting wire. This arrangement is needed in order to maximize the force imposed on the conductor by introducing the *longest possible* conductor into the

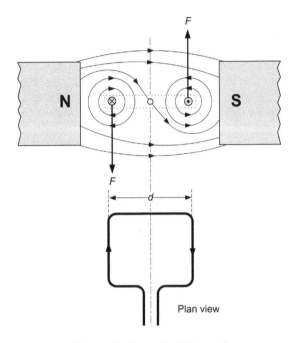

Torque, *T* = Force, *F* × Distance, *d*

Figure 4.10 Torque on a current-carrying loop suspended within a permanent magnetic field

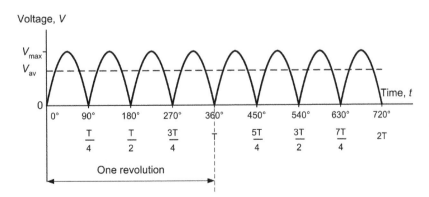

Figure 4.9 E.m.f. generated (compare with Fig. 4.7)

magnetic field. Also, from the relationship $F = BIl$ it can be seen that the force used to provide the torque in a motor is directly proportional to the size of the magnetic flux, B. Instead of using a permanent magnet to produce this flux, in a real motor, an electromagnet is used. Here an electromagnetic field is set up using the **solenoid** principle (Fig. 4.12). A long length of conductor is wound into a coil consisting of many turns and a current passed through it. This arrangement constitutes a **field winding** and each of the turns in the field winding assists each of the other turns in order to produce a strong magnetic field, as shown in Fig. 4.12.

As in the case of the DC generator, this field may be intensified by inserting a ferromagnetic core inside the coil. Once the current is applied to the conducting coil, the core is magnetized and all the time the current is on it acts in combination with the coil to produce a permanent magnet, having its own N–S poles.

Now returning to the simple motor illustrated in Fig. 4.11, we know that when current is supplied to the rotating armature (**rotor**) a torque is produced. In order to produce continuous rotary motion, this torque (turning moment) must always act in the same direction.

Therefore, the current in each of the armature conductors must be reversed as the conductor passes between the north and south magnetic field poles. The commutator acts like a rotating switch, reversing the current in each armature conductor at the appropriate time to achieve this continuous rotary motion. Without the presence of a commutator in a DC motor, only a half turn of movement is possible!

In Fig. 4.13(a) the rotation of the armature conductor is given by Fleming's left-hand rule (see Fig. 4.14). When the coil reaches a position mid-way between the poles (Fig. 4.13(b)), no rotational torque is produced in the coil. At this stage the commutator reverses the current in the coil. Finally (Fig. 4.13(c)) with the current reversed, the motor torque now continues to rotate the coil in its original direction.

Key point

The torque produced by a DC motor is directly proportional to the product of the current flowing in the rotating armature winding.

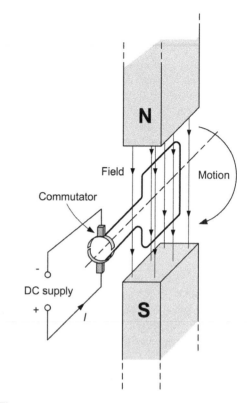

Figure 4.11 Simple electric motor with commutator

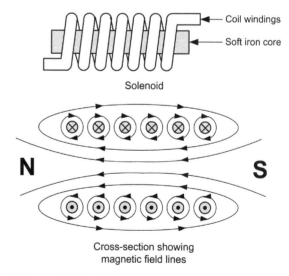

Figure 4.12 Magnetic field produced by a solenoid

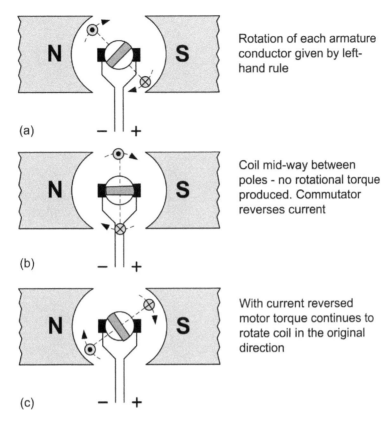

(a)   − ‖ +

Rotation of each armature conductor given by left-hand rule

Coil mid-way between poles - no rotational torque produced. Commutator reverses current

(b)   − ‖ +

With current reversed motor torque continues to rotate coil in the original direction

(c)   − ‖ +

Figure 4.13   Action of the commutator

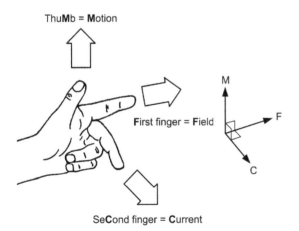

Figure 4.14   Fleming's left-hand rule

4.1.4 Field connections

The field winding of a DC motor can be connected in various different ways according to the application

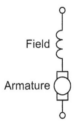

Figure 4.15   Series-wound DC motor

envisaged for the motor in question. The following configurations are possible:

- series-wound
- shunt-wound
- compound-wound (where both series and shunt windings are present).

In the **series-wound** DC motor the field winding is connected in series with the armature and the full armature current flows through the field winding (see Fig. 4.15). This arrangement results in a DC motor

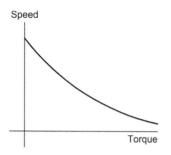

Figure 4.16 Typical torque and speed characteristic for a series-wound DC motor

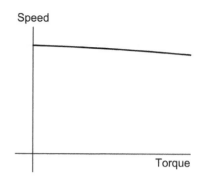

Figure 4.18 Typical torque and speed characteristic for a shunt-wound DC motor

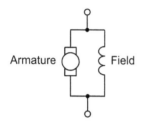

Figure 4.17   Shunt-wound DC motor

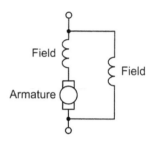

Figure 4.19   Compound-wound DC motor

that produces a large starting torque at slow speeds. This type of motor is ideal for applications where a heavy load is applied from rest. The disadvantage of this type of motor is that on light loads the motor speed may become excessively high. For this reason this type of motor should not be used in situations where the load may be accidentally removed. A typical set of torque and speed characteristics for a series-wound DC motor is shown in Fig. 4.16.

In the **shunt-wound** DC motor the field winding is connected in parallel with the armature and thus the supply current is divided between the armature and the field winding (see Fig. 4.17). This arrangement results in a DC motor that runs at a reasonably constant speed over a wide variation of loads but does not perform well when heavily loaded. A typical set of torque and speed characteristics for a shunt-wound DC motor is shown in Fig. 4.18.

The **compound-wound** DC motor has both series and shunt field windings (see Fig. 4.19) and is therefore able to combine some of the advantages of each type of motor. A typical set of torque and speed characteristics for a compound-wound DC motor is shown in Fig. 4.20.

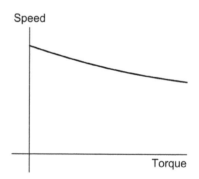

Figure 4.20 Typical torque and speed characteristic for a compound-wound DC motor

---

Key point

In order to avoid the need for a large permanent magnet in a DC machine (i.e. a motor or generator), a separate field winding is used. This winding is energized with direct current. In the case of a DC generator, this current can be derived from the output of the generator (in which case it is referred to as a self-excited generator) or it can be energized from a separate DC supply.

Key point

There is no real difference between the construction of a DC generator and that of a DC motor. The only significant distinction is that in a generator rotational mechanical energy is being converted to electrical energy whilst in the case of a motor the converse applies.

Key point

The output of a generator can be regulated by controlling the intensity of its field: increasing the field current will produce a corresponding increase in output voltage whereas reducing the field current will produce a corresponding reduction in output.

4.1.5 Starter-generators

Some small aircraft starter-generators eliminate the need for separate engine starter motors and DC generators. They usually have separate field windings (one for the starter motor and one for the generator) together with a common armature winding. When used for starting, the starter-generator is connected as a series-wound DC motor capable of producing a very high starting torque. However, when used as a generator the connections are changed so that the unit operates as a shunt-wound generator producing reasonably constant current over a wide range of speed.

In the start condition, the low-resistance starter field and common armature windings of the starter-generator are connected in series across the DC supply via a set of contactors. This arrangement ensures that a torque is produced that is sufficient to start an aircraft's turbine engine.

When the engine reaches self-sustaining speed, the current is broken through the first set of contactors and a second set of contactors operate, removing the external DC power supply from the starter-generator and reconnecting the arrangement so that

the generated armature voltage is fed to the higher-resistance shunt field and the aircraft's main voltage regulator.

The advantage of this arrangement is not only that the starter-generator replaces two individual machines (i.e. a starter and a generator) with consequent savings in size and weight, but additionally that only a single mechanical drive is required between the engine and the starter-generator unit. The disadvantage of this arrangement is that the generator output is difficult to maintain at low engine r.p.m. and therefore starter-generators are mainly found on turbine-powered aircraft that maintain a relatively high engine r.p.m.

Test your understanding 4.1

1. In relation to a simple DC machine, explain the meaning of each of the following terms:
   (a) slip-rings
   (b) brushes
   (c) field winding.
2. Explain the advantages and disadvantages of series-wound DC motors compared with shunt-wound DC motors.

4.2 AC generators

AC generators, or **alternators**, are based on the principles that relate to the simple AC generator that we met earlier in Section 4.1.1. However, in a practical AC generator the magnetic field is rotated rather than the conductors from which the output is taken. Furthermore, the magnetic field is usually produced by a rotating electromagnet (the *rotor*) rather than a permanent magnet. There are a number of reasons for this including:

(a) the conductors are generally lighter in weight than the magnetic field system and are thus more easily rotated
(b) the conductors are more easily insulated if they are stationary
(c) the currents which are required to produce the rotating magnetic field are very much smaller than those which are produced by the conductors. Hence the slip-rings are smaller and more reliable.

Figure 4.22 shows the simplified construction of a **single-phase AC generator**. The *stator* consists of five coils of insulated heavy gauge wire located in

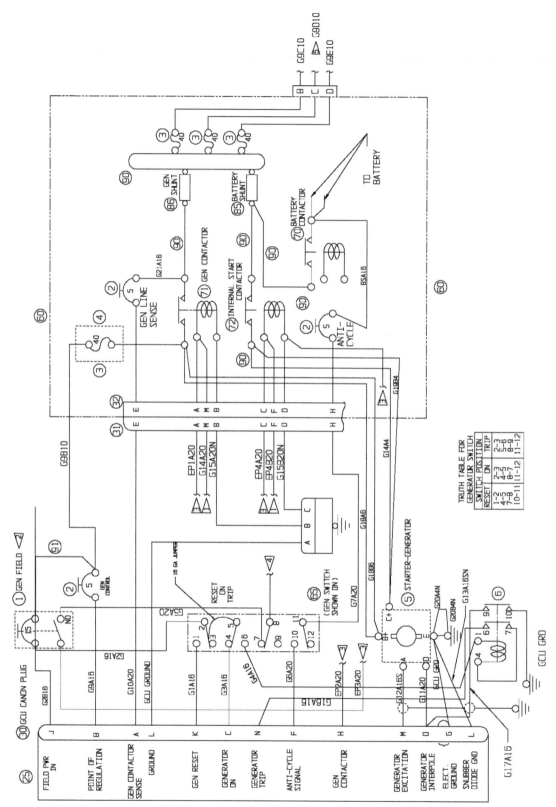

Figure 4.21 Starter-generator circuit showing changeover contactors

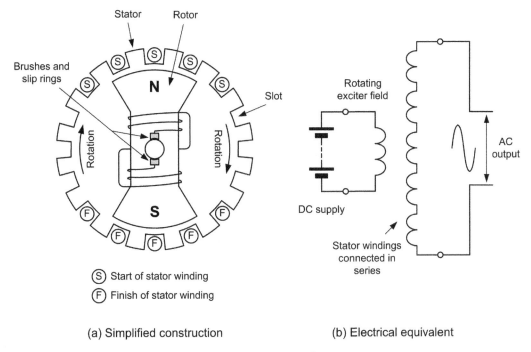

Stator   Rotor

Brushes and
slip rings

Rotation

Rotation

N

S

Slot

Rotating
exciter field

DC supply

Stator windings
connected in
series

AC
output

(S) Start of stator winding
(F) Finish of stator winding

(a) Simplified construction   (b) Electrical equivalent

Figure 4.22   Simplified construction of a single-phase AC generator

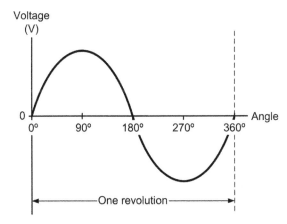

Voltage
(V)

0

0°   90°   180°   270°   360°   Angle

One revolution

Figure 4.23   Output voltage produced by the single-phase AC generator shown in Fig. 4.22

slots in the high-permeability laminated core. These coils are connected in series to make a single stator winding from which the output voltage is derived.

The two-pole rotor comprises a field winding that is connected to a DC field supply via a set of slip-rings and brushes. As the rotor moves through one complete revolution the output voltage will complete one full cycle of a sine wave, as shown in Fig. 4.23.

By adding more pairs of poles to the arrangement shown in Fig. 4.22 it is possible to produce several cycles of output voltage for one single revolution of the rotor. The frequency of the output voltage produced by an AC generator is given by:

$$f = \frac{pN}{60}$$

where f is the frequency of the induced e.m.f. (in Hz), p is the number of pole pairs, and N is the rotational speed (in rev/min).

Example 4.1

An alternator is to produce an output at a frequency of 60 Hz. If it uses a 4-pole rotor, determine the shaft speed at which it must be driven.

Re-arranging $f = \dfrac{pN}{60}$ to make N the subject gives:

$$N = \frac{60f}{p}$$

A 4-pole machine has 2 pairs of poles, thus $p = 2$ and:

$$N = \frac{60 \times 60}{2} = 1,800 \text{ rev/min}$$

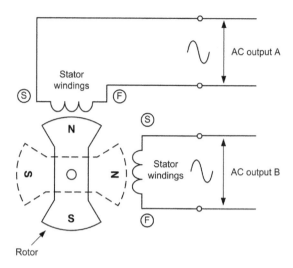

Figure 4.24 Simplified construction of a two-phase AC generator

Key point

In a practical AC generator, the magnetic field excitation is produced by the moving rotor whilst the conductors from which the output is taken are stationary and form part of the stator.

4.2.1 Two-phase AC generators

By adding a second stator winding to the single-phase AC generator shown in Fig. 4.22, we can produce an alternator that produces two separate output voltages which will differ in phase by 90°. This arrangement is known as a two-phase AC generator.

When compared with a single-phase AC generator of similar size, a two-phase AC generator can produce more power. The reason for this is attributable to the fact that the two-phase AC generator will produce two positive and two negative pulses per cycle whereas the single-phase generator will only produce one positive and one negative pulse. Thus, over a period of time, a multi-phase supply will transmit a more evenly distributed power and this, in turn, results in a higher overall efficiency.

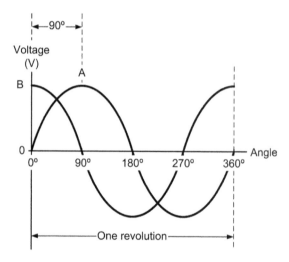

Figure 4.25 Output voltage produced by the two-phase AC generator shown in Fig. 4.24

next section. In a practical three-phase system the three output voltages are identified by the colours red, yellow, and blue or by letters A, B, and C respectively.

Key point

Three-phase AC generators are more efficient and produce more constant output than comparable single-phase AC generators.

4.2.2 Three-phase AC generators

The three-phase AC generator has three individual stator windings, as shown in Fig. 4.26. The output voltages produced by the three-phase AC generator are spaced by 120° as shown in Fig. 4.27. Each phase can be used independently to supply a different load or the generator outputs can be used with a three-phase distribution system like those described in the

Test your understanding 4.2

An alternator uses a 12-pole rotor and is to operate at a frequency of 400 Hz. At what speed must it be driven?

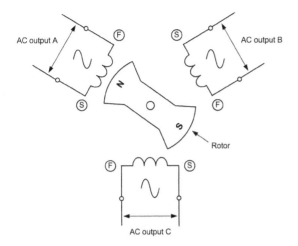

Figure 4.26 Simplified construction of a three-phase AC generator

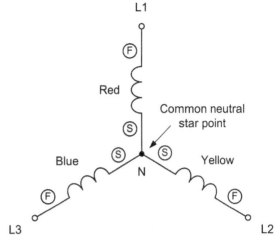

Figure 4.28 Star connection

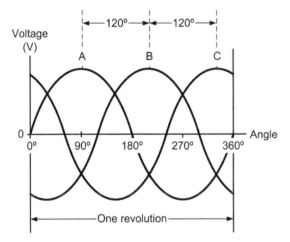

Figure 4.27 Output voltage produced by the three-phase AC generator shown in Fig. 4.26

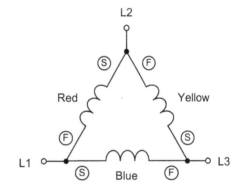

Figure 4.29 Delta connection

4.3 Three-phase generation and distribution

When three-phase supplies are distributed there are two basic methods of connection:

● star (as shown in Fig. 4.28)
● delta (as shown in Fig. 4.29)

A complete star-connected three-phase distribution system is shown in Fig. 4.30. This shows a three-phase AC generator connected to a three-phase load. Ideally, the load will be **balanced** in which case all three load resistances (or impedances) will be identical.

The relationship between the line and phase voltages shown in Fig. 4.30 can be determined from the **phasor diagram** shown in Fig. 4.31. This diagram shows the relative directions of the three alternating phase voltages (V_P) and the voltages between the lines (V_L). From this diagram it is important to note that three line voltages are 120° apart and that the line voltages lead the phase voltages by 30°. In order to obtain the relationship between the line voltage, V_L, and the phase voltage, V_P, we need to resolve any one of the triangles, from which we find that:

$$V_L = 2(V_P \times \cos 30°)$$

Now

$$\cos 30° = \frac{\sqrt{3}}{2}$$

and hence:

$$V_L = 2\left(V_P \times \frac{\sqrt{3}}{2}\right)$$

from which:

$$V_L = \sqrt{3}\, V_P$$

Note also that the phase current is the same as the line current, hence:

$$I_P = I_L$$

An alternative, delta-connected three-phase distribution system is shown in Fig. 4.32. Once again this shows a three-phase AC generator connected to a three-phase load. Here again, the load will ideally be balanced in which case all three load resistances (or impedances) will be identical.

In this arrangement the three line currents are 120° apart and lag the phase currents by 30°. Using a similar phasor diagram to that which we used earlier, we can show that:

$$I_L = \sqrt{3}\, I_P$$

It should also be obvious that:

$$V_P = V_L$$

Example 4.2

In a star-connected three-phase system the phase voltage is 105 V. Determine the line voltage.

$$V_L = \sqrt{3}\, V_P = \sqrt{3} \times 115 = 199.2\,V$$

Example 4.3

In a delta-connected three-phase system the line current is 6 A. Determine the phase current.

$$I_L = \sqrt{3}\, I_P$$

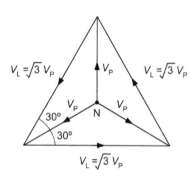

Figure 4.31 Phasor diagram for the three-phase star-connected system

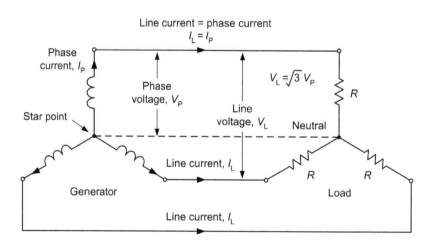

Figure 4.30 A complete star-connected three-phase distribution system

from which:

$$I_P = \frac{I_L}{\sqrt{3}} = \frac{6}{1.732} = 3.46\,A$$

4.3.1 Power in a three-phase system

In an unbalanced three-phase system the total power will be the sum of the individual phase powers. Hence:

$$P = P_1 + P_2 + P_3$$

or

$$P = V_1 I_1 \cos \phi_1 + V_2 I_2 \cos \phi_2 + V_3 I_3 \cos \phi_3$$

However, in the balanced condition the power is simply:

$$P = 3\,V_P I_P \cos \phi$$

where V_P and I_P are the phase voltage and phase current respectively and ϕ is the phase angle.

Using the relationships that we derived earlier, we can show that, for both the star- and delta-connected systems the total power is given by:

$$P = \sqrt{3}\,V_L I_L \cos \phi$$

Example 4.4

In a three-phase system the line voltage is 110 V and the line current is 12 A. If the power factor is 0.8 determine the total power supplied.

Here it is important to remember (see Section 1.7.7) that:

$$\text{Power factor} = \cos \phi$$

and hence:

$$\begin{aligned}P &= \sqrt{3}\,V_L I_L \times \text{power factor} \\ &= \sqrt{3} \times 110 \times 12 \times 0.8 = 1829 = 1.829\ \text{kW}\end{aligned}$$

Key point

The total power in a three-phase system is the sum of the power present in each of the three phases.

Test your understanding 4.3

1. The phase voltage in a star-connected AC system is 220 V. What will the line voltage be?
2. The phase current in a delta-connected system is 12 A. What will the line current be?
3. A three-phase system delivers power to a load consisting of three 12 Ω resistors. If a current of 8 A is supplied to each load, determine the total power supplied by the system.
4. In a three-phase system the line voltage is 105 V and the line current is 8 A. If the power factor is 0.75 determine the total power supplied.

4.4 AC motors

AC motors offer significant advantages over their DC counterparts. AC motors can, in most cases,

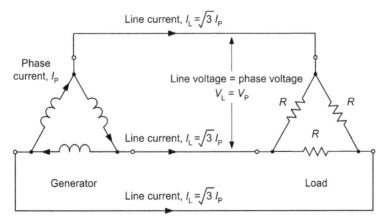

Figure 4.32 A complete delta-connected three-phase distribution system

duplicate the operation of DC motors and they are significantly more reliable. The main reason for this is that the commutator arrangements (i.e. brushes and slip-rings) fitted to DC motors are inherently troublesome. Because the speed of an AC motor is determined by the frequency of the AC supply that is applied to it, AC motors are well suited to constant-speed applications.

AC motors are generally classified into two types:

- synchronous motors
- induction motors.

The synchronous motor is effectively an AC generator (i.e. an alternator) operated as a motor. In this machine, AC is applied to the stator and DC is applied to the rotor. The induction motor is different in that no source of AC or DC power is connected to the rotor. Of these two types of AC motor, the induction motor is by far the most commonly used.

Key point

The principle of all AC motors is based on the generation of a rotating magnetic field. It is this rotating field that causes the motor's rotor to turn.

4.4.1 Producing a rotating magnetic field

Before we go any further it's important to understand how a rotating magnetic field is produced. Take a look at Fig. 4.33 which shows a three-phase stator to which three-phase AC is applied. The windings are connected in delta configuration, as shown in Fig. 4.34. It is important to note that the two windings for each phase (diametrically opposite to one another) are wound in the *same* direction.

At any instant the magnetic field generated by one particular phase depends on the current through that phase. If the current is zero, the magnetic field is zero. If the current is a maximum, the magnetic field is a maximum. Since the currents in the three windings are 120° out of phase, the magnetic fields generated will also be 120° out of phase.

The three magnetic fields that exist at any instant will combine to produce one field that acts on the rotor. The magnetic fields inside the motor will combine to produce a moving magnetic field and, at the end of one complete cycle of the applied current, the

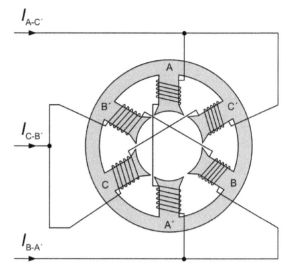

Figure 4.33 Arrangement of the field windings of a three-phase AC motor

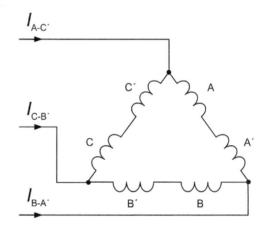

Figure 4.34 AC motor as a delta-connected load

magnetic field will have shifted through 360° (or one complete revolution).

Figure 4.35 shows the three current waveforms applied to the field system. These waveforms are 120° out of phase with each other. The waveforms can represent either the three alternating magnetic fields generated by the three phases, or the currents in the phases.

We can consider the direction of the magnetic field at regular intervals over a cycle of the applied current (i.e. every 60°). To make life simple we take the times at which one of the three current waveforms passes

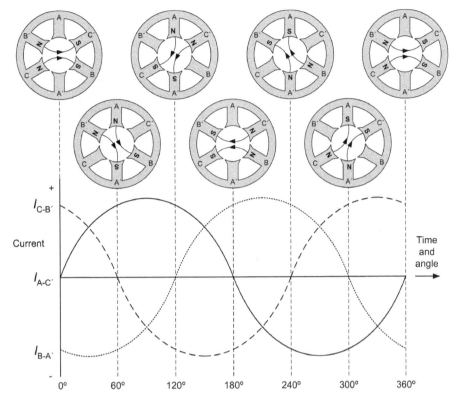

Figure 4.35   AC waveforms and magnetic field direction

through zero (i.e. the point at which there will be no current and therefore no field produced by one pair of field windings). For the purpose of this exercise we will use the current applied to A and C' as our reference waveform (i.e. this will be the waveform that starts at 0° on our graph).

At 0°, waveform C–B' is positive and waveform B–A' is negative. This means that the current flows in opposite directions through phases B and C, and so establishes the magnetic polarity of phases B and C. The polarity is shown in Fig. 4.35. Note that B' is a north pole and B is a south pole, and that C is a north pole and C' is a south pole.

Since at 0° there is no current flowing through phase A, its magnetic field is zero. The magnetic fields leaving poles B' and C will move towards the nearest south poles C' and B. Since the magnetic fields of B and C are equal in amplitude, the resultant magnetic field will lie between the two fields, and will have the direction shown.

At the next point, 60° later, the current waveforms to phases A and B are equal and opposite, and

waveform C is zero. The resultant magnetic field has rotated through 60°. At point 120°, waveform B is zero and the resultant magnetic field has rotated through another 60°. From successive points (corresponding to one cycle of AC), you will note that the resultant magnetic field rotates through one revolution for every cycle of applied current. Hence, by applying a three-phase alternating current to the three windings we have been able to produce a rotating magnetic field.

Key point

If three windings are placed round a stator frame, and three-phase AC is applied to the windings, the magnetic fields generated in each of the three windings will combine into a magnetic field that rotates. At any given instance, these fields combine together in order to produce a resultant field that acts on the rotor. The rotor turns because the magnetic field rotates.

4.4.2 Synchronous motors

We have already shown how a rotating magnetic field is produced when a three-phase alternating current is applied to the field coils of a stator arrangement. If the rotor winding is energized with DC, it will act like a bar magnet and it will rotate in sympathy with the rotating field. The speed of rotation of the magnetic field depends on the frequency of the three-phase AC supply and, provided that the supply frequency remains constant, the rotor will turn at a constant speed. Furthermore, the speed of rotation will remain constant regardless of the load applied. For many applications this is a desirable characteristic; however, one of the disadvantages of a synchronous motor is that it cannot be started from a standstill by simply applying three-phase AC to the stator. The reason for this is that the instant AC is applied to the stator, a high-speed rotating field appears. This rotating field moves past the rotor poles so quickly that the rotor does not have a chance to get started. Instead, it is repelled first in one direction and then in the other.

Another way of putting this is simply that a synchronous motor (in its pure form) has no starting torque. Instead, it is usually started with the help of a small induction motor (or with windings equivalent to this incorporated in the synchronous motor). When the rotor has been brought near to synchronous speed by the starting device, the rotor is energized by connecting it to a DC voltage source. The rotor then falls into step with the rotating field. The requirement to have an external DC voltage source as well as the AC field excitation makes this type of motor somewhat unattractive!

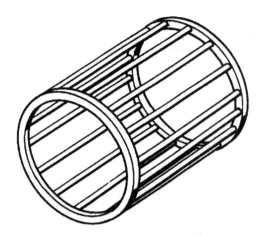

Figure 4.36   Squirrel cage rotor construction

4.4.3 Three-phase induction motors

The induction motor derives its name from the fact that AC currents are induced in the rotor circuit by the rotating magnetic field in the stator. The stator construction of the induction motor and of the synchronous motor are almost identical, but their rotors are completely different.

The induction motor rotor is a laminated cylinder with slots in its surface. The windings in these slots are one of two types. The most common uses so-called **squirrel cage** construction (see Fig. 4.36) that is made up of heavy copper bars connected together at either end by a metal ring made of copper or brass. No insulation is required between the core and the bars because of the very low voltages generated in the rotor bars. The air gap between the rotor and stator is kept very small so as to obtain maximum field strength.

The other type of winding contains coils placed in the rotor slots. The rotor is then called a **wound rotor**. Just as the rotor usually has more than one conductor, the stator usually has more than one pair of poles per coil, as shown in Fig. 4.37.

Key point

The synchronous motor is so called because its rotor is synchronized with the rotating field set up by the stator. Its construction is essentially the same as that of a simple AC generator (alternator).

Key point

Synchronous motors are not self-starting and must be brought up to near synchronous speed before they can continue rotating by themselves. In effect, the rotor becomes 'frozen' by virtue of its inability to respond to the changing field!

Key point

The induction motor is the most commonly used AC motor because of its simplicity, its robust construction and its relatively low cost. These advantages arise from the fact that the rotor of an induction motor is a self-contained component that is *not actually electrically connected to an external source of voltage.*

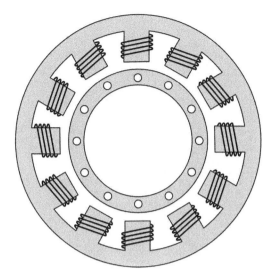

Figure 4.37 Typical stator construction

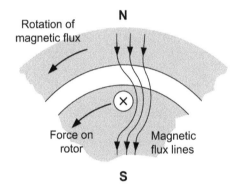

Figure 4.38 Force on the rotor of an induction motor

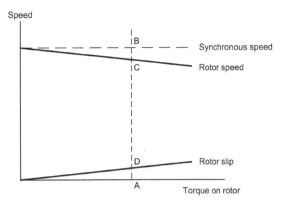

Figure 4.39 Relationship between torque and slip

cancel out the continuous motion of the stator field. Hence the rotor will move in the same direction as the stator field and will attempt to align with it. In practice, it gets as close to the moving stator field but never quite aligns perfectly with it!

Key point

The induction motor has the same stator as the synchronous motor. The rotor is different in that it does not require an external source of power. Current is induced in the rotor by the action of the rotating field cutting through the rotor conductors. This rotor current generates a magnetic field which interacts with the stator field, resulting in a torque being exerted on the rotor and causing it to rotate.

4.4.4 Slip, torque and speed

We have already said that the rotor of an induction motor is unable to turn in sympathy with the rotating field and, in practice, a small difference always exists. In fact, if the speeds were exactly the same, no relative motion would exist between the two, and so no e.m.f. would be induced in the rotor. For this reason the rotor operates at a lower speed than that of the rotating magnetic field. This phenomenon is known as **slip** and it becomes more significant as the rotor develops increased torque, as shown in Fig. 4.39.

From Fig. 4.39, for a torque of A the rotor speed will be represented by the distance AC whilst the slip will be represented by distance AD. Now:

$$AD = AB - AC = CB$$

Regardless of whether a squirrel cage or wound rotor is used, the basic principle of operation of an induction motor is the same. The rotating magnetic field generated in the stator induces an e.m.f. in the rotor. The current in the rotor circuit caused by this induced e.m.f. sets up a magnetic field. The two fields interact, and cause the rotor to turn. Fig. 4.38 shows how the rotor moves in the same direction as the rotating magnetic flux generated by the stator.

From **Lenz's law** (see Section 1.6.2) we know that an induced current opposes the changing field which induces it. In the case of an induction motor, the changing field is the rotating stator field and so the force exerted on the rotor (caused by the interaction between the rotor and the stator fields) attempts to

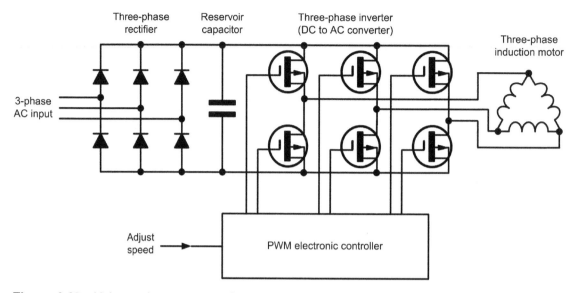

Figure 4.40 Using an inverter to produce a variable output speed from an AC induction motor

For values of torque within the working range of the motor (i.e. over the linear range of the graph shown in Fig. 4.39), the slip directly proportional to the torque and the **per-unit slip** is given by:

$$\text{Per-unit slip} = \frac{\text{slip}}{\text{synchronous speed}} = \frac{AD}{AB}$$

Now since AD = AB − BC,

$$\text{slip} = (\text{synchronous speed}) - (\text{rotor speed})$$

thus:

$$\text{Per-unit slip} = \frac{(\text{synchronous speed}) - (\text{rotor speed})}{\text{synchronous speed}}$$
$$= \frac{AB - BC}{AB}$$

The percentage slip is given by:

$$\text{Percentage slip} = \frac{(\text{synchronous speed}) - (\text{rotor speed})}{\text{synchronous speed}}$$
$$\times 100\% = \frac{AB - BC}{AB} \times 100\%$$

The actual value of slip tends to vary from about 6% for a small motor to around 2% for a large machine. Hence, for most purposes the induction motor can be considered to provide a constant speed (determined by the frequency of the current applied to its stator); however, one of its principal disadvantages is the fact that it is not easy to vary the speed of such a motor!

Note that, in general, it is not easy to control the speed of an AC motor unless a variable frequency AC supply is available. The speed of a motor with a wound rotor can be controlled by varying the current induced in the rotor but such an arrangement is not very practical as some means of making contact with the rotor windings is required. For this reason, DC motors are usually preferred in applications where the speed must be varied. However, where it is essential to be able to adjust the speed of an AC motor, the motor is invariably powered by an **inverter**. This consists of an electronic switching unit which produces a high-current three-phase **pulse-width-modulated** (**PWM**) output voltage from a DC supply, as shown in Fig. 4.40.

Key point

The rotor of an induction motor rotates at less than synchronous speed, in order that the rotating field can cut through the rotor conductors and induce a current flow in them. This percentage difference between the synchronous speed and the rotor speed is known as slip. Slip varies very little with normal load changes, and the induction motor is therefore considered to be a constant-speed motor.

Example 4.5

An induction motor has a synchronous speed of 3600 r.p.m. and its actual speed of rotation is measured as 3450 r.p.m. Determine (a) the per-unit slip and (b) the percentage slip.

(a) The per-unit slip is found from:

$$\text{Per-unit slip} = \frac{3{,}600 - 3{,}450}{3{,}600} = \frac{150}{3{,}600} = 0.042$$

(b) The percentage slip is given by:

$$\text{Percentage slip} = \frac{3{,}600 - 3{,}450}{3{,}450} \times 100\%$$
$$= \frac{150}{3{,}450} \times 100\% = 4.2\%$$

Inside an induction motor, the speed of the rotating flux, N, is given by the relationship:

$$N = \frac{f}{p}$$

where N is the speed of the flux (in revolutions per second), f is the frequency of the applied AC (Hz) and p is the number of pole pairs.

Now the per-unit slip, s, is given by:

$$s = \frac{AB - BC}{AB} = \frac{N - N_r}{N}$$

where N is the speed of the flux (in revolutions per second) and N_r is the rotor speed.

Now:

$$sN = N - N_r$$

from which:

$$N_r = N - sN = N(1 - s)$$

and:

$$N_r = N(1 - s) = \frac{f}{p}(1 - s)$$

where N_r is the speed of the rotor (in revolutions per second), f is the frequency of the applied AC (Hz) and s is the per-unit slip.

Example 4.6

An induction motor has four poles and is operated from a 400 Hz AC supply. If the motor operates with a slip of 2.5% determine the speed of the output rotor.

Now:

$$N_r = \frac{f}{p}(1 - s) = \frac{400}{2}(1 - 0.025)$$
$$= 200 \times 0.975 = 195$$

Thus the rotor has a speed of 195 revolutions per second (or 11,700 r.p.m.).

Example 4.7

An induction motor has four poles and is operated from a 60 Hz AC supply. If the rotor speed is 1700 r.p.m. determine the percentage slip.

Now:

$$N_r = \frac{f}{p}(1 - s)$$

from which:

$$s = 1 - \frac{N_r p}{f} = 1 - \frac{\left(\dfrac{1{,}700}{60}\right) \times 2}{60}$$
$$= 1 - \frac{56.7}{60} = 1 - 0.944 = 0.056$$

Expressed as a percentage, this is 5.6%

4.4.5 Single- and two-phase induction motors

In the case of a two-phase induction motor, two windings are placed at right angles to each other. By exciting these windings with current which is 90° out of phase, a rotating magnetic field can be created. A single-phase induction motor, on the other hand, has only one phase. This type of motor is extensively used in applications which require small low-output motors. The advantage gained by using single-phase motors is that in small sizes they are less expensive to manufacture than other types. Also they eliminate the need for a three-phase supply. Single-phase motors are used in communication equipment, fans, portable power tools, etc. Since the field due to the single-phase AC voltage applied to the stator winding is pulsating, single-phase AC induction motors develop a pulsating torque. They are therefore less efficient than three-phase or two-phase motors, in which the torque is more uniform.

Single-phase induction motors have only one stator winding. This winding generates a field which can be said to alternate along the axis of the single winding, rather than to rotate. Series motors, on the other hand, resemble DC machines in that they have commutators and brushes.

When the rotor is stationary, the expanding and collapsing stator field induces currents in the rotor which generate a rotor field. The opposition of these fields exerts a force on the rotor, which tries to turn it 180° degrees from its position. However, this force is exerted through the centre of the rotor and the rotor will not turn unless a force is applied in order to assist it. Hence some means of **starting** is required for all single-phase induction motors.

Key point

Induction motors are available that are designed for three-phase, two-phase and single-phase operation. The three-phase stator is exactly the same as the three-phase stator of the synchronous motor. The two-phase stator generates a rotating field by having two windings positioned at right angles to each other. If the voltages applied to the two windings are 90° out of phase, a rotating field will be generated.

Key point

A synchronous motor uses a single- or three-phase stator to generate a rotating magnetic field, and an electromagnetic rotor that is supplied with DC. The rotor acts like a magnet and is attracted by the rotating stator field. This attraction will exert a torque on the rotor and cause it to rotate with the field.

Key point

A single-phase induction motor has only one stator winding; therefore the magnetic field generated does not rotate. A single-phase induction motor with only one winding cannot start rotating by itself.

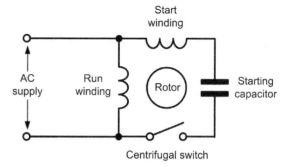

Figure 4.41   Capacitor starting arrangement

Once the rotor is started rotating, however, it will continue to rotate and come up to speed. A field is set up in the rotating rotor that is 90° out of phase with the stator field. These two fields together produce a rotating field that keeps the rotor in motion.

4.4.6 Capacitor starting

In an induction motor designed for capacitor starting, the stator consists of the main winding together with a starting winding which is connected in parallel with the main winding and spaced at right angles to it. A phase difference between the current in the two windings is obtained by connecting a capacitor in series with the auxiliary winding. A switch is included solely for the purposes of applying current to the auxiliary winding in order to start the rotor (see Fig. 4.41).

On starting, the switch is closed, placing the capacitor in series with the auxiliary winding. The capacitor is of such a value that the auxiliary winding is effectively a resistive–capacitive circuit in which the current leads the line voltage by approximately 45°. The main winding has enough inductance to cause the current to lag the line voltage by approximately 45°. The two field currents are therefore approximately 90° out of phase. Consequently the fields generated are also at an angle of 90°. The result is a revolving field that is sufficient to start the rotor turning.

After a brief period (when the motor is running at a speed which is close to its normal speed) the switch opens and breaks the current flowing in the auxiliary winding. At this point, the motor runs as an ordinary single-phase induction motor. However, since the two-phase induction motor is more efficient than a single-phase motor, it can be desirable to maintain the current in the auxiliary winding so that motor runs as a two-phase induction motor.

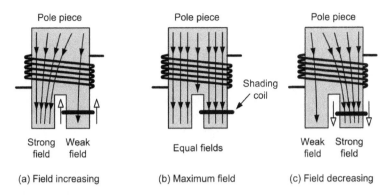

Figure 4.42   Action of a shaded pole

In some types of motor a more complicated arrangement is used with more than one capacitor switched into the auxiliary circuit. For example, a large value of capacitor could be used in order to ensure sufficient torque for starting a heavy load and then, once the motor has reached its operating speed, the capacitor value can be reduced in order to reduce the current in the auxiliary winding. A motor that employs such an arrangement, where two different capacitors are used (one for *starting* and one for *running*) is often referred to as **capacitor-start, capacitor-run** induction motor. Finally, note that, since phase shift can also be produced by an inductor, it is possible to use an inductor instead of a capacitor. Capacitors tend to be less expensive and more compact than comparable inductors and therefore are more frequently used.

Since the current and voltage in an inductor are also 90° out of phase, inductor starting is also possible. Once again, a starting winding is added to the stator. If this starting winding is placed in series with an inductor across the same supply as the running winding, the current in the starting winding will be out of phase with the current in the running winding. A rotating magnetic field will therefore be generated, and the rotor will rotate.

up to speed, the current in the auxiliary winding can be switched-off, and the motor will continue running as a single-phase motor.

4.4.7 Shaded pole motors

A different method of starting a single-phase induction motor is based on a **shaded-pole**. In this type of motor, a moving magnetic field is produced by constructing the stator in a particular way. The motor has projecting pole pieces just like DC machines; and part of the pole surface is surrounded by a copper strap or **shading coil**.

As the magnetic field in the core builds, the field flows effortlessly through the unshaded segment. This field is coupled into the shading coil which effectively constitutes a short-circuited loop. A large current momentarily flows in this loop and an opposing field is generated as a consequence. The result is simply that the unshaded segment initially experiences a larger magnetic field than does the shaded segment. At some later time, the fields in the two segments become equal. Later still, as the magnetic field in the unshaded segment declines, the field in the shaded segment strengthens. This is illustrated in Fig. 4.42.

Key point

In order to make a single-phase motor self-starting, a starting winding is added to the stator. If this starting winding is placed in series with a capacitor across the same supply as the running winding, the current in the starting winding will be out of phase with the current in the running winding. A rotating magnetic field will therefore be generated, and the rotor will rotate. Once the rotor comes

Key point

In the shaded pole induction motor, a section of each pole face in the stator is shorted out by a metal strap. This has the effect of moving the magnetic field back and forth across the pole face. The moving magnetic field has the same effect as a rotating field, and the motor is self-starting when switched on.

Figure 4.43 Main three-phase AC generator mounted in the engine compartment of a large passenger aircraft

Test your understanding 4.4

1. Explain the difference between synchronous AC motors and induction motors.
2. Explain the main disadvantage of the synchronous motor.
3. An induction motor has a synchronous speed of 7200 r.p.m. and its actual speed of rotation is measured as 7000 r.p.m. Determine (a) the per-unit slip and (b) the percentage slip.
4. An induction motor has four poles and is operated from a 400 Hz AC supply. If the motor operates with a slip of 1.8% determine the speed of the output rotor.
5. An induction motor has four poles and is operated from a 60 Hz AC supply. If the rotor speed is 1675 r.p.m. determine the percentage slip.
6. Explain why a single-phase induction motor requires a means of starting.

4.5 Practical aircraft generating systems

Generators are a primary source of power in an aircraft and can either produce direct or alternating current (DC or AC) as required. They are driven by a belt

drive (in smaller aircraft), or engine/APU accessory gearbox in larger aircraft. Generators will have sufficient output to supply all specified loads and charge the battery(s). Most avionic equipment requires a regulated and stable power supply depending on its function, e.g. in the case of lighting, it would be inconvenient if the intensity of lighting varied with engine speed. Generator output is affected by internal heat and this has to be dissipated. Cooling methods can include natural radiation from the casing, however this is inadequate for high-output devices where ram-air is directed from a scoop and directed into the generator's brush-gear and commutator. In some installations, e.g. helicopters, a fan is installed to provide cooling when the aircraft is hovering. We know from basic theory that a generator's output will vary depending on the input shaft speed. A means of regulating the generator's output to the bus is required as is a means of overload protection.

4.5.1 DC generators

DC generators are less common on modern aircraft due to their low power-to-weight ratio, poor performance at low r.p.m. and high servicing costs. The latter is due to the need for inspection and servicing of brushes and commutators since they have irregular

Figure 4.44 Tail-mounted auxiliary power unit (APU) with three-phase AC generator (top of picture)

Figure 4.45 Overhead electrical panel fitted with main and APU generator indicators and displays. This panel provides the crew with an instant overview of the status of the electrical generating system and the AC and DC distribution buses

surfaces/contact area and conduct the entire load current. Carbon brushes are porous and will absorb substances including moisture; this provides an amount of inherent lubrication. At altitude, the atmosphere is dryer and this leads to higher brush wear. Without any lubrication, arcing occurs and static charges build up; brush erosion is accelerated. Additives can be incorporated into the brushes that deposit a lubricating film on the commutator; this needs time to build up a sufficient protection; brushes need to be run in for several hours before the protective layer forms (this is often mistaken for contamination). The alternative is an in-built lubrication that is consumed as part of the natural brush wear, i.e. no film is deposited.

Key maintenance point

Automotive style alternators are normally installed on general aviation and light aircraft to overcome the shortfalls of DC generators. Larger aircraft use brushless AC generators.

4.5.2 Alternators

Automotive style alternators comprise a rotor, stator and rectifier pack. The rotor contains the field coil arranged in six sections around the shaft. Each section forms a pole piece that is supplied via slip-rings and brushes. The alternator has no residual magnetism, its field has to be excited by a DC supply (e.g. the battery). When energised, the rotor's pole pieces

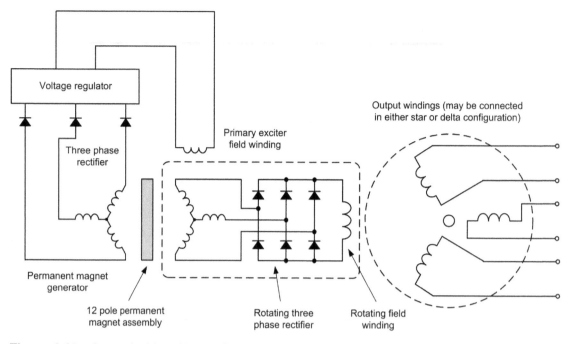

Figure 4.46 A practical brushless AC generator arrangement

produce north and south poles. As these poles are rotated they induce currents in the stator windings; these are wound at 120° and this produces three-phase AC. The AC output is fed to a diode rectifier pack comprising six high-current diodes, see Fig. 4.46 which produces a DC output. This has to be regulated before connecting to the various aircraft systems. Voltage regulators used with alternators on general aviation aircraft can be electromechanical or electronic. There are two types of electromechanical regulators: sensing coil with contacts and carbon-pile. Modern solid-state electronic regulators are more reliable as they use no mechanical parts.

The alternators previously described rely on slip-rings and brushes, albeit with reduced current loading. Slip-rings and brushes require maintenance in the workshop thereby incurring an associated cost burden. The **brushless generator** is a more complex device but has significantly increased reliability coupled with reduced maintenance requirements. A schematic diagram for the brushless generator is shown in Fig. 4.46; the device can be divided into three main sections:

- permanent magnet generator
- rotating field
- three-phase output.

The AC generator uses a brushless arrangement based on a rotating rectifier and **permanent magnet generator** (**PMG**). The output of the PMG rectifier is fed to the voltage regulator which provides current for the primary exciter field winding. The primary exciter field induces current into a three-phase rotor winding. The output of this winding is fed to the shaft-mounted rectifier diodes which produce a pulsating DC output which is fed to the rotating field winding. It is important to note that the excitation system is an integral part of the rotor and that there is no direct electrical connection between the stator and rotor.

The output of the main three-phase generator is supplied via **current transformers** (one for each phase) that monitor the load current in each line. An additional current transformer can also be present in the neutral line to detect an **out-of-balance** condition (when the load is unbalanced an appreciable current will flow in the generator's neutral connection).

The generator output is fed to the various aircraft systems and a solid-state regulator. This rectifies the output and sends a regulated direct current to the stator exciter field of the PMG. The regulator maintains the output of the generator at 115 V AC and is normally contained within a **generator control unit** (**GCU**); this unit is further described in Chapter 8.

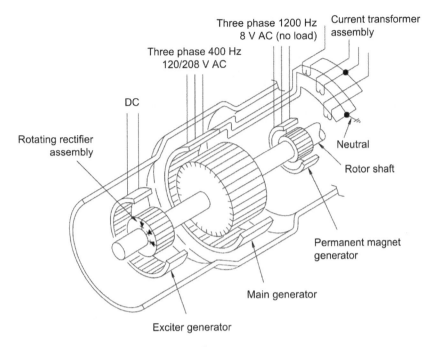

Three phase 1200 Hz
8 V AC (no load)

Current transformer
assembly

Three phase 400 Hz
120/208 V AC

DC

Rotating rectifier
assembly

Neutral

Rotor shaft

Permanent magnet
generator

Main generator

Exciter generator

Figure 4.47   Layout of a typical brushless AC generator

Although the regulator controls the output voltage of the generator, its frequency will vary depending on the speed of shaft rotation. Variable frequency power supplies (sometimes called **frequency wild**) are acceptable for resistive loads, e.g. de-icing, but they are not suitable for many induction motor loads that need to run at constant speed, e.g. fuel pumps and gyroscopic instruments. Furthermore, certain loads are designed for optimum efficiency at the specified frequency of 400 Hz, e.g. cooling fans. Some larger multi-engine aircraft operate the generators in parallel; it is essential that each generator is operating at the same frequency. Constant frequency can be achieved in one of two ways: controlling the shaft speed by electromechanical methods using a **constant speed drive** (**CSD**) or by controlling the generator output frequency electronically (**variable speed constant frequency: VSCF**).

Key point

A three-phase AC generator can be made brushless by incorporating an integral excitation system in which the field current is derived from a rotor-mounted rectifier arrangement. In this type of generator the coupling is entirely magnetic and no brushes and slip-rings are required.

4.5.3 Constant speed drive/integrated drive generator

The CSD is an electromechanical device installed on each engine. The input shaft is connected to the engine gearbox; the output shaft is connected to the generator. The CSD is based on a variable ratio drive employing a series of hydraulic pumps and differential gears. CSDs can be disconnected from the engine via a clutch, either manually or automatically. Note that it is only possible to reconnect the clutch on the ground. Modern commercial aircraft employ a combined CSD and brushless AC generator, in one item – the integrated drive generator (IDG). Typical characteristics are a variable input speed of 4500/9000 r.p.m. and a constant output speed of 12,000 ± 150 r.p.m. The IDG on a large commercial aircraft is oil-cooled and produces a 115/200 V 400 Hz three-phase, 90 m kVA output.

4.5.4 Variable speed constant frequency (VSCF)

Both the constant speed drive and integrated drive generator are complex and very expensive electromechanical devices. Advances in semiconductor technology has facilitated the development of solid-state products that can convert variable frequencies into

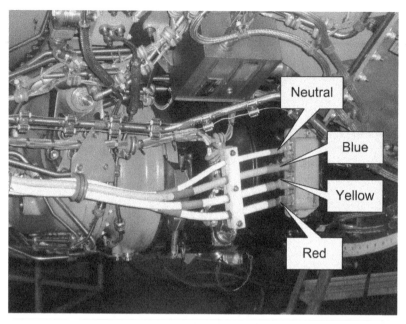

Figure 4.48   Typical main generator showing three-phase output cabling

Figure 4.49   Typical generator control unit (GCU)

115/200 V AC, 400 Hz three-phase power supplies. Variable speed constant frequency (VSCF) systems comprise a generator and power converter.

A brushless AC generator is mounted onto the engine accessory gearbox as before; its output voltage and frequency varies in accordance with engine speed. The gearbox increases the generator speed by a ratio of 1:3, producing a variable output frequency between 1300 and 2500 r.p.m. The three-phase output of VSCF is full-wave rectified to produce a 270 V DC output. This direct current output is smoothed by large capacitors, filtered and fed into an inverter that produces a square wave output. The inverter converts the DC level into a three phase, pulse-width-modulated waveform. This is then converted in the sinusoidal output voltage. These outputs are then converted into AC through electronic circuits. The final output stage is monitored by a **current transformer** and **electromagnetic interference** filter (**CT/EMI**). Within the **generator control unit** (**GCU**) a **generator control relay** (**GCR**) energizes the field. This circuit can be interrupted by the pilot or automatically under fault conditions.

The VSCF **generator conversion control unit** (**GCCU**) can either be integrated with the generator as a single engine mounted device (weighing typically 65 kg) or it can be located in the airframe. The latter arrangement has the advantage of making the engine accessories smaller; this means a lower profile nacelle. In addition, the electronics can be located in a zone with reduced temperature and vibration.

VSCF systems are more reliable compared with constant speed drive and integrated drive generators since there are fewer moving parts. The VSCF system's moving parts consist of the generator's rotor and an oil pump used for cooling. The VSCF can be used for both primary and secondary power supplies; outputs of 110 kVA are achievable. Enabling technology for VSCF are the power transistors and diodes capable of handling currents in excess of 500 A. These diodes and transistors form the core of the rectifier and conversion circuits of the GCCU. The VSCF contains an oil pump mounted on the generator shaft that circulates oil through the system; this oil is passed through a heat exchanger. Oil temperatures and pressures are closely monitored; warnings are given to the crew in the event of malfunctions. Oil level can be checked during ground servicing through a sight glass.

4.6 Multiple choice questions

1. The slip-rings in an AC generator provide a means of:
   (a) connecting an external circuit to a rotating armature winding
   (b) supporting a rotating armature without the need for bearings
   (c) periodically reversing the current produced by an armature winding.

2. Decreasing the current in the field coil of a DC generator will:
   (a) decrease the output voltage
   (b) increase the output voltage
   (c) increase the output frequency.

3. The rotor of an AC induction motor consists of:
   (a) a laminated iron core inside a 'squirrel cage' made from copper or aluminium
   (b) a series of coil windings on a laminated iron core with connections via slip-rings
   (c) a single copper loop which rotates inside the field created by a permanent magnet.

4. The slip speed of an AC induction motor is the difference between:
   (a) the synchronous speed and the rotor speed
   (b) the frequency of the supply and the rotor speed
   (c) the maximum speed and the minimum speed.

5. When compared with three-phase induction motors, single-phase induction motors:
   (a) are not inherently 'self starting'
   (b) have more complicated stator windings
   (c) are significantly more efficient.

6. A three-phase induction motor has three pairs of poles and is operated from a 60 Hz supply. Which one of the following gives the motor's synchronous speed?
   (a) 1200 r.p.m.
   (b) 1800 r.p.m.
   (c) 3600 r.p.m.

7. In a star-connected three-phase system, the line voltage is found to be 200 V. Which one of the following gives the approximate value of phase voltage?
   (a) 67 V
   (b) 115 V
   (c) 346 V.

8. A single-phase AC generator has twelve poles and it runs at 600 r.p.m. Which one of the following gives the output frequency of the generator?
   (a) 50 Hz
   (b) 60 Hz
   (c) 120 Hz.

9. In a balanced star-connected three-phase system the line current is 2 A and the line voltage is

200 V. If the power factor is 0.75 which one of the following gives the total power in the load?
(a) 300 W
(b) 520 W
(c) 900 W.

10. The commutator in a DC generator is used to:
    (a) provide a means of connecting an external field current supply
    (b) periodically reverse the connections to the rotating coil winding
    (c) disconnect the coil winding when the induced current reaches a maximum value.

11. The brushes fitted to a DC motor/generator should have:
    (a) low coefficient of friction and low contact resistance
    (b) high coefficient of friction and low contact resistance
    (c) low coefficient of friction and high contact resistance.

12. Self-excited generators derive their field current from:
    (a) the current produced by the armature
    (b) a separate field current supply
    (c) an external power source.

13. In a shunt-wound generator:
    (a) none of the armature current flows through the field
    (b) some of the armature current flows through the field
    (c) all of the armature current flows through the field.

14. When combined with a CSD, a brushless three-phase AC generator is often referred to as:
    (a) a compound generator
    (b) a 'frequency wild' generator
    (c) an IDG.

15. An out-of-balance condition in an AC three-phase system can be detected by means of:
    (a) voltage sensors connected across each output line
    (b) a dedicated field coil monitoring circuit
    (c) a current transformer connected in the neutral line.

Chapter 5 Batteries

Batteries are primary sources of electrical power found on most aircraft, delivering direct current. (Secondary sources of power are described in Chapter 6.) There are several types of battery used on aircraft, defined by the types of materials used in their construction; these include lead-acid and nickel-cadmium batteries. The choice of battery type depends mainly on performance and cost. Other types of battery are being considered for primary power on aircraft; these include lithium and nickel-metal hydride. Electrical power delivered by batteries is used for a variety of applications, e.g. lights, radios, instruments, and motors. This chapter reviews the battery types used on aircraft, typical applications and how they are installed and maintained.

5.1 Overview

The main aircraft battery is a primary source of electrical power; its use can be controlled by the pilot or by automatic means. The main battery provides autonomous starting for the engine(s) or auxiliary power unit (APU) when external ground power is not available. Typical current requirement during APU starting is 1000A, albeit for a short period of time. Batteries also supply **essential loads** in the event of generator failure. It is an airworthiness requirement that the main battery(s) supplies essential services for a specified period of time. Other aircraft systems are supplied with their own dedicated batteries, e.g. aircraft emergency lights. Individual computers use their own battery sources to provide non-volatile memory. Battery type and maintenance requirements have to be understood by the aircraft engineer to ensure safe and reliable operation and availability.

The battery is constructed from a number of individual cells; generic cell features consist of two electrodes (the **anode** and **cathode**) and electrolyte contained within a casing. Cell materials vary depending on the type of battery performance required for a given cost. The simple primary cell (Fig. 5.1) causes an electron flow from the cathode (negative) through the external load to the anode (positive). The materials used refer to the two types of battery cell in widespread use on aircraft for the primary source of power: **lead-acid** or **nickel-cadmium**. These are maintained on the aircraft and treated as line replaceable units; a full description of these two battery types is provided in this chapter. Cells used within other aircraft equipment or systems are typically made from **lithium** or **nickel-metal hydride** materials. These are not maintained as individual items on the aircraft, they are installed/removed as part of the equipment that they are fitted into; in this case only a brief description is provided.

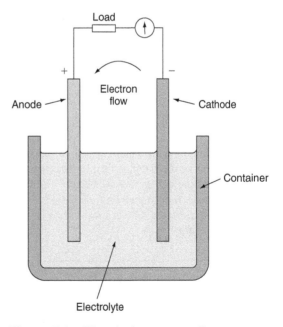

Figure 5.1 Electrical storage cell

Key point

The main aircraft battery is a primary source of electrical power; its use can be controlled by the pilot or by automatic means.

5.2 Storage cells

The basic function of any electrical cell is the conversion of chemical energy into electrical energy. The cells can be considered as a chemical means of storing electrical energy. Electrons are removed from the (**positive**) cathode and deposited on the (**negative**) anode. The electrolyte is the physical means of migration between the cathode/anode. The attraction of electrons between cathode/anode creates a potential difference across the cell; the cathode/anode are attached to external terminals for connection to the equipment or system. Material types used for the cathode/anode and electrolyte will determine the cell voltage.

Key maintenance point

Different battery types possess different characteristics both in terms of what they are used for, and in terms of how they should be maintained; always refer to maintenance manual instructions for servicing.

Cells are categorized as either primary (where they can only be used once) or secondary (where they can be recharged). In the primary cell, the chemical activity occurs only once, i.e. during discharge. By applying current through a secondary cell in the opposite direction to that of discharging, the chemical reaction is reversed and the cell can be used again. The cathode/anode are returned to their original charged form; the cell therefore becomes a chemical means of storing electrical energy.

Key maintenance point

It can be dangerous to attempt charging a primary cell. In the secondary cell, the chemical activity is reversible.

The energy storage **capacity** of a cell is determined by the amount of material available for chemical reaction. To maximize the storage capacity, the physical areas of the cathode and anode are made as large as possible, normally by constructing them as plates. Capacity is stated in ampere-hours; batteries are rated with low or high discharge rates, either 10 hours or 1 hour. The battery's capacity will gradually deteriorate over time depending on usage, in particular the charge and discharge rates. For aircraft maintenance purposes, we need to define the acceptable capacity of the main battery(s); this is the ratio of actual capacity and rated capacity, expressed as a percentage. Actual capacity must not fall below 80% of the rated capacity; therefore testing is required on a periodic basis. **Memory effect** is observed in some secondary cells that cause them to hold less charge; cells gradually lose their maximum capacity if they are repeatedly recharged before being fully discharged. The net result is the cell appears to retain less charge than specified.

All secondary cells have a finite life and will gradually lose their capacity over time due to secondary chemical reactions; this occurs whether the cell is used or not. They also have a finite number of charge and discharge cycles since they lose a very small amount of storage capacity during each cycle. Secondary cells can be damaged by repeated deep discharge or repeated over-charging.

Storage cells have **internal resistance**; this is usually very small but it has the effect of limiting the amount of current that the cell can supply and also reducing the amount of electromotive force (e.m.f.) available when connected to a load. Internal resistance varies significantly with the distance between plates. For this reason, the gap is made as small as practicably possible. This internal resistance is sometimes shown as a series resistor within the cell for design purposes, but it is normally omitted in circuit diagrams used in maintenance and wiring diagram manuals. Internal resistance is affected by temperature and this leads to practical issues for certain cell types.

Test your understanding 5.1

Describe the process whereby secondary cells gradually lose their maximum capacity (memory effect).

A number of cells are linked together in series to form a **battery**. The total battery terminal voltage is

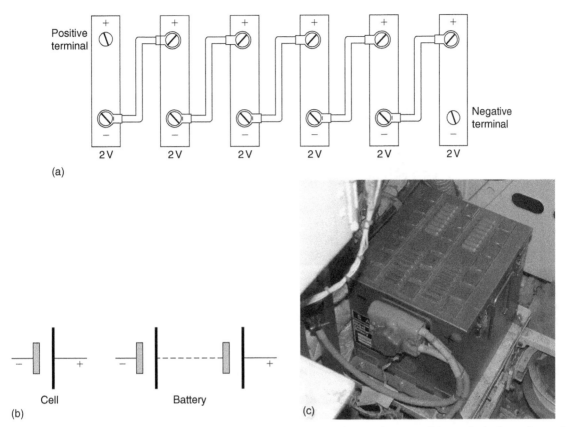

Figure 5.2 Cells and batteries: (a) connection of cells to form a battery; (b) symbols for cells and a battery; (c) typical battery casing

the sum of individual cell voltages, see Fig. 5.2(a). In this illustration, six cells are connected in series to form a 12 V battery. The circuit symbols for individual cells and a battery are shown in Fig. 5.2(b). All of the individual cells are contained within a battery case, see Fig. 5.2(c).

5.3 Lead-acid batteries

Developed in 1859, this is the oldest secondary cell technology in aircraft use today. Despite advances in alternative technologies, lead-acid batteries have retained market share (particularly in general aviation) due to the relatively low cost and mature technology. This type of battery has widespread applications on general aviation fixed and rotary wing aircraft due to the high current available for engine start and relatively low manufacturing cost (compared with nickel-cadmium batteries). The surface area of the plates, strength of the electrolyte and temperature determine

the actual capacity of a lead-acid cell. There are two types of lead-acid battery used in aircraft: **flooded** (wet-cell) and **sealed**. The disadvantages of flooded batteries are that they require regular maintenance, they liberate gas during charging and the electrolyte can be spilt or leak. Spillage and/or leakage of the electrolyte requires immediate clean-up to avoid corrosion. These problems are overcome with sealed lead-acid batteries. Although lead-acid batteries remain popular with GA aircraft, this battery technology will eventually be phased out due to **environmental** issues.

5.3.1 Construction

Flooded cells are housed within an impact- and acid-resistant casing made from polystyrene-based materials. The casing retains the two terminals and includes a vent cap to prevent gas pressure build-up whilst not allowing the electrolyte to escape. A single battery cell contains a number of positive and negative **plate groups** constructed as illustrated in Fig. 5.3. The

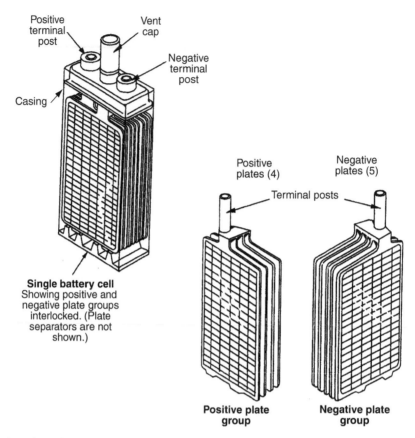

Figure 5.3 Lead-acid cell construction

individual plates are separated by a porous material to prevent short circuit through physical contact; there is space below the plates to allow any material shed from the plates to accumulate without shorting the plates. Flooded cells can be accessed on an individual basis for checking the content and condition of the electrolyte.

Key point

Since each positive plate is always positioned between two negative plates, there is actually one more negative plate than the positives.

Each positive plate is a cast lead/antimony frame formed as a grid; this is impregnated with a paste of lead dioxide (PbO_2). The negative plate is a similar frame containing lead (Pb); this is sometimes referred to as '**spongy lead**'. In practice, a typical cell is constructed with several plates in order to get the required current output. Positive plates distort when chemical

reactions take place on only one side; for this reason, there are always an even number of positive plates sandwiched between an odd number of negative plates. All positive plates are connected together as are all the negatives. The plates are interlaced and separated by a porous separator that allows free circulation of the electrolyte at the plate surfaces; the plates are all stacked within the cell container. The **electrolyte** is sulphuric acid (H_2SO_4) diluted with distilled (pure) water.

Test your understanding 5.2

What are the advantages and disadvantages of flooded (wet-cell) batteries?

5.3.2 Charging/discharging

When fully charged, each cell has a potential difference of 2.5 V (falling to 2.2 V after a period of approximately one hour) at its terminals; when discharged,

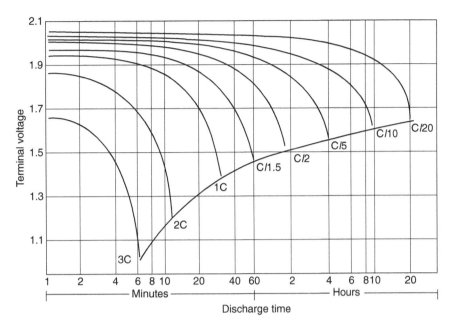

Figure 5.4 Lead-acid cell discharge characteristics

this potential difference is 1.8 V. A six-cell battery would produce 13.2 V fully charged, and 10.8 V DC when discharged. A twelve-cell battery would produce 26.4 V DC fully charged, and 21.6 V DC when discharged. During normal use of lead-acid cells, the terminal voltage stays around 2 V for a long period of cell life, this is referred to as the cell's **nominal voltage**. When fully charged, the positive plate is lead dioxide (PbO_2) and the negative plate is lead (Pb).

Connecting an external load to the battery completes the electrical circuit, electrons are transferred from the negative plate and the battery starts to discharge. The chemical reaction that takes place during discharge changes each of the plates into lead sulphate ($PbSO_4$). Molecules of water are formed, thereby diluting the electrolyte. For a given battery capacity, a steady discharge rating forms part of the battery specification, e.g. a 20 hour rate produces a constant current for 20 hours until the cell is discharged. Figure 5.4 illustrates typical lead-acid battery characteristics at different discharge currents. The discharge current, in amps (A), is expressed as a fraction of the numerical value of C. For example, 0.1 C means C/10 A, and discharging will take approximately 10 hours. If the battery capacity was 35 Ah, a discharge current of 3.5 A can be expressed as 0.1 C (or C/10). This means that batteries of different sizes can be compared by a single set of graphs. Since a battery may be rated for different discharge times, its rated capacity will normally be an indication of current used. With a 20-hour discharge capacity, the chart shows that C/20 will discharge the battery at 1 A current in 20 hours.

The condition of each cell can be determined by the **specific gravity** (SG) of its electrolyte. When the battery is charged, the above process is reversed. The lead sulphate on the positive plate is returned to lead peroxide. The negative plate is returned to lead, and the electrolyte is restored to its original specific gravity; SG ranges will be from 1.25–1.3 (charged) down to 1.15–1.2 when discharged. Table 5.1 summarizes the chemical aspects of a charged and discharged lead-acid cell.

5.3.3 Maintenance

Flooded lead-acid batteries are susceptible to damage at low temperatures due to freezing of the electrolyte causing plate damage. The point at which the electrolyte freezes depends on its specific gravity; at a specific gravity of 1.15 (discharged) the freezing point is −15°C. To prevent freezing, the specific gravity should be maintained at higher levels; at a specific gravity of 1.275 (charged) the freezing point

Table 5.1 Chemical aspects of a charged and discharged lead-acid cell.

State	Positive plate	Negative plate	Electrolyte
Charged	Lead dioxide (PbO_2)	Lead (Pb)	Concentrated sulphuric acid (H_2SO_4)
Discharged	Lead sulphate ($PbSO_4$)	Lead sulphate ($PbSO_4$)	Weak sulphuric acid (H_2SO_4)

is $-62°C$. Although this guards against freezing, the consequence of maintaining a battery in this condition is that it will gradually self-discharge.

Lead-acid batteries require a three-month **capacity check**, and have approximately 18–24 months' life. The condition of a fully charged lead-acid battery can be confirmed by three factors:

- The terminal voltage remains at its maximum level
- There is a free discharge of gas
- The SG is in the range 1.25–1.3.

The **specific gravity** of the electrolyte provides the definitive means of checking the charged condition of a lead-acid cell; this must be checked with a hydrometer on a periodic basis. (Specific gravity of a fluid is the relative density, or ratio of fluid's weight compared to pure water.) The electrolyte must always cover the plates; it can be topped up with distilled water. Differences of specific gravity readings between cells indicates that the battery is reaching the end of its useful life.

Key maintenance point

Specific gravity is temperature-dependent, and correction factors must be applied when taking measurements (refer to maintenance manual instructions).

Key maintenance point

When taking specific gravity readings from lead-acid cells, the acid sample should always be returned into the same cell.

Charging of lead-acid batteries should be from a constant voltage source. Excessive charging rates can lead to boiling of the electrolyte; fumes containing droplets of electrolyte can escape the battery. These fumes can become noxious unless the battery is properly ventilated. The voltage per cell during charging should not exceed 2.35 V.

Sulphation occurs when an excess of lead sulphate builds up on the plates. This happens with a fully charged battery over a period of several weeks when the battery self-discharges. To prevent this, the battery should be re-charged in accordance with the maintenance manual instructions. Sulphation can eventually occur on a permanent basis and the sulphate will not go back into solution when charged. Over time, the lead sulphate gradually occupies more space on the plates thereby reducing capacity. This can be removed by drawing a heavy charge current causing particles to be removed from the plates and subsequently accumulated at the bottom of the cell. Eventually the plates become uneven in cross-section and distorted, leading to cracks being formed. Particles will accumulate at the bottom of the cell and this can lead to shorting of the plates. Sulphating is accelerated by small (trickle) discharging/charging together with incorrect electrolyte strength and levels.

In the event of electrolyte **spillage/leaks**, (always refer to the aircraft maintenance manual for any specific requirements) the following generic actions should be taken:

1. Report the incident.
2. Mop-up the electrolyte with a damp rag or sponge.
3. Brush the affected area with a dilute solution of sodium bicarbonate.
4. Sponge the area with clean water; dry thoroughly.
5. Press a moist piece of blue litmus paper on the affected area; a change of colour to red indicates the presence of acid (repeat steps 3–4 until the acid is removed).
6. Leave for 24 hours, and then check for any evidence of corrosion.
7. Restore any protective finish to the aircraft structure.

5.3.4 Sealed batteries

Maintenance and servicing costs associated with flooded cells can be overcome with sealed lead-acid batteries. This technology was developed in the 1970s and has been in place since the 1980s and is known as **valve-regulated lead-acid** (VRLA); the sealed lead-acid (SLA) effectively provides maintenance-free lead-acid batteries.

Cell plates are made from lead calcium; the electrolyte is sulphuric acid diluted with distilled water. Plates are separated by an **absorbent glass mat** (AGM) that absorbs gases liberated from the plates during charging. The lead plates are purer (99.99%) than flooded cell materials since they do not have to support their own weight. The electrolyte is absorbed between the plates and immobilized by a very fine fibreglass mat. This glass mat absorbs and immobilizes the acid while keeping the electrolyte in contact with the plates. This allows a fast reaction between and electrolyte and plate material during charge/discharge. There is no disintegration of the active materials leading to a short circuit.

Key maintenance point

It is not possible to check the SG of the electrolyte; the battery can only be checked by measuring the terminal voltage.

The **internal resistance** of sealed lead-acid cells is lower than flooded cells, they can handle higher temperatures, and self-discharge more slowly. They are also more tolerant to the attitude of the aircraft. The product is inherently safer than the flooded cell due to reduced risk of spillage, leakage and gassing. Maintenance requirements are for a capacity check only. The overall capacity-to-weight ratio of sealed lead-acid batteries is superior to flooded lead-acid batteries. Since they are sealed, they can be shipped as non-hazardous material via ground or air.

5.4 Nickel-cadmium batteries

Nickel-cadmium battery technology became commercially available for aircraft applications in the 1950s. At that time the major sources of batteries for aircraft were either vented lead-acid or silver-zinc technology. The nickel-cadmium (Ni-Cd) battery (pronounced 'nye-cad') eventually became the preferred battery type for larger aircraft since it can withstand higher charge/discharge rates and has a longer life. Ni-Cd cells are able to maintain a relatively steady voltage during high discharge conditions. The disadvantages of nickel-cadmium batteries are that they are more

expensive (than lead-acid batteries) and have a lower voltage output per cell (hence their physical volume is larger than a lead-acid battery).

5.4.1 Construction

Plates are formed from a nickel mesh on which a nickel powder is sintered. The sintering process (where powdered material is formed into a solid) is used to form the porous base-plates (called **plaques**). This process maximizes the available quantity of active material. The plaques are vacuum impregnated with nickel or cadmium salts, electrochemically deposited with the pores of the plaques. Nickel tabs are spot-welded onto the plates and formed into the terminals; these plates are then stacked and separated by a porous plastic in a similar fashion to the lead-acid battery. The electrolyte is potassium hydroxide (KOH) diluted in distilled water giving a specific gravity of between 1.24 and 1.3. Both the plates and electrolyte are sealed in a plastic container.

5.4.2 Charging

During charging, there is an exchange of ions between plates. Oxygen is removed from the negative plate, and transferred to the positive plate. This transfer takes place for as long as charging current exists, until all the oxygen is driven out of the negative plate (leaving metallic cadmium) and the positive plate becomes nickel oxide. The electrolyte acts as an ionized conductor and it does not react with the plates in any way. There is virtually no chemical change taking place in the electrolyte during charging or discharging, therefore its condition does not provide an indication of cell condition. Towards the end of charging, gassing occurs as a result of **electrolysis** and the water content of the electrolyte is reduced. Gas emitted by decomposition of water molecules is converted into hydrogen at the negative plate and oxygen at the positive plate. This gassing leads to the loss of some water; the amount of gas released is a function of electrolyte temperature and charging voltage. When fully charged, each cell has a potential difference of between 1.2 and 1.3 V across its terminals. This reduces to 1.1 V when discharged. An aircraft battery containing 19 cells at 1.3 V therefore produces a battery of 24.7 V. Charging voltage depends on the design and construction, but will be in the order of 1.4/1.5 V per cell.

5.4.3 Discharging

This is a reverse chemical activity of the charging process; the positive plate gradually loses oxygen and the negative plate gradually regains oxygen. No gassing takes place during a normal discharge; the electrolyte is absorbed into the plates and may not be visible over the plates. When fully charged, the volume of electrolyte is high; this is the only time that water should be added to a Ni-Cd battery. Ni-Cd battery electrolyte freezes at approximately –60°C and is therefore less susceptible to freezing compared to lead-acid. The formation of white crystals of potassium carbonate indicates the possibility that overcharging has occurred.

Referring to Fig. 5.5, the nickel-cadmium cell voltage remains relatively constant at approximately 1.2 V through to the end of discharge, at which point there is a steep voltage drop. The discharge characteristics of a cell are affected by the:

* discharge rate
* discharge time
* depth of discharge
* cell temperature
* charge rate and overcharge rate
* charge time, and rest period after charge
* previous cycling history.

Every nickel-cadmium cell (and hence a battery) has a specific:

* rated capacity
* discharge voltage
* effective resistance.

Individual cells are rated at a nominal 1.2 V, and voltage for battery voltages are multiples of the individual cell nominal voltage. Five cells connected in series would therefore result in a 6 V battery. It can be seen from Fig. 5.6 that the discharge voltage will exceed 1.2 V for some portion of the discharge period. Cell capacity is normally rated by stating a conservative estimate of the amount of capacity that can be discharged from a relatively new, fully charged cell. The cell rating in ampere-hours (or milliampere-hours) is therefore quoted by most manufacturers to a voltage of 0.9 V at 5 hour discharge rate.

Figure 5.6 shows that when rates of discharge are reduced, the available capacity becomes less dependent on the discharge rate. When rates of discharge rates increase, the available capacity decreases.

Charging of nickel-cadmium batteries needs specific methods since they can suffer from an effect called **thermal runaway**. This occurs at high temperatures and if the battery is connected to a constant charging voltage that can deliver high currents. Thermal runaway causes an increase in temperature and lower internal resistance, causing more current to flow into the battery. In extreme cases sufficient heat may be generated to destroy the battery. Dedicated battery charges (either on the aircraft or in the workshop) are designed to take this into account by regulating the charging current. Temperature sensors are installed in the batteries to detect if a runaway condition is occurring. Table 5.2 summarizes the chemical aspects of a charged and discharged nickel-cadmium cell.

5.4.4 Maintenance

Since there is virtually no chemical change taking place during nickel-cadmium cell charging or discharging, the condition of the electrolyte does not provide an indication of the battery's condition. Cell terminal voltage does not provide an indication of charge since it remains relatively constant. The only accurate and practical way to determine the condition of the nickel-cadmium battery is with a **measured discharge** in the workshop. The fully charged battery is tested after a two hour 'resting' period, after which the electrolyte is topped up using distilled or demineralized water. Note that since the electrolyte level depends on the state of charge, water should never be added to the battery on the aircraft. This could lead to the electrolyte overflowing when the battery discharges, leading to corrosion and self-discharging (both of which could lead to premature failure of the battery). Ni-Cd batteries emit gas near the end of the charging process and during overcharging. This is an explosive mixture and must

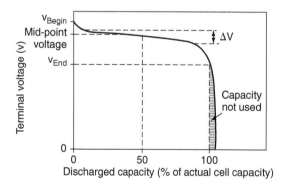

Figure 5.5 Nickel-cadmium cell discharge characteristics

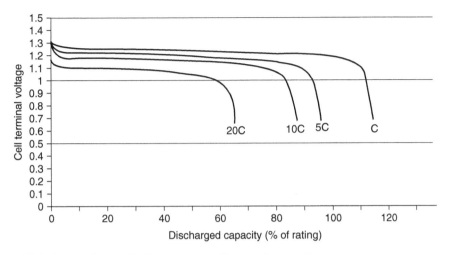

Figure 5.6 Nickel-cadmium cell discharge profiles and capacity

Table 5.2 Chemical aspects of a charged and discharged nickel-cadmium cell

State	Positive plate	Negative plate	Electrolyte
Charged	Nickel oxides (Ni_2O_2 Ni_3O_3)	Cadmium (Cd)	Potassium hydroxide (KOH)
Discharged	Nickel hydroxide $Ni(OH_2)$	Cadmium hydroxide $Cd(OH_2)$	Potassium hydroxide (KOH)

be prevented from accumulating; maintenance of the venting system is essential.

In the event of electrolyte **spillage/leaks** (always refer to the aircraft maintenance manual for specific details):

- report incident
- mop electrolyte with damp rag or sponge
- cover the area with a dilute solution of acetic acid, 5% solution of chromic acid, or 10% solution of boric acid
- press moist piece of red litmus paper on affected area; change of colour to blue indicates presence of alkaline
- leave for a minumum of 24 hours, check for corrosion
- restore protective finish.

In addition to providing primary power, Ni-Cd batteries are also used in aircraft for emergency equipment, e.g. lighting. This type of cell is sealed and the electrolyte cannot be topped up. Extreme care must be taken with how these batteries are charged.

Key maintenance point

Servicing equipment used for lead-acid batteries must not be used for nickel-cadmium batteries; sulphuric acid is detrimental to the operation of nickel-cadmium batteries.

Key maintenance point

If Ni-Cd batteries are replacing lead-acid batteries, always neutralize the battery compartment.

Key maintenance point

Electrolyte used in lead-acid and nickel cadmium batteries is actively corrosive.

Key maintenance point

Main batteries need to be kept upright in the aircraft to avoid spilling any of the electrolyte.

5.5 Lithium batteries

Lithium batteries include a family of over 20 different products with many types of anodes, cathodes and electrolytes. The type of materials selected depends on many factors, e.g. cost, capacity, temperature, life, etc.; these are all driven by what the application requirements are.

Applications range from consumer products (accounting for the largest market requirement) through to specialist applications including communications and medical equipment. Aircraft are often equipped with systems requiring an autonomous source of energy, e.g. emergency locator beacons, life rafts and life jackets. Lithium (Li) is one of the alkali group of **reactive metals**; it is one of the lightest elements, giving it an immediate advantage for aircraft applications. It has a single valence electron with low combining power, therefore readily becoming a positive ion. The materials used in these cells are:

- electrolyte: lithium-ion
- cathode: cobalt
- anode: graphite.

Lithium-ion is a fast-growing and promising battery technology. This type of battery is often found in consumer products (mobile phones and laptop computers) because they have very high energy-to-weight ratios, no memory effect, and a slow discharge charge rate when not in use. They are being introduced for aircraft applications (e.g. in smoke detectors) on a cautious basis because they are significantly more susceptible to thermal runaway. Applications on aircraft now include engine start and emergency back-up power, the first such application of the devices in the business aviation sector. In the longer term, they are being developed for main battery applications. They offer several advantages compared to lead-acid and nickel-cadmium products, including:

- longer life
- less weight
- low maintenance
- reduced charging time.

Disadvantages are the higher product cost and the fact that the electrolyte is extremely flammable. They can lose up to 10% of their storage capacity every year from when they are manufactured, irrespective of usage. The rate at which the ageing process occurs is subject to temperature; higher temperatures results in faster ageing.

The lithium-ion main aircraft battery will not be a 'drop-in' replacement for main battery applications. Safety features are required within the aircraft as well as in the battery. These features include protection circuits and hardware to maintain voltage and current within safe limits. The nominal cell voltage is 3.6 V, charging requires a constant voltage of 4.2 V with associated current limiting. When the cell voltage reaches 4.2 V, and the current drops to approximately 7% of the initial charging current, the cell is fully recharged. Figure 5.7 illustrates the typical discharge curve of a lithium-ion cell when discharged at the 0.2 C rate. Lithium-ion cells have a very flat discharge curve, and cell voltage cannot be used to determine the state of charge. The effective capacity of the lithium-ion cell is increased with low discharge rates and reduced if the cell is discharged at higher rates.

Software-based monitoring and alarms are needed for safe operation during charging. Specific design and maintenance considerations for these batteries in aircraft include:

- maintaining safe cell temperatures and pressures
- mitigating against explosion
- preventing the electrolyte escaping from the battery
- disconnecting the charging source in the event of over-temperature
- providing a low battery charge warning.

The next-generation of power solutions from True Blue Power® (a division of Mid-Continent Instrument Co., Inc.) is based on Nanophosphate® lithium-ion cell technology for aviation applications. Current production includes Universal Serial Bus (USB) charging ports, inverters, converters, emergency power supplies and advanced lithium-ion batteries. Ideal for the piston, turbine, and emergency power market, the TB17 advanced lithium-ion features superior energy density – Nanophosphate® lithium-ion cells offer three times the energy per kilogram, resulting in a battery that is 45% lighter than lead-acid or nickel-cadmium alternatives.

5.6 Nickel-metal hydride batteries

Nickel-metal hydride (Ni-MH) is a secondary battery technology, similar to the sealed nickel-cadmium product. Ni-MH batteries provide a constant voltage during discharge, excellent long-term storage and long cycle life (over 500 charge–discharge

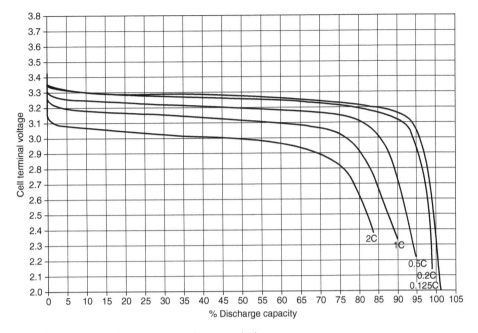

Figure 5.7   Lithium-ion cell discharge characteristics

cycles). No maintenance is required on this type of battery; however, care must be taken in charging and discharging. The evolution of Ni-MH technology is being driven by the need for environmentally friendly materials and higher energy efficiency. The materials used in the Ni-MH battery technology are:

- anode: nickel and lanthanum
- cathode: nickel hydroxide
- electrolyte: potassium hydroxide.

The charging voltage is in the range of 1.4/1.6 V per cell. A fully charged cell measures between 1.35 and 1.4 V (unloaded), and supplies a nominal 1.2 V per cell during use, reducing to approximately 1 volt per cell (further discharge may cause permanent damage). The Ni-MH cell requires a complex charging algorithm, and hence dedicated charger equipment.

Figure 5.8 illustrates the voltage profile of a metal-hydride cell, discharged at the 5-hour rate (0.2 C rate). This profile is affected by temperature and discharge rate; however under most conditions, the cell voltage retains a flat plateau that is ideal for electronics applications. As with nickel-cadmium cells, the nickel-metal hydride cell exhibits a sharp 'knee' at the end of the discharge where the voltage drops rapidly.

Key maintenance point

A Ni-Cad charger should not be used as a substitute for a Ni-MH charger.

A new generation of nickel-metal hydride 12 V batteries has been designed by Advanced Technological Systems International Limited (ATSI) as a direct

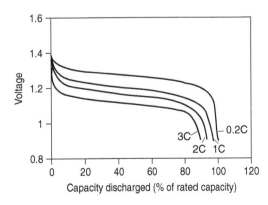

Figure 5.8 Metal-hydride cell discharge characteristics

Table 5.3 Metal-hydride aircraft battery: technical specifications (courtesy of ATSI)

Dimensions	$65 \times 95 \times 150$ mm
Weight	1.9 kg
Nominal voltage	12 V
Capacity	10 Ah
Maximum discharge	5 A
Fuse	Self setting internal
Charge time	8 hours flat to full
Minimum charge cycles	600
Operating temperatures	$-20/+60°C$

replacement for the conventional sealed lead-acid battery typically used in gliders. It delivers more than twice the power of its lead-acid counterpart whilst having the same base footprint and lower weight. The integral advanced electronics guarantees that it will always deliver maximum output up to the point of total discharge. Unlike sealed lead-acid batteries, it does not suffer any loss of performance even after many deep discharge cycles, or storage in a discharged state, making it one of the most advanced batteries in the world today. The new battery type will be longer lasting than the equivalent sealed lead-acid battery and requires a purpose-designed charging unit.

5.7 Battery locations

An aircraft is fitted with one or two main batteries depending on its size and role. The battery is located as close as possible to its point of distribution; this is to reduce IR losses through heavy-duty cables. In smaller general aviation (GA) aircraft, the battery can be located in the engine compartment, alternatively behind the luggage compartment in the rear fuselage, see Fig. 5.9(a). On some larger GA aircraft the battery is located in the leading edge of the wing, see Fig. 5.9(b). Other locations include the nose equipment bay on medium size helicopters (Fig. 5.9(c)) or attached to the **external airframe**, see Fig. 5.9(d). For larger aircraft, e.g. the Boeing 747, one battery is located in the **flight compartment**; the other is located in the **auxiliary power unit** (APU) bay at the rear of the aircraft. Batteries are installed in a dedicated box or compartment designed to retain it in position and provide ventilation. The battery compartment is usually fitted with a tray to collect any spilt electrolyte and protect the airframe. Tray material will

be resistant to corrosion and non-absorbent. The structure around the battery compartment will be treated to reduce any damage from corrosion resulting from any spilt electrolyte or fumes given off during charging. Batteries must be secured to prevent them from becoming detached during aircraft manoeuvres; they are a **fire risk** if they become detached from their tray.

Key maintenance point

When installing batteries in the aircraft, extreme care must be taken not to directly connect (or 'short circuit') the terminals. This could lead to a high discharge of electrical energy causing personal harm and/or damage to the aircraft.

Key maintenance point

The battery must be secured without causing any deformation of the casing which could lead to plate buckling and internal shorting.

5.8 Battery venting

Main battery installations must be vented to allow gases to escape, and accommodate electrolyte spillage. Rubber or other non-corroding pipes are used as ventilation lines which direct the gases overboard, usually terminating at the fuselage skin. On pressurized aircraft the differential pressures between cabin and atmosphere are used to draw air through the venting system. Some installations contain traps to retain harmful gases and vapours. Figure 5.10 illustrates battery venting, acid traps and how pressurized cabin air is used to ventilate the battery.

Key maintenance points

- Avoid personal contact with battery electrolyte (fluid and fumes).
- Observe safety precautions for the protection of hands and eyes.
- Always use personal protective equipment (goggles, rubber gloves, aprons) when handling electrolyte to prevent serious burns.
- Seek first aid in the event of electrolyte contact.
- When mixing electrolyte, acid is always added to the water. (Adding water to acid is very dangerous.)

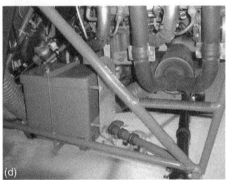

Figure 5.9 Typical battery locations: (a) battery compartment (GA aircraft); (b) wing leading edge (Beech King Air); (c) nose equipment bay (medium helicopter); (d) externally mounted (small helicopter)

5.9 Battery connections

These depend on the type of battery and aircraft installation. On smaller aircraft the cable connections simply fit over the terminal lugs and are secured with a nut, bolt and washers. On larger aircraft, the main batteries have **quick-release** connectors, see Fig. 5.11. These provide protection for the terminals and cable connections, the aircraft connector is a plastic housing with two shrouded spring-loaded terminals (for connecting the battery cables) and a hand-wheel with lead-screw. The battery connection is a plastic housing integrated into the casing; it contains two shrouded pins and a female lead screw. When the two halves are

engaged, the lead screws are pulled together and eventually form a lock. This mechanism provides good contact pressure and a low-resistance connection. The main battery(s) is connected into the aircraft distribution system; this is described in Chapter 8.

Key maintenance point

Batteries must not be exposed to temperatures or charging currents in excess of their specified values. This will result in the electrolyte boiling, rapid deterioration of the cell(s) eventually leading to battery failure.

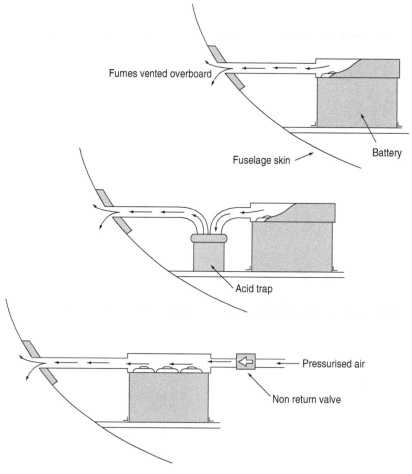

Fumes vented overboard

Fuselage skin

Battery

Acid trap

Pressurised air

Non return valve

Figure 5.10   Battery venting

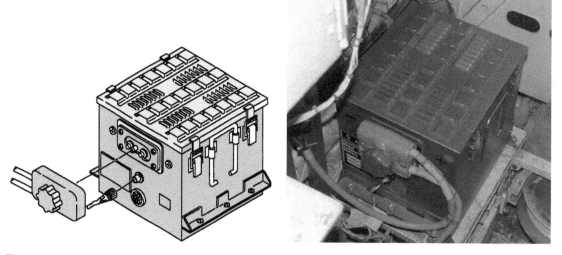

Figure 5.11   Battery connections

Key maintenance point

Removal of the aircraft battery can result in loss of power to any clocks that are electrically powered. It will usually be necessary to check and reset the clocks on the flight deck when battery power is eventually restored.

Key maintenance point

Some aircraft main batteries are heavy and may require a hoist for removal/installation into the aircraft.

5.10 Multiple choice questions

1. In a simple cell, electrons are:
   (a) removed from the (positive) cathode and deposited on the (negative) anode
   (b) removed from the (negative) cathode and deposited on the (positive) anode
   (c) removed from the (negative) anode and deposited on the (positive) cathode.

2. The energy storage capacity of a cell is determined by the:
   (a) terminal voltage
   (b) electrolyte specific gravity
   (c) amount of material available for chemical reaction.

3. When mixing electrolyte:
   (a) acid is always added to the water
   (b) water is always added to the acid
   (c) it is not important how water and acid are mixed.

4. Servicing equipment used for lead-acid batteries:
   (a) can also be used for nickel-cadmium batteries
   (b) must not be used for nickel-cadmium batteries
   (c) must be disposed of after use.

5. Battery capacity is measured in:
   (a) volts
   (b) amperes
   (c) ampere-hours.

6. Lead-acid batteries are recharged by constant:
   (a) voltage
   (b) current
   (c) ampere-hours.

7. The only accurate and practical way to determine the condition of the nickel-cadmium battery is with a:
   (a) specific gravity check of the electrolyte
   (b) measured discharge in the workshop
   (c) check of the terminal voltage.

8. A cell that can only be charged once is called a:
   (a) secondary cell
   (b) metal-hydride cell
   (c) primary cell.

9. The only time that water should be added to a Ni-Cd battery is:
   (a) when fully charged, and the volume of electrolyte is high
   (b) when fully discharged, and the volume of electrolyte is high
   (c) when fully charged, and the volume of electrolyte is low.

10. The only accurate and practical way to determine the condition of the lead-acid battery is with a:
    (a) specific gravity check of the electrolyte
    (b) measured discharge in the workshop
    (c) check of the terminal voltage.

11. Referring to Fig. 5.12 (page 132), the reason for having an even number of positive plates in a lead-acid battery is because:
    (a) positive plates distort when chemical reactions take place on both sides
    (b) positive plates distort when chemical reactions take place on one side
    (c) negative plates distort when chemical reactions take place on both sides.

12. Referring to Fig. 5.13 (page 132), a 20 Ah battery, when discharging at 2A, will be fully discharged in approximately:
    (a) two hours
    (b) 15 minutes
    (c) ten hours.

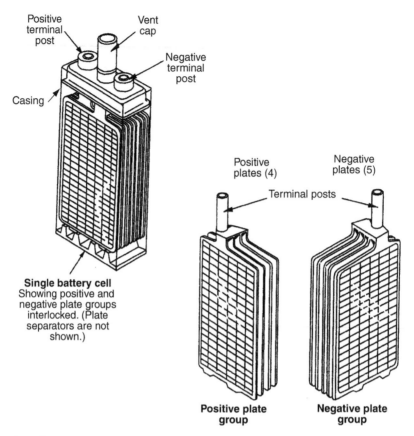

Figure 5.12   See Question 11

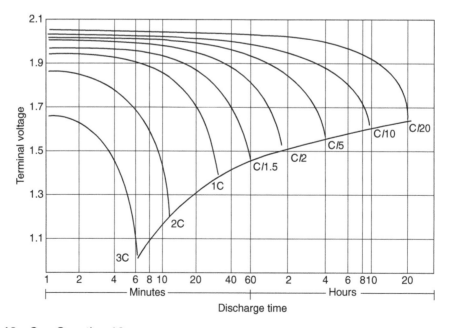

Figure 5.13   See Question 12

Chapter 6

Power supplies

Aircraft electrical power can be derived from a variety of sources; these are categorized as either primary or secondary supplies. Batteries and generators are primary sources of electrical power; inverters and transformer rectifier units (TRUs) are secondary sources of power. This power is either in the form of direct or alternating current depending on system requirements. In addition to onboard equipment, most aircraft have the facility to be connected to an external power source during servicing or maintenance. The basic power source found on most aircraft is the battery, delivering direct current (DC). Generators can supply either direct or alternating current; the outputs of generators need to be regulated. Alternating current generators are also referred to as alternators.

Inverters are used to convert DC (usually from the battery) into alternating current (AC). Transformer rectifier units (TRUs) convert AC into DC; these are often used to charge batteries from AC generators. In some installations transformers (as described in Chapter 1) are used to convert AC into AC, typically for stepping down from 115 to 26 V AC. An auxiliary power unit (APU) is normally used for starting the aircraft's main engines via the air distribution system. While the aircraft is on the ground, the APU can also provide electrical power. In the event of generator failure(s), continuous power can be provided by a ram air turbine (RAT). In this chapter, we review the various sources of electrical power used on aircraft and their typical applications.

6.1 Regulators

We know from basic theory that a generator's output will vary depending on the input shaft speed. A means of regulating the generator's output is therefore required.

6.1.1 Vibrating contact regulator

This device comprises voltage and current regulators as shown in Fig. 6.1. They are used on small general

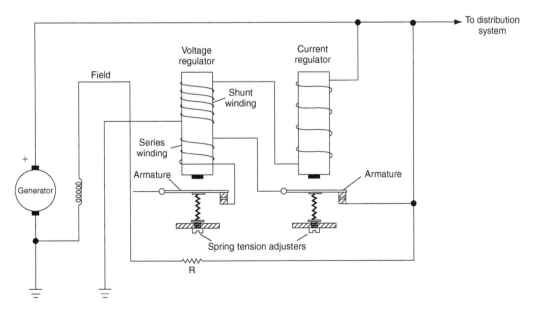

Figure 6.1 Vibrating contact regulator schematic

aviation (GA) aircraft that have relatively low generator power outputs. When the engine starts, the alternator output voltage builds up rapidly to the nominal aircraft level (either 14 or 28 V DC). Contacts of both regulators remain closed to allow current to flow into the field windings. When the generator output **voltage** increases beyond 14/28 V, the voltage coil contacts open and this introduces the resistor into the field windings, thereby reducing the field excitation current, and subsequently reduces the generator output. Once the output voltage drops to below 14/28 V, the contacts close (by a spring mechanism) and the resistor is bypassed, allowing full excitation current back into the field. The on/off cycle repeats between 50 and 200 times per second, or 50–200 Hz. This process regulates the generator output to a mean level, typically 14 ± 0.5 V (or 28 ± 1 volt).

Current regulation is achieved in a similar way, i.e. by controlling the field current. When loads are high, the voltage output may be insufficient to open the contacts. The result is that the output will continue to increase until the maximum rated current is reached. At this point, the current regulator contacts open and the resistor is connected into the field windings.

The accuracy of this type of regulation depends on the resistor value and spring tensions. In the event of high rotor speed and low electrical load on the generator, the output could exceed the specified system voltage despite the field being supplied via the resistor. In this event, the contact is pulled to ground, thereby reducing the output to below the regulated mean level. Although simple, this type of regulator has the disadvantage of contact wear; a typical vibrating contact regulator product is shown in Fig. 6.2.

Key maintenance point

The accuracy of the vibrating contact regulator depends on the resistor value and spring tension.

6.1.2 Carbon-pile regulator

Another type of electromechanical regulator is the carbon-pile device. This type of regulator is used in generator systems with outputs in excess of 50 A and provides smoother regulation compared with the vibrating contact regulator. Carbon-pile regulators

Figure 6.2   Vibrating contact regulator overview

consists of a variable resistance in series with the generator's shunt-wound field coil. The variable resistance is achieved with a stack (or pile) of **carbon discs** (washers). These are retained by a ceramic tube that keeps the discs aligned. Figure 6.3 shows the main features of the regulator in cross-section. The surface of each disc is relatively rough; applying pressure to the discs creates more surface contact, thereby reducing the resistance of the pile. When pressure is reduced, the reverse process happens, and the resistance through the pile increases. Pressure is applied to the pile by a spring plate. This compression is opposed by the action of an electromagnet connected to the generator output; the strength of the electromagnet's flux varies in proportion with generator output voltage.

Higher generator output increases the current in the electromagnet; this attracts the steel centre of the spring, which reduces compression on the pile, thereby increasing its resistance. Less field current reduces the generator output voltage; the current in the voltage coil reduces electromagnetic effect and the spring compresses the pile, reducing its resistance. The varying force applied by the electromagnet and spring thereby controls the pile's resistance to control field current and maintains a constant generator output voltage. The regulator is contained within a cylinder (typically three inches in diameter and six

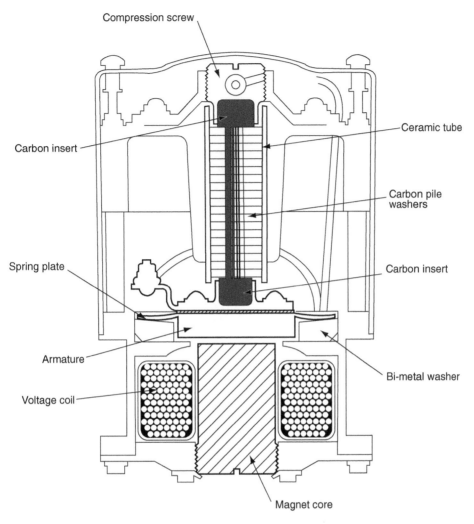

Figure 6.3 Carbon-pile regulator – cross section

inches in length) with cooling fins. Functions of each component are as follows:

- **Compression screw**: the means of setting up compression on the pile and compensating for erosion of the pile during its life.
- **Spring plate and armature**: this compresses the pile to its minimum resistance position.
- **Voltage coil**: contains a large number of turns of copper wire and, with the core screw, forms an electromagnet when connected across the generator output.
- **Magnet core**: concentrates the coil flux; it is also used for voltage adjustment during servicing.

- **Bi-metallic washers**: providing temperature compensation.

Figure 6.4 shows the carbon-pile regulator connected into the generator's regulating circuit. The ballast resistor has a low-temperature coefficient and minimizes the effects of temperature on the voltage coil. The trimmer resistors (in series with the ballast resistor) allow the generator output voltage to be trimmed on the aircraft. The **boost** resistor is normally shorted out; if the switch is opened it allows a slight increase in generator output to meet short-term increases in loading. This is achieved by temporarily reducing the current through the voltage coil. The boost resistor can either be located in the regulator and/or at a remote location for easy access during maintenance.

Key point

Higher generator output increases the current in the carbon pile regulator, which reduces compression on the pile thereby increasing its resistance.

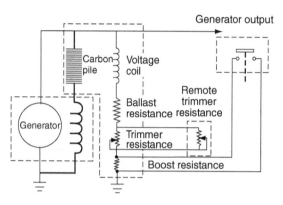

Figure 6.4 Carbon-pile regulator – schematic

Test your understanding 6.1

Explain the purpose of the carbon pile regulator voltage coil.

6.1.3 Electronic voltage regulator

There are many types and configurations of electronic voltage regulators. A representative type is illustrated in Fig. 6.5. The **alternator master switch** used in AC systems energizes the field relay and applies current to the base of TR_2 and the resistor network of R_1, R_2, R_{V1}. This network, together with the Zener diode (Z) is used to establish the nominal operating voltage. Current flows through the alternator's field coil via transistors TR_2 and TR_3, allowing the generator's output to increase. When the output reaches its specified value (14 or 28 V DC depending on the installation) Zener diode Z conducts which turns on transistor TR_1, shorting out transistor TR_2 and TR_3. The generator voltage falls and Zener diode Z stops conducting, thereby turning off transistor TR_1. This turns transistors TR_2 and TR_3 back on, allowing the generator output to increase again. This operation is repeated

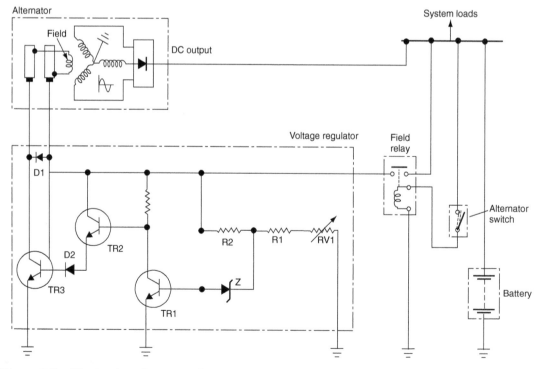

Figure 6.5 Electronic voltage regulator

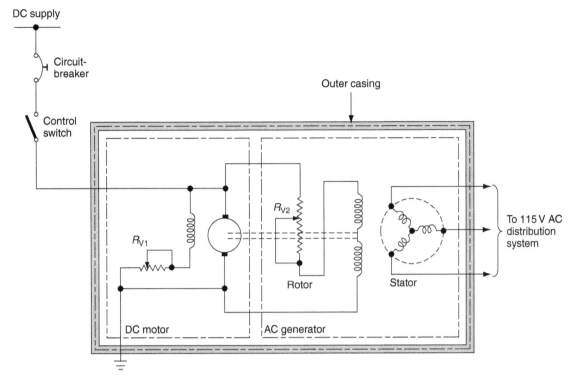

Figure 6.6   Rotary inverter schematic

many times per second as with the vibrating contact regulator; the difference being that electronic circuits have no moving parts and do not suffer from arcing across contacts. Diode D_1 provides protection against the back e.m.f. induced in the field each time TR_3 is switched. The trimming resistor R_{V1} can be used to adjust the nominal voltage output of the regulator.

6.2 External power

In addition to the onboard equipment that has been described, most aircraft have the facility to be connected to an external power source during servicing or maintenance. This allows systems to be operated without having to start the engines or use the battery. The external ground power can either be from a battery pack, a ground power unit (that has a diesel engine and generator) or from industrial power converters connected to the national grid. More details of external power and how it is used is given in Chapter 8.

6.2.1 Power conversion

Equipment used on aircraft to provide secondary power supplies include:

- inverters
- transformer rectifier units (TRUs)
- transformers.

6.3 Inverters

Inverters are used to convert direct current into alternating current. The input is typically from the battery; the output can be a low voltage (26 V AC) for use in instruments, or high voltage (115 V AC single or three phase) for driving loads such as pumps. Older **rotary inverter** technology uses a DC motor to drive an AC generator, see Fig. 6.6. A typical rotary inverter has a four-pole compound DC motor driving a star-wound AC generator. The outputs can be single- or three-phase; 26 V AC, or 115 V AC. The desired output frequency of 400 Hz is determined by

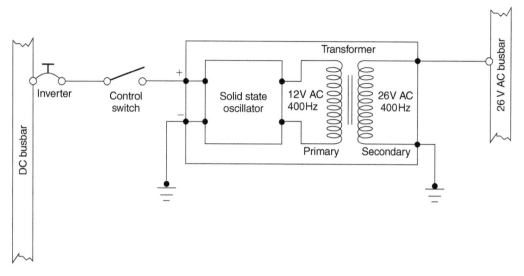

Figure 6.7 Static inverter schematic

the DC input voltage. Various regulation methods are employed, e.g. a trimming resistor (R_v) connected in series with the DC motor field sets the correct speed when connected to the 14 or 28 V DC supply.

Key point

The desired output frequency of a rotary inverter is determined by the DC input voltage.

Modern aircraft equipment is based on the **static inverter**; it is solid state, i.e. it has no moving parts (see Fig. 6.7). The DC power supply is connected to an oscillator; this produces a low-voltage 400 Hz output. This output is stepped up to the desired AC output voltage via a transformer.

The static inverter can either be used as the sole source of AC power or to supply specific equipment in the event that the main generator has failed. Alternatively they are used to provide power for passenger use, e.g. lap-top computers. The DC input voltage is applied to an oscillator that produces a sinusoidal output voltage. This output is connected to a transformer that provides the required output voltage. Frequency and voltage controls are usually integrated within the static inverter; it therefore has no external means of adjustment.

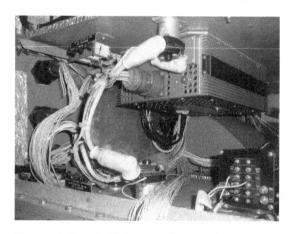

Figure 6.8 Static inverter installation

Key point

Inverters are used to convert direct current into alternating current.

A typical inverter used on a large commercial aircraft can produce 1 kVA. Static inverters are located in an electrical equipment bay; a remote on/off switch in the flight compartment is used to isolate the inverter if required. Figure 6.8 shows an inverter installation in a general aviation aircraft.

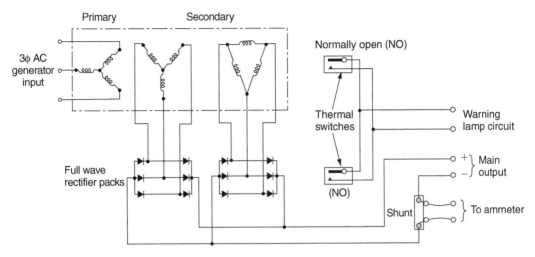

Figure 6.9 Transformer rectifier unit (TRU) schematic

This particular inverter has the following features:

Input:	28 V DC	39 A
Outputs:	115 V AC	6.5 A
	26 V AC	5.8 A
	400 Hz 750 VA	
	continuous	
Power factor:	0.8 to 0.95	
Weight:	15.6 lb	
Dimensions:	$270 \times 220 \times 100$ mm	

6.4 Transformer rectifier units

Transformer rectifier units (TRUs) convert AC into DC; these are often used to charge batteries from AC generators. A schematic diagram for a TRU is shown in Fig. 6.9. The three-phase 115/200 V 400 Hz input is connected to star-wound primary windings of a transformer. The dual secondary windings are wound in star and delta configuration. Outputs from each of the secondary windings are rectified and connected to the main output terminals. A series (**shunt**) resistor is used to derive the current output of the TRU. Overheat warnings are provided by locating thermal switches at key points within the TRU.

Test your understanding 6.2

Explain the applications of inverters and transformer rectifier units (TRUs).

6.5 Transformers

Transformers are devices that convert (or transfer) electrical energy from one circuit to another through inductively coupled electrical conductors. The transformer used as a power supply source can be considered as having an input (the primary conductors, or windings) and output (the secondary conductors, or windings). A changing current in the primary windings creates a changing magnetic field; this magnetic field induces a changing voltage in the secondary windings. By connecting a load in series with the secondary windings, current flows in the transformer. The output voltage of the transformer (secondary windings) is determined by the input voltage on the primary and ratio of turns on the primary and secondary windings. In practical applications, we convert high voltages into low voltages or vice versa; this conversion is termed step down or step up. (More transformer theory is given in Chapter 1.)

Circuits needing only small step-up/down ratios employ **auto-transformers**. These are formed from single winding, tapped in a specific way to form primary and secondary windings. Referring to Fig. 6.10(a), when an alternating voltage is applied to the primary (P_1–P_2) the magnetic field produces links with all turns on the windings and an EMF is induced in each turn. The output voltage is developed across the secondary turns (S_1–S_2) which can be connected for either step-up or step-down ratios. In practice, auto-transformers are smaller in size and weight than conventional transformers. Their disadvantage is that,

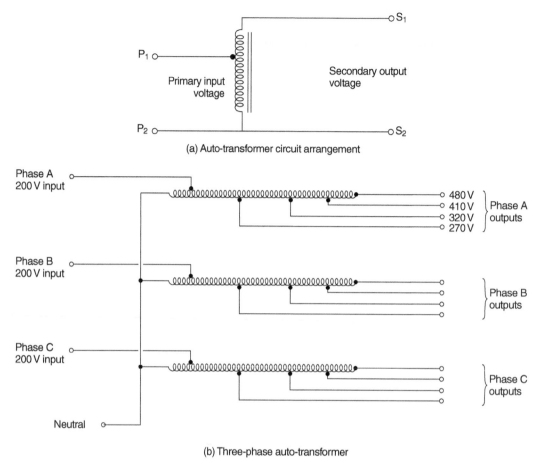

(a) Auto-transformer circuit arrangement

(b) Three-phase auto-transformer

Figure 6.10   (a) Auto-transformer principles; (b) three-phase auto-transformer

since the primary and secondary windings are physically connected, a breakdown in insulation places the full primary e.m.f. onto the secondary winding.

The arrangement for a three-phase auto-transformer is shown in Fig. 6.10(b). This is a **star-connected** step-up configuration. Primary input voltage is the 200 V AC from the aircraft alternator; multiple outputs are derived from the secondary tappings: 270, 320, 410 and 480 V AC. Applications for this type of arrangement include **windscreen heating**.

Test your understanding 6.3

Explain the difference between conventional and auto-transformers.

6.6 Auxiliary power unit (APU)

An APU is a relatively small gas turbine engine, typically located in the tail cone of the aircraft. The APU is a two-stage **centrifugal compressor** with a single turbine. Bleed air is tapped from the compressor and connected into the aircraft's air distribution system. Once started (see Chapter 10) the APU runs at constant speed, i.e. there is no throttle control. The APU shuts down automatically in the event of malfunction.

APUs are used for starting the aircraft's main engines via the air distribution system. While the aircraft is on the ground, the APU can also provide:

- electrical power
- hydraulic pressure
- air conditioning.

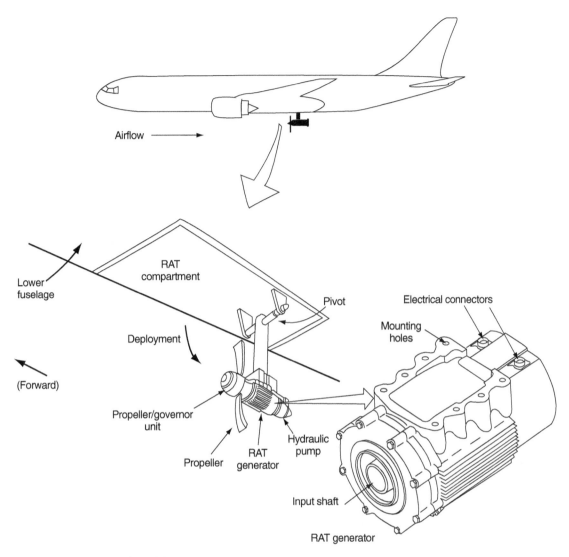

Figure 6.11   Ram air turbine

The APU itself is started from the main aircraft battery. In some aircraft, the APU can also provide electrical power in the air in the event of main generator failure. The Boeing 787 aircraft has more electrical systems and less pneumatic systems than aircraft it is replacing. In this case the APU delivers only electrical power.

APUs fitted to extended-range twin-engine operations aircraft (ETOPS) are critical to the continued safe flight of the aircraft since they supply electrical power, hydraulic pressure and an air supply in the event of a failed main engine generator or engine.

Some APUs on larger four-engined aircraft are not certified for use while the aircraft is in flight.

Key maintenance point

It is essential to remove electrical power from the relevant busbar (or in some cases the entire aircraft) before removing or installing electrical components. Failure to observe this precaution can result in electric shock as well as damage to components and equipment.

Key maintenance point

It is essential that electrical power is removed from an external power cable before connecting the cable to the aircraft. Failure to observe this precaution can result in electric shock as well as damage to components and equipment.

6.7 Emergency power

In the event of generator failure, continuous power can be provided by a ram air turbine (RAT). Also referred to as an air-driven generator, this is an emergency source of power that can be called upon when normal power sources are not available. The RAT is an air-driven device that is stowed in the wing or fuselage and deployed in the event that the aircraft loses normal power. When deployed, it derives energy from the airflow, see Fig. 6.11. RATs typically comprises a two-bladed fan, or propeller that drives the generator shaft via a governor unit and gearbox; the gear ratios increase the generator shaft speed.

The RAT can be deployed between aircraft speeds of 120 to 430 knots; some RATs feature variable pitch blades operated by a hydraulic motor to maintain the device at typical speeds of 4,800 r.p.m. Typical RAT generators produce an AC output of 7.5 kVA to a TRU. Heaters are installed in the RAT generator to prevent ice formation. RATs can weigh up to 400 lbs on very large transport aircraft, with blade diameters of between 40 and 60 inches depending on power requirements.

6.8 Multiple choice questions

1. Circuits needing only small step-up/down ratios employ:
   (a) auto-transformers
   (b) transformer rectifier units
   (c) inverters.

2. Inverters are used to convert what forms of electrical power?
   (a) AC into DC
   (b) AC into AC
   (c) DC into AC.

3. The accuracy of the vibrating contact regulator depends on the:
   (a) input voltage
   (b) generator output current
   (c) resistor value and spring tension.

4. The voltage coil of a carbon pile regulator contains a:
   (a) large number of copper wire turns connected across the generator output
   (b) low number of copper wire turns connected across the generator output
   (c) large number of copper wire turns connected in series with the generator output.

5. Transformer rectifier units (TRUs) are often used to:
   (a) convert battery power into AC power
   (b) charge batteries from AC generators
   (c) connect batteries in series.

6. Stepping a power supply down from 115 to 26 V AC would normally be achieved by a:
   (a) transformer rectifier unit
   (b) inverter
   (c) transformer.

7. The desired output frequency of a rotary inverter is determined by the:
   (a) AC input voltage
   (b) input frequency
   (c) DC input voltage.

8. Inverters and transformer rectifier units (TRUs) are sources of:
   (a) emergency power
   (b) secondary power
   (c) primary power.

9. Higher generator output increases the current in the carbon pile regulator, this:
   (a) reduces compression on the pile thereby increasing its resistance
   (b) increases compression on the pile thereby increasing its resistance
   (c) reduces compression on the pile thereby decreasing its resistance.

10. TRUs are used to convert what forms of electrical power:
    (a) AC into DC
    (b) AC into AC
    (c) DC into AC.

Wiring and circuit protection

The safe and economic operation of an aircraft is becoming ever more dependent on electrical and electronic systems. These systems are all interconnected with wires and cables, and take many forms. Electrical wires and cables have to be treated as an integral part of the aircraft requiring careful installation; this is followed by direct ongoing inspection and maintenance requirements for continued airworthiness. Wire and cable installations cannot be considered (or treated) as 'fit and forget'. System reliability will be seriously affected by wiring that has not been correctly installed or maintained. Legislation is being proposed to introduce a new term: **electrical wire interconnection system** (EWIS); this will acknowledge the fact that wiring is just one of many components installed on the aircraft. EWIS relates to any wire, wiring device, or combination of these, including termination devices, installed in the aircraft for transmitting electrical energy between two or more termination points.

We need to distribute the sources of electrical power safely and efficiently and control its use on the aircraft. Once installed, the wires and cables must be protected from overload conditions that could lead to overheating, causing the release of toxic fumes, possibly leading to fire. This chapter describes the physical construction of wires and cables together with how they are protected from overload conditions before power is distributed to the various loads on the aircraft. (Practical installation considerations for wiring are addressed in Chapter 20.)

7.1 Overview

In the first instance, we need to make a distinction between **wires** and **cables**. Wires are formed from a single solid conductor or stranded conductors, contained within insulation and protective sheath materials. Cables can be defined as:

- two or more separate wires within the same insulation and protective sheath

- two or more wires twisted together
- any number of wires covered by a metallic braid, or sheath
- a single insulated conductor covered by a metallic outer conductor (co-axial cable).

The terms wires and cables are often interchanged. In this book, reference will be made to 'wiring' in the all-embracing generic sense; cables will be referred to in specific terms as and when required.

Key maintenance point

Wire and cable installations cannot be considered (or treated) as 'fit and forget'. System reliability will be seriously affected by wiring that has not been correctly installed or maintained.

7.1.1 Types of wire and cable

There is a vast range of wire and cable types used in an aircraft; these can be categorized with their specific uses and applications:

- airframe wires and cables
- equipment wires and cables
- ignition system cables
- thermocouple cables
- data bus cables
- radio-frequency (RF) cables.

This chapter addresses airframe wire and cable applications in detail. (Wiring inside of equipment is not specifically addressed.) The application of ignition and thermocouple cables are described in Chapter 10 (engine systems). Wires or cables designed for small signals, e.g. data bus cables carrying digital data, are screened to prevent their signals from being affected by electromagnet interference. Wires and cables that carry high power and/or high frequencies are also shielded to prevent them being the cause of electromagnetic interference. **Screening** and **shielding** to protect against electromagnetic interference (EMI) and electromagnetic compatibility (EMC) are described in this chapter; the subject of EMI/EMC protection is discussed in more detail in Chapter 19. Radio-frequency (RF) cables (or feeders) are specialized types of screened cables. (These are described in more detail in a related book in the series, *Aircraft Communications and Navigation Systems*.)

7.1.2 Operating environment

Wiring installed in aircraft has to operate in a harsher environment than that found in cars, buildings or industrial applications. In addition to conducting current (often at high voltages), aircraft wiring will be exposed to a variety of environmental and in-service conditions including contaminants, for example:

- hydraulic fluid
- fuel and/or oil
- temperature extremes
- abrasion
- vibration

General-purpose wiring used in the airframe comprises the majority of the installed material. There are also special considerations for wiring, e.g. when routed through fire zones. The materials used for the conductor and insulation must take these factors into account. Each application has to be planned from the initial design, taking into account the inspection and maintenance required for continued airworthiness of the wiring.

Test your understanding 7.1

Why are cables shielded?

7.2 Construction and materials

Aircraft wiring needs to be physically flexible to allow it to be installed, and then to withstand the vibration of the aircraft that will cause the wires to flex. **Multistranding** of the conductor increases the flexibility of the wire or cable, making it easier to install and withstand vibration of the aircraft. The insulating material has to be able to withstand the applied voltage; the sheath material needs to be able to withstand the

specified contaminants. Conductors need to be able to carry the required current without overheating or burning; they must also have low insulation resistance to minimize voltage drops. From **Ohm's law**, we know that (at a given temperature) the voltage across a resistance is proportional to the current. For a given current (I) and resistance (R), the voltage drop is quoted in terms of *IR* **losses**. For these reasons, most aircraft conductors are constructed from copper or aluminium contained within man-made insulating material(s).

Aluminium conductors are sometimes used in aircraft; however, the majority of installations are copper. The choice of conductor material is a trade-off; the first consideration is the material's resistance over a given length (given the term **resistivity**, symbol ρ, measured in ohm-metres, abbreviated Ωm). Annealed copper at 20°C has a resistivity of 1.725×10^{-8} Ωm; copper is more ductile than aluminium, and can be easily soldered. Aluminium has a resistivity value of 2.8×10^{-8} Ωm, it is 60% lighter than copper but it is more expensive.

A major consideration for using aluminium is that it is **self-oxidizing**; this reduces manufacturing costs (no plating required) but extra precautions are necessary for terminating the conductors due to the increased termination resistance. The quantity and gauge of these strands depends on the current-carrying capacity (or rating) and degree of flexibility required. Individual strands of copper need to be coated to prevent oxidation. The choice of coating for the strands depends on the operating temperature of the wire. In general terms, three types of coating are used: tin, silver or nickel, giving temperature ratings of 135°C, 200°C and 260°C respectively.

To summarize, copper conductors are used extensively on aircraft due to the material's:

- low resistivity
- high ductility
- high tensile strength
- ease of soldering.

Test your understanding 7.2

Explain why cables and wires are multi-stranded.

Conductors must be insulated to prevent short-circuits between adjacent circuits and the airframe. Power supplies of 12 or 28V do not pose a threat of electrical shock, but the wires must be insulated to prevent arcing, loss of system integrity and equipment failure. The combined effects of insulation damage and fluid contamination gives rise to **wet arc tracking**. This phenomenon can occur when insulating surfaces are contaminated with any material containing free ions; the surface then behaves as an electrically conductive medium (an **electrolyte**). Leakage currents are sufficiently high to vaporize the contamination; this drives away the electrolyte and results in the formation of localized dry areas. These areas now offer a higher resistance to the current flow. In turn, high voltages will develop across these areas and result in small surface discharges. Initially, these discharges will emit flashes of light at the insulation surface, and produce localized temperatures in the order of 1000°C. These high temperatures cause degradation of the insulation material. The ability of aircraft wiring to resist wet arc tracking is highly dependent on the wire insulation material. The conductivity level of the electrolyte will influence the **failure mode** resulting from this wet arc tracking. Higher power supply voltages are potentially lethal and the insulation provides a level of protection against this. The insulation material and its thickness depend mainly on the operating temperature and system voltage; examples of insulating and sheath materials include:

- ethylene tetrafluoroethylene (ETFE): this is a fluorocarbon-based polymer (fluoropolymer) in the form of plastic material
- polytetrafluoroethylene (PTFE): a synthetic fluoropolymer
- fluorinated ethylene-propylene (FEP): this retains the properties of PTFE, but is easier to form
- polyvinylidene fluoride (PVF_2 or PVDF): has good abrasion and chemical resistance, and (like most fluoropolymers) is inherently flame-retardant.

Test your understanding 7.3

Explain the term wet arc tracking.

7.3 Specifications

These have become more complex over the years to address the higher performance needed from wires and cables. This has been driven largely from in-service experience with electrical fires and the drive to reduce weight as more and more avionics equipment is introduced onto the aircraft. Reduced weight for a given length and diameter of wire is achieved through reducing the wall thickness of the insulation. Typical specifications used for wires and cables are contained in the US military specification **MIL-W-M22759E**. This specification covers fluoropolymer-insulated single conductor electrical wires manufactured with copper or copper alloy conductors coated with either tin, silver or nickel. The fluoropolymer insulation of these wires can either be PTFE, PVF_2, FEP or ETFE. Wires manufactured to this specification are given a part number using the following format: M22759/x-xx-x.

- *M22759/x*-xx-x determines the specific wire type (insulation, sheath and wire coating) from a table in the specification
- M22759/x-*xx*-x determines the wire size from a table in the specification
- M22759/x-xx-*x* determines the insulation colour from another specification (MIL-STD-681).

Examples of wires are shown in Fig. 7.1. The single-walled construction in Fig. 7.1(a) has a composite insulator and protective sheath to reduce cost and weight. The twin-walled construction, Fig. 7.1(b), illustrates the conductor, insulator and separate protective sheath. Outer sheaths on either type of wire can crack over long periods of time; in the single-wall-type wire, moisture ingress can migrate into the wire through capillary action. This can lead to tracking inside the insulation, leading to overheating; twin-walled wiring is more resistant to this effect.

Key maintenance point

The combined effects of insulation damage and fluid contamination gives rise to wet arc tracking.

7.3.1 Wire size

Wire sizes used on aircraft are defined in accordance with the **American Wire Gauge** or **Gage** (AWG). For a given AWG, the wire will have a specified diameter and hence a known conductance (the reciprocal of resistance). Cable size relates to the conductor's diameter; the overall wire or cable diameter is therefore larger due to the insulation. The largest wire size is 0000 AWG; the smallest is 40 AWG. The range of AWG wire/cable sizes is detailed in the Appendices. Selection of wire size depends on the specified current to be conducted. Larger diameter wires add

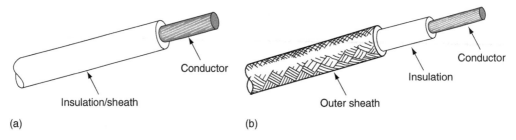

(a)   (b)

Figure 7.1   Typical aircraft wires and cables: (a) single-walled, (b) twin-walled

weight, but offer less voltage drop for a given length and lower heating effect due to I^2R losses.

Wire identification is printed on the outer surface of the sheath at intervals between 6 and 60 inches. This includes the wire part number and manufacturers' commercial and government entity (**CAGE**) designation. The printing is either green or white (depending on the actual colour of the wire).

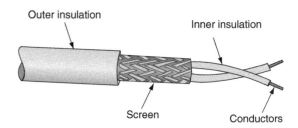

Figure 7.2   Screened cable

Key point

Conductors must have low insulation resistance to minimize IR losses.

7.3.2 Performance requirements

Wires manufactured to MIL-W-M22759E have to demonstrate compliance with dimensions and construction, together with criteria to meet the following requirements:

- ease of removing the insulation
- ease of soldering
- dielectric testing
- flexibility
- elongation and tensile strength
- wicking
- high- and low-temperature testing
- flammability
- life cycle testing
- fluid immersion
- humidity
- smoke emission.

7.4 Shielding/screening

Shielded or screened wiring either prevents **radiation** from circuits switching high currents or protects

susceptible circuits (see Fig. 7.2). The inner conductor carries the system current; the screen provides a low-resistance path for coupling of electromagnetic fields. These fields are coupled into the shield and are dissipated to ground. (Further details on the theory of electromagnetic fields and shielding are given in Chapter 19.) Typical applications where shielding is used includes wiring installed near generators, ignition systems or contacts that are switching high currents. Shielded wires can be formed with single, twin, triple or quadruple cores. The classic example of screened wiring occurs with the Arinc 429 data bus which uses twisted screened pairs of cable to transmit digital data. This is being transmitted at either 12.5–14.5 (low speed) or 100 (high speed) K bits per second with a differential voltage of 10 V between the pair of wires. Examples of screened wire and cable specifications are found in M27500-22TG1T-14.

Key point

Wires and cables that carry high power, and/or high frequencies are shielded to prevent electromagnetic interference.

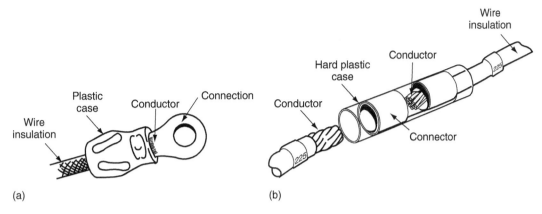

Figure 7.3   Wire termination and splicing: (a) ring tongue terminal, (b) in-line splice crimping

7.4.1 Crimps and splices

Individual wires and cables can be terminated or connected using crimps and splices. Care must be taken when stripping shielded cables; both the inner conductor and outer screen must be exposed in order to make the connection. The outer shield is formed into pigtails and terminated with a **crimp**, or ring-tongue terminal, see Fig. 7.3(a). Alternatively, individual wires can be joined with an in-line **splice**, e.g. if a system is being modified with additional wiring, see Fig. 7.3(b). Crimps and splices are formed over the exposed conductor and insulating material. The entire crimp or splice termination is then protected mechanically with a hard plastic case.

Test your understanding 7.4

Give some examples of wire/cable performance specifications.

7.4.2 Coaxial cables

A specialized version of the shielded wire is the coaxial cable. The inner conductor is solid or stranded; it can be plain copper or plated. The outer conductor forms a shield and is a single wire **braid** made from fine strands of copper or steel. The inner and outer conductors are separated by a solid insulation, forming a **dielectric**. The outer sheath or jacket provides protection against fluid contaminants. Coaxial cables are normally used to guide radio-frequency (RF) energy between antennas and receivers or transmitters. The inner conductor is shielded by the braiding

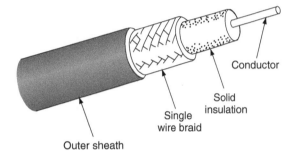

Figure 7.4   Coaxial cable overview

from electric and magnetic fields; the conductor's own field is contained within the same shield. The net result is that fields from the inner and outer conductors cancel each other out. In most practical RF applications, coaxial cable radiation and susceptibility are virtually eliminated.

The typical construction of coaxial cable is shown in Fig. 7.4; the salient features are:

- **inner conductor:** silver plated copper
- **solid insulation:** (dielectric) foamed FEP
- **single wire braid:** (screen) tin plated copper
- **outer insulation:** (jacket) FEP.

(Further details on coaxial cables and antennas used in radio-frequency (RF) applications can be found in *Aircraft Communications and Navigation Systems*.) Other applications for coaxial cables include fuel quantity systems (see Chapter 11).

Coaxial cables are terminated with **bayonet** or **thread-nut couplings** (BNCs and TNCs respectively). BNC has the advantage of a quick-release connection. TNC performs well under vibration conditions since

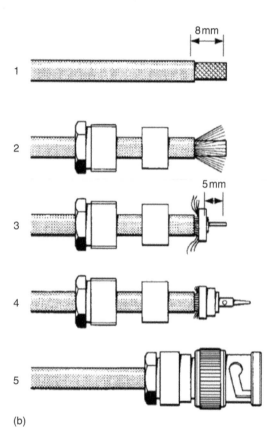

Figure 7.5 Coaxial cable connectors: (a) typical end fittings, (b) terminating the coax cable

they have reduced axial movement. Figure 7.5 gives an illustration of typical coaxial cable connectors and how they are fitted to the cable.

Test your understanding 7.5

What is the difference between shielded and coaxial cables?

7.5 Circuit protection

The current-carrying capacity of a wire or cable is determined by its length and cross sectional area; heat dissipation is determined by I^2R losses. When the circuit or system is designed, the wire size is selected to safely carry this current. Wires and cables are subjected to abrasion during the normal service life of the aircraft; this can lead to the conductor being exposed. This exposure could lead to a low-resistance path between the conductor and the airframe and/or an adjacent conductor. Faulty equipment, low-resistance paths or overloading from additional circuits will cause the current to increase and this might exceed the current-carrying limit of the conductor. Heat will build up in the wire leading to fumes, smoke and ultimately fire. It is vital that we protect against this whilst allowing for transients; the methods used in aircraft are selected from the following devices:

- fuse
- circuit-breaker
- limiting resistor.

7.5.1 Fuses

Fuses are links of wire that are connected in series with the circuit. Their current-carrying capacity is predetermined and they will heat up and melt when this is exceeded, thereby interrupting and isolating the circuit. Materials used for the **fusible link** include lead, tin-bismuth alloy, copper or silver alloys. Referring to Fig. 7.6(a), the fuse wire is contained within a glass or ceramic casing (or cartridge) to prevent any particles of hot metal escaping which could cause secondary damage. End-caps provide a connection for the fuse wire and make contact with the circuit wiring. Fuse holders consist of terminals and a panel clamp-nut.

Some fuse holders have an indication of the fuse condition, i.e. if the fuse has blown. The **indicating cap** is black with an integrated coloured light. When the fuse has blown, the cap illuminates; different colours indicate different power supply voltages. Heavy-duty fuses (typically protecting circuits with up to 50 A current) are constructed with a ceramic body and terminals, see Fig. 7.6(c). Fuses are either clipped into position on a terminal board, see Fig. 7.7, or screwed into a panel, see Fig. 7.8.

Fuses are relatively low cost items, but they can only be used once. In some applications, the fuse

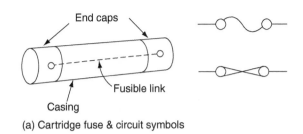

(a) Cartridge fuse & circuit symbols

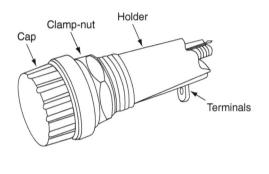

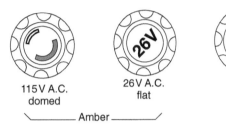

115 V A.C.
domed

26 V A.C.
flat

26 V D.C.
flat
White

Amber

(b) Fuse holders (left) and indicating caps (right)

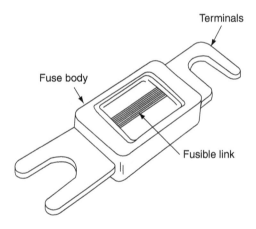

(c) Heavy duty fuse construction

Figure 7.6   Aircraft fuses

material and physical construction is designed to have a time delay; the so-called **slow-blow** fuse, or **current limiter**. This is made from a copper alloy that has a higher melting point than lead/tin. It has a single strip of material waisted into a narrow cross-section to provide the fusing point. Heavy-duty fuses are used at power distribution points. They have multi-strands of parallel elements and are rated up to 500 A. This type of fuse is fitted with a packing medium to contain the debris following rupture. Materials used include quartz, magnesium oxide, kieselguhr or calcium carbonate (chalk).

Fuses have a rating that determines the maximum current it can carry without melting. The fuse will also have a minimum fusing current that is affected by ageing; a process that occurs when the fuse is

Figure 7.7 Fuse clipped directly into a terminal board

Figure 7.8 Fuses located in a panel

operated at the minimum fuse rating for prolonged periods of time. Ambient temperature affects the current rating and response time of a fuse. They must be located close to power source to minimize the length of unprotected wire; at the same time they have to be accessible for replacement. Spare fuses must be carried on the aircraft and be accessible to the flight crew. Typical requirements are to carry 50% of each rating as spares, e.g. if the aircraft is fitted with four 10 A and five 15 A fuses, then two 10 A and three 15 A fuses should be carried as spares.

Key maintenance points

1. Fuses can rupture, or 'blow', when exposed to vibration and thermal cycling as well as during overload conditions.
2. Determine the reason that caused the fuse to blow before replacing.
3. Never assume that the blown fuse is to be replaced with the same rating; always check the aircraft documents.

7.5.2 Circuit-breakers

Circuit-breakers are electromechanical devices that interrupt and isolate a circuit in the event of excessive current. Unlike fuses, circuit-breakers can be reset (assuming that the fault condition has cleared). There are two circuit-breaker principles: electromagnetic and thermal.

An **electromagnetic** circuit-breaker is essentially a relay with current flowing through a coil; the resulting

magnetic field attracts an armature mechanism. The current is normally a proportion of the main load current; this increases in proportion to the main load current. The armature mechanism is linked to a pair of contacts that carry the main load current. These contacts are opened when the current through the coil exceeds a certain limit.

Thermal-type circuit-breakers consist of a bimetallic thermal element, switch contacts and mechanical latch. The internal schematic of a thermal circuit-breaker and its external features are illustrated in Fig. 7.9. The thermal element is a **bimetallic** spring that heats up as current passes through it; this eventually distorts and trips the mechanism when the rated level is exceeded. The mechanism is linked to the main switch contacts, when the circuit-breaker '**trips**' the contacts open, thereby disconnecting power from the circuit. When the contacts open, a **button** is pushed out of the circuit-breaker. This button is used to manually reset the contacts; a **white collar** just below the button provides visual indication that the circuit-breaker is closed or tripped.

Some circuit-breakers use a large collar grip to identify specific systems. They can be locked open if required, e.g. if the system is installed but not certified for operation. As with fuses, the circuit-breaker should be located as close as possible to source of power; they are often arranged on the panels in groups.

Circuit-breakers can also be used to conveniently isolate circuits, e.g. during maintenance. Certain circuit-breakers are fitted with removable collars so that they can be readily identified. The circuit-breaker current rating is engraved on the end of the button (see Fig. 7.10). Circuit-breakers can be single- or multipole devices (poles are defined as the number of links

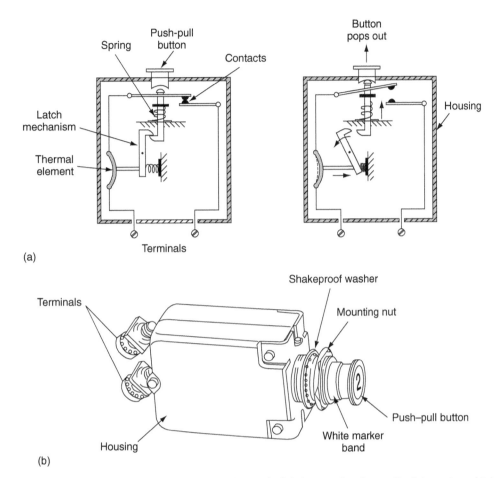

(a)

(b)

Figure 7.9 Aircraft circuit-breakers (thermal type): (a) internal schematic (closed and tripped), (b) external features

Figure 7.10 Circuit-breaker panel

that a switching device contains). Multi-pole devices are used in three-phase AC circuits.

Various configurations of circuit-breaker are installed on aircraft, including:

1. automatic reset
2. automatic trip/push to reset
3. switch types
4. trip free.

The circuit symbols for each of these types is illustrated in Fig. 7.11. **Trip-free** circuit-breaker contacts cannot be closed whilst a fault exists. This is the preferred type of circuit-breaker on aircraft, especially on new installations.

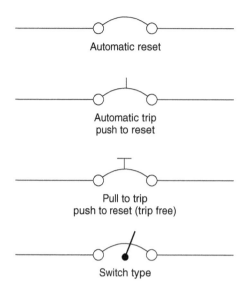

Figure 7.11 Circuit-breaker symbols/type

7.5.3 Limiting resistors

These are used to limit current surges, primarily in DC circuits, where the initial current surge is large. When these circuits are switched they create large current flows that can be harmful to other components and reduce the power supply voltage for a period of time (determined by the time constants of the circuit). Limiting resistors are connected in series with such circuits and then automatically shorted out once the circuit current has stabilized. Typical applications of limiting resistors are found in engine starting circuits and voltage regulators. Limiting resistors are also used in fire extinguishing systems (see also Chapter 16). Fire extinguishers are activated by applying direct current (DC) through current-limiting resistors to the associated squib; this ruptures a disc that allows extinguishing agent to be expelled under pressure. The limiting resistors prevent inadvertent operation of the squib. In electronics, limiting resistors are used to protect devices such as diodes.

7.6 Multiple choice questions

1. Wires and cables that carry high power and/or high frequencies are shielded to prevent:
   (a) the wire or cable being the cause of electromagnetic interference
   (b) the wire or cable being affected by electromagnetic interference
   (c) wet arc tracking.

2. Two or more separate wires within the same insulation and protective sheath is referred to as a:
   (a) screened wire
   (b) coaxial cable
   (c) cable.

3. Fuses have a rating that determines the:
   (a) maximum current it can carry without opening
   (b) minimum current it can carry without opening
   (c) time taken for the fuse to rupture.

4. Conductors must have low insulation resistance to minimize:
   (a) current surges
   (b) electromagnetic interference
   (c) IR losses.

5. The combined effects of insulation damage and fluid contamination gives rise to:
   (a) electromagnetic interference
   (b) wet arc tracking
   (c) current surges

6. For a given AWG, the wire will have a specified:
   (a) diameter and hence a known conductance
   (b) length
   (c) screening.

7. Coaxial cables are normally used for:
   (a) digital signals
   (b) motors and generators
   (c) radio-frequency (RF) signals.

8. Trip-free circuit-breaker contacts:
   (a) can always be closed whilst a fault exists
   (b) cannot be closed whilst a fault exists
   (c) are only used during maintenance.

9. Limiting resistors are connected:
   (a) momentarily in series
   (b) momentarily in parallel
   (c) permanently in series.

10. The white collar just below a circuit-breaker button provides visual indication of the:
    (a) circuit-breaker trip current
    (b) system being protected
    (c) circuit-breaker being closed or tripped.

Chapter 8

Distribution of power supplies

Electrical power is supplied to the various loads in the aircraft via common points called busbars. The electrical power distribution system is based on one or more busbar(s); these provide pre-determined routes to circuits and components throughout the aircraft. The nature and complexity of the distribution system depends on the size and role of the aircraft, ranging from single-engine general aviation through to multi-engine passenger transport aircraft. For any aircraft type, the distribution system will comprise the following items:

- busbar
- protection
- control
- wiring
- loads.

In this chapter, we will focus on busbar configurations and how these are arranged for the protection and management of the various power supply sources available on the aircraft.

The word 'bus' (as used in electrical systems) is derived from the Latin word *omnibus* meaning 'for all'. Busbars are often formed from thick strips of copper, hence the name 'bar'. These have holes at appropriative intervals for attaching one side of the protection device (circuit-breaker or fuse). Alternatively, they are made from heavy gauge wire. The busbar can be supplied from one or more of the power sources previously described (generator, inverter, transformer rectifier unit or battery). Protection devices, whether fuses or circuit-breakers, are connected in series with a specific system; they will remove the power from that system if an overload condition arises. There also needs to be a means of protecting the power source and feeder lines to the busbar, i.e. before the individual circuit protection devices. The various methods used for aircraft power supply distribution for a wide range of aircraft types are described in this chapter.

8.1 Single engine/general aviation aircraft

A simple busbar distribution system is fed by the battery and controlled by a **battery master switch**, see Fig. 8.1. Since the current supplied from the battery

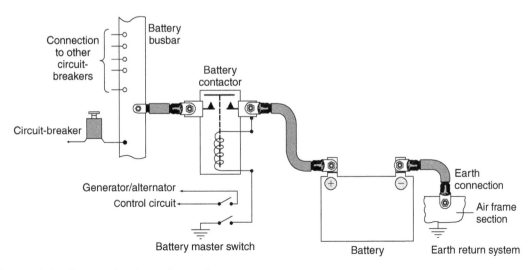

Figure 8.1 Battery busbar schematic

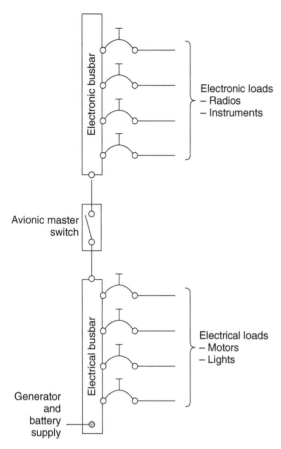

Figure 8.2   Split avionic bus schematic

to the bus will be very high, a heavy-duty relay or contactor is often used; the battery master switch activates the contactor when power is required on the bus. The same switch is also used for generator control.

For general aviation aircraft the battery is normally used for an engine start. The connection for this is taken before the bus to minimize voltage drop (IR) losses. A variation of the simple battery bus is the **split avionic bus**, see Fig. 8.2. This arrangement isolates electronic devices from heavy surges during engine start, or when using external power to start the engine. The arrangement for an AC generator system is shown schematically in Fig. 8.3. This illustrates the voltage regulator and battery master switch used for alternator control. It is normal practice for the battery charge/discharge current to be monitored via an ammeter.

For certification purposes, it is often a requirement that no single failure of any component should cause

loss of power to the radios. One way of achieving this is supplying the bus from the main contactor and an emergency switched contact.

8.1.1 Reverse current relay

This circuit arrangement is needed on any DC generation system to prevent the battery from feeding excess current back through to the generator's armature. Without this protection, the battery would discharge when the generator output voltage is less than battery terminal voltage, e.g. during engine shutdown. An automatic method is needed to disconnect the generator from the battery under these conditions. Many circuits have been developed, some are standalone, some are integrated with other generator functions within the same unit e.g. voltage regulation. To illustrate the principles of the reverse current relay, an electromechanical device is described.

Referring to Fig. 8.4, voltage and current coils are wound onto the same soft-iron core. The voltage coil comprises many turns of fine gauge wire; this is connected in parallel with the generator output. The current coil is made of a few turns of thick gauge wire; it is connected in series with the generator output. The combined coils and core form an electromagnet.

A pair of high-current-rated switch contacts are held open by the force of a spring; the contacts can be closed by an armature that is controlled by the influence of the electromagnet's field.

When the generator output is higher than the terminal voltage of the battery, the contacts are held closed by the magnetic field created by the current in both the coils. (These are wound such that the magnetic fields assist each other.) If the engine is slowed down to the point where the generator output is less than the terminal voltage of the battery, the contacts are opened since current is now flowing through the current coil in the opposite direction (the electromagnetic field is weakened as a result). With the contacts open, the battery is effectively disconnected from the generator.

8.1.2 Current limiter

This device reduces the output from a DC generator under fault conditions, e.g. a short circuit on the busbar, to protect the generator from overheating and burning out its windings. The limiter reduces generator output if its maximum safe load is exceeded; if the fault condition is removed, the generator automatically becomes

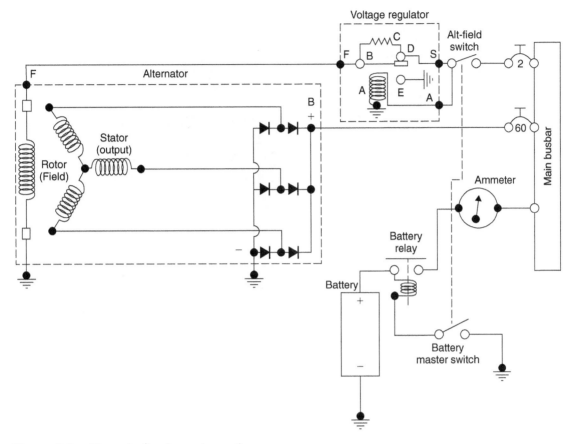

Figure 8.3   Alternator/busbar schematic

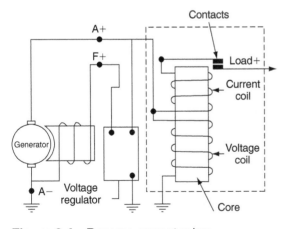

Figure 8.4   Reverse current relay

available again. It operates on a similar principle to the electromechanical voltage regulator; as with the reverse current relay, it could be a stand-alone device or be built into a generator control unit. To illustrate the

principles of the current limiter, an electromechanical device is described.

Referring to Fig. 8.5, the current coil is connected in series with the output of the generator. If the output exceeds a pre-determined limit, the coil's magnetic field opens a pair of switch contacts; this puts a resistor in series with the generator's field thereby reducing the generator's output. This decrease in generator output weakens the field of the current coil and the contacts close again (shorting out the resistor). If the fault condition has been removed, the generator remains connected to the bus. Should a permanent overload condition exist, the contacts open/close on a repetitive basis to keep the output within safe limits.

Test your understanding 8.1

Explain the different functions of a reverse current relay and current limiter.

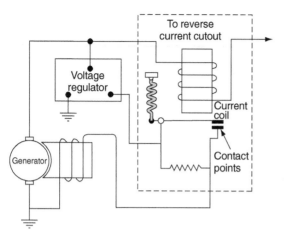

Figure 8.5 Current limiter

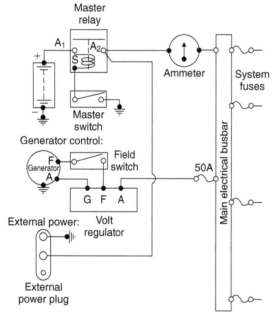

Figure 8.6 External power supply schematic

Figure 8.7 External ground power from a battery pack

8.1.3 External power

The aircraft battery provides an autonomous means of starting the engine. Certain types of operation, e.g. cold weather and repeated starts, could lead to excessive demands, resulting in a battery that is not fully charged. An external power supply system schematic (as illustrated in Fig. 8.6) provides power to the aircraft, even when the battery is flat, or not installed.

External power can be from a ground power unit or simply from a battery pack as shown in Fig. 8.7. On larger aircraft installations, a connector with three sockets supplies external power. These sockets connect with three pins on the aircraft fuselage as shown in Fig. 8.8. Different-size pins are used on the connectors to prevent a reverse polarity voltage being applied.

Some aircraft installations have a **ground power relay**. Power can only be supplied into the aircraft via the main pins when the third (shorter) pin makes contact. The third pin is used to energize the ground power relay; this additional relay prevents arcing on the power connector as illustrated in Fig. 8.9. External AC power is applied in a similar way, except that the three phases have to be connected via individual circuits.

8.1.4 Battery charging: single engine

Battery charging on the ground is achieved by an external power supply (it would be unsafe to run the

(a)

(b)

Figure 8.8   External ground power: (a) external ground power 3-pin connector, (b) external ground power connected

engine in the hangar). External power in the form of another battery, a diesel-powered generator or adapted mains supply provides a source of energy. With external power available and switched on, power is now available at the battery relay output terminal via the external power relay. Power is also supplied to the closing circuit (the resistor and diode reduces the

charging voltage). When the battery master switch is selected on, this closes the battery relay thus providing full charging current. Protection against a short circuit is from a fuse.

Test your understanding 8.2

What kind of instrument would be used to measure battery charging/discharging current?

8.2 Twin engine general aviation aircraft

In the basic configuration, each engine drives its own generator, and the outputs are supplied to a common battery busbar. Engine speed variations mean that one generator could be supplying more output than the other generator and so an equalizing system, using carbon pile regulators, is employed to balance the outputs. This balancing circuit provides automatic adjustment so that each generator delivers an equal output.

Referring to Fig. 8.10, the left and right carbon pile regulators (items 3 and 4) each have equalizing coils (items 7 and 8) wound on the same core as the voltage regulator coils (items 5 and 6). Low value resistors (items 1 and 2, typically $0.01\,\Omega$) are fitted in each generator ground connection; these develop voltage drops in proportion to the generator output currents.

Both equalizing coils are connected in series with the top end of the resistor; the system senses and adjusts the generator outputs to balance both generators at normal engine speeds. The generator outputs are connected through the reverse current relays (items 11 and 12) to the battery busbar. Equalizing circuits connect each resistor via equalizing coils and equalizing switches (items 13 and 14). When each generator has the same output, e.g. 50 A, the voltage dropped across each resistor is therefore $IR = 50 \times 0.01 = 0.5\,V$; there is no voltage difference and no current flows in the equalizing circuit, the circuit is now balanced. Over time, variations in output develop due to the mechanical differences in the regulators, e.g.

- carbon pile wear
- contact surface resistance
- mechanical variations in the armature.

For illustration purposes, assume that the left generator is regulating at an output less than the right generator. The voltage dropped across each of the resistors is

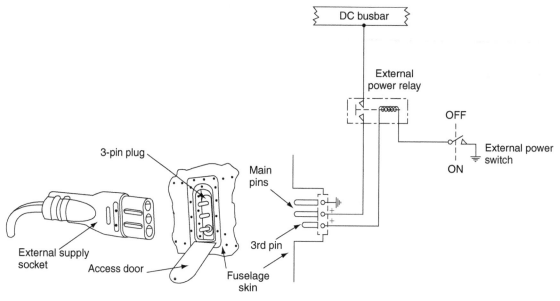

Figure 8.9 External ground power 3-pin connector schematic

now different since the left generator is supplying less current. To illustrate this effect, if the left generator is supplying 40 A and the right generator is supplying 50 A, the voltage drops across the respective resistors are:

$$V_{R1} = 40 \times 0.01 = 0.4 \text{ V}$$

$$V_{R2} = 50 \times 0.01 = 0.5 \text{ V}$$

This difference in voltage (0.1 volt) causes a current to flow from R2 through the right equalizing coil (8) into the left equalizing coil (7). This relaxes the right carbon pile regulator thereby decreasing the generator output. The left regulator coil compresses thereby increasing the left generator output. In practice, this circuit can only compensate for relatively small variations in output; the generators need to be trimmed on a periodic basis.

8.3 Larger aircraft systems

Larger (commuter, business and passenger) aircraft have many more electrical systems compared with general aviation aircraft; there is a requirement for a comprehensive approach to account for potential failures of generators, wiring, etc. The management of potential failures is addressed by categorizing the various loads and then disconnecting them in accordance with a predetermined sequence. The process of

switching loads off the bus is called **load-shedding**; this can be achieved by automatic or manual control. These loads are connected onto specific busbars that fulfil a specific function. These can be categorized into a hierarchy as illustrated in Fig. 8.11. Connections between busbars are via heavy-duty contactors, or breakers. Aircraft types vary, however the following categories are typical for many installations:

Main bus: this is sometimes called the non-essential, generator, or load bus. It will include loads such as the galleys, in-flight entertainment (IFE) and main cabin lights. These loads can be disconnected and isolated in flight without affecting the safe operation of the aircraft.

Essential bus: this is sometimes called the vital or safety bus. It will include equipment and instruments required for the continued safe operation of the aircraft.

Battery bus: this is sometimes called the standby, or emergency bus. It supplies the equipment required for the safe landing of the aircraft, e.g. radios, fuel control, landing gear and fire protection.

There are three main types of distribution system architecture used on aircraft to fulfil the above:

- split bus system
- parallel system
- split parallel system.

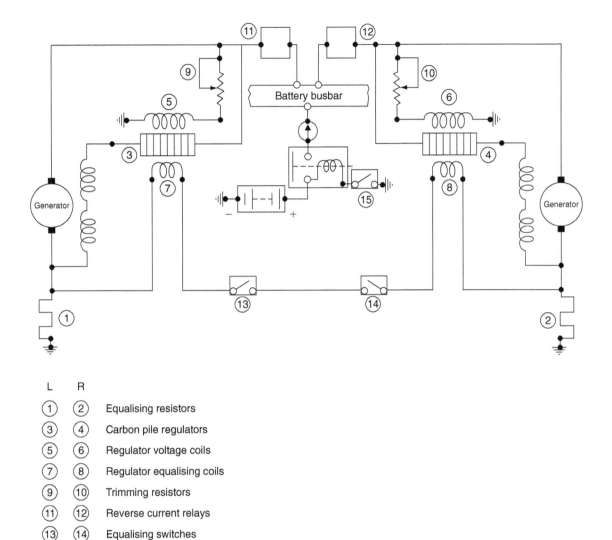

L	R	
①	②	Equalising resistors
③	④	Carbon pile regulators
⑤	⑥	Regulator voltage coils
⑦	⑧	Regulator equalising coils
⑨	⑩	Trimming resistors
⑪	⑫	Reverse current relays
⑬	⑭	Equalising switches
	⑮	Battery switch

Figure 8.10   Simple twin engine equalizing system schematic

8.4 Split bus system

This is a completely isolated twin generation system, sometimes called a **non-parallel** system used on twin-engine aircraft, see Fig. 8.12. **Primary power** is based on two main AC integrated drive generators (typically 40 kVA on each engine). An APU generator (40 kVA) is used as back-up in the event of a main integrated drive generator (IDG) failure. Note that the APU is normally a constant speed device in its own right; therefore an IDG is not required. The advantage of a split-bus system is that the generators do not need

to be operating at exactly the same frequency and can be running out of phase with each other. **Secondary power** is derived from step-down transformers to provide 26 V AC; transformer rectifier units (TRUs) provide 28 V DC for the DC busbars and battery charging.

Referring to Fig. 8.12, the right and left generators feed their own busbars to which specific loads are connected. Each generator bus is connected to a transfer bus via transfer relays. In the event of a generator failure, the remaining generator (engine or APU) supplies essential loads. Control of the system

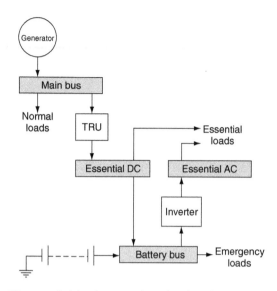

Figure 8.11 Large aircraft electrical power hierarchy

is via a number of flight compartment switches, control breakers and relays arranged to connect and disconnect the generators and busbars. Typical control panel features for a split bus system are shown in Fig. 8.13; the features of this panel are:

Ammeters for the main generators to indicate load current

Ground power available (blue) when external power supply is connected

Ground power on/off switch to select ground power onto the aircraft

Transfer bus off (amber) when the transfer relay is de-energized (either normal or transfer)

Bus off (amber) both respective generator circuit-breakers (GCBs) and bus tie breakers (BTBs) open

Generator bus off (blue) if the respective GCB is open

APU generator bus off (blue) APU running at >95% RPM, no power from generator.

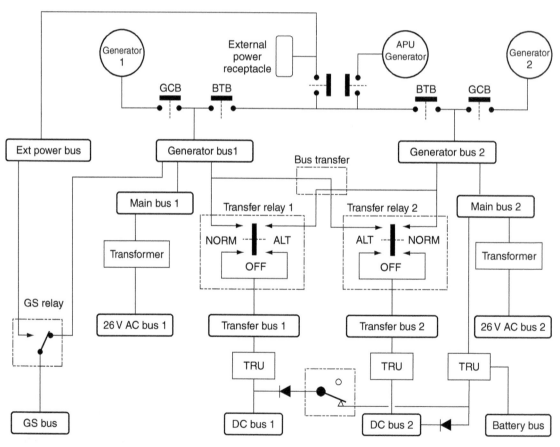

Figure 8.12 Split bus system

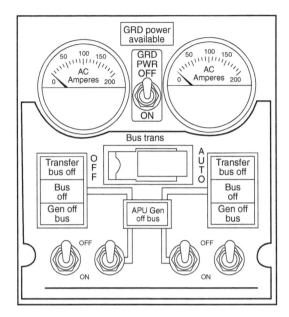

Figure 8.13 Typical electrical power control panel features

Key maintenance point

External AC power cannot be paralleled with the aircraft generators.

8.5 Parallel bus system

The electrical distribution system on larger passenger aircraft (with three or four engines) are based on a parallel load distribution system, see Fig. 8.14. In this configuration, all generators are connected to their own AC load bus and a distribution bus; any generator can supply any load bus to provide equal load-sharing. All generator voltages, frequencies and phase relationships must be controlled to very close tolerances. Any attempt to connect generators in parallel before these conditions are met could result in loss of generator power due to large circulating currents. Referring to Fig. 8.14, when all four GCBs and BTBs are closed, all four generators are **synchronized** and connected to the tie (or synchronized) distribution busbar.

If one generator fails, its GCB is opened; this isolates the generator from its own load busbar. That busbar is now powered from the remaining generators. If this bus becomes overloaded, opening its GCB

and BTB isolates it. With more than two generator failures, **load-shedding** is introduced. External power can be made available by one or two power supply units (or carts). The APU can also be connected onto the distribution bus.

8.6 Split/parallel bus system

This is a flexible load distribution system for large passenger aircraft; it provides the advantages of the parallel system and maintains isolation when needed. Primary power supply features include: one IDG per engine, two APU generators and two external power connections. A split system breaker links left and right sides of distribution system. Any generator can supply any load busbar; any combination of generators can operate in parallel.

8.7 Standby and essential power

Essential 115 V AC power is provided as a single phase supply, see Fig. 8.15; this is selected from one phase of a main busbar. (On a four-engined aircraft, the normal source is the main AC bus number four.) If this normal source fails, a supply is maintained by selecting a different main bus from the remaining systems. The standby AC bus is normally powered from the essential AC bus; if this source fails, a static inverter (supplied from the main battery) is selected to provide standby AC power.

Some aircraft distinguish between **battery busbars** (that can be disconnected from the battery) and **hot battery busbars** that are connected directly to the battery, i.e. without any switching; this is illustrated in Fig. 8.15. This arrangement splits the battery bus with a direct connection (the 'hot' battery bus), and a switched battery bus controlled by the battery switch. Essential DC power is from a transformer rectifier unit (TRU) powered from the essential AC bus. Standby AC and DC bus power is normally from the respective essential supplies; when selected ON, standby AC and DC power is from an inverter and the battery respectively.

Test your understanding 8.3

Explain the difference between main, essential and battery busbars.

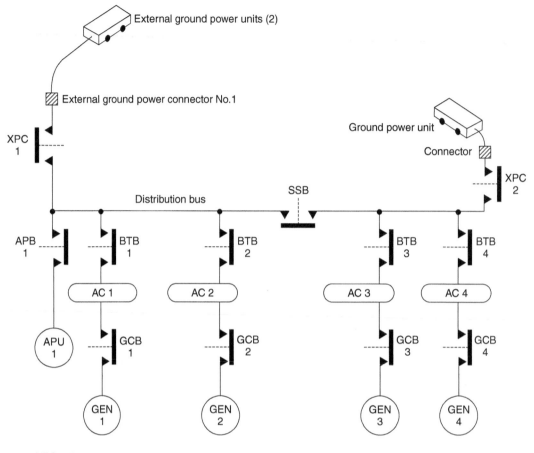

Figure 8.14 Parallel load distribution system

8.8 Battery charging

The battery charger on a large aircraft operates from 115-volt, three-phase, 400 Hz AC power supplied by the AC ground service bus. A completely discharged nickel-cadmium battery can be recharged in approximately 60–90 minutes. The battery charger operates in one of two modes depending on whether the aircraft is being supplied by external ground power or not.

With the aircraft on the ground, the charger is powered from the ground services bus and provides a constant current to the battery. When the battery's terminal voltage reaches a pre-defined level (adjusted for battery temperature), charging is automatically disconnected. When external power is not available, the battery is charged from a transformer rectifier unit (TRU) that provides 28 V DC to maintain battery charge, and supply loads on the battery bus. The battery's temperature sensor forms part of the charging system to prevent battery damage. If the battery temperature is outside a predefined range (typically −18°C to 60°C), the charging circuit is disconnected.

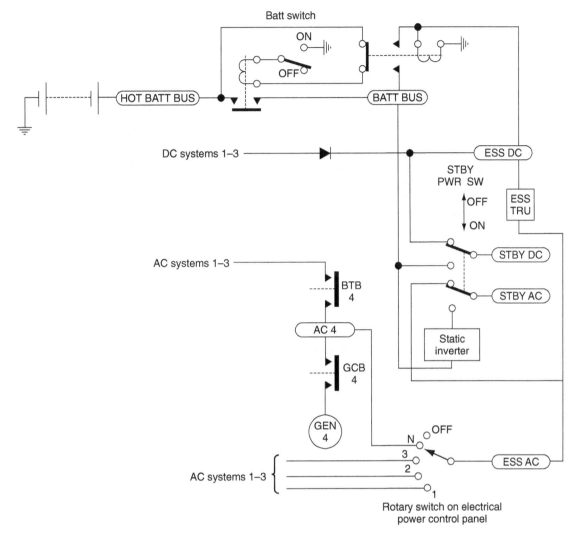

Figure 8.15   AC/DC standby and essential power

8.9 Control and protection

Various components are used for both control and
protection of the power distribution system:

- current transformers
- differential current protection
- phase protection
- breakers/contactors.

8.9.1 Current transformers

These are used to sense current for control, protection and indication applications. The primary winding

is the main heavy-duty AC feeder cable being monitored; the secondary winding is contained within a
housing, see Fig. 8.16. The secondary windings are in
the form of inductive pick-up coils. When current I_P
flows in the feeder cable, the corresponding magnetic
field induces current I_S into the secondary windings;
this is the output signal that is used by a control, protection or indication device.

8.9.2 Differential current protection

This circuit detects short-circuits in AC generator
feeder lines or busbars; it is a method of protecting the generator from overheating and burning out.

Assuming a three-phase AC generator is installed, each phase has its own protection circuit. For illustration purposes, the circuit for a single phase is described. Two control transformers (CTs) are located at either end of the distribution system, see Fig. 8.17. CT_1 is located in the negative (earthed) connection of the generator's output. CT_2 is located at the output from the busbar performing a monitoring function in a **generator control unit** (GCU).

If a fault were to develop between the generator and busbar, a current I_F flows to ground. The net current received at the busbar is therefore the total generator output current I_T minus the fault current (I_F). The fault current flows back through the earth return system through CT_1 and back into the generator; the

remaining current ($I_T - I_F$) flows through CT_2 and into the loads. Current transformer CT_1 therefore detects ($I_T - I_F$) + I_F which is the total generator current. Current transformer CT_2 detects ($I_T - I_F$); the difference between control transformer outputs is therefore I_F. At a pre-determined differential current, the generator control relay (GCR) is automatically tripped by the GCU and this opens the generator field.

8.9.3 Phase protection (Merz Price circuit)

This circuit protects against faults between phases, or from individual phase to ground faults. Connections are shown for protection of a single phase in Fig. 8.18; a three-phase system would require the same circuit per phase. Two current transformers (CTs) are located at each end of the feeder distribution line:

* CT_1 monitors the current output from the generator
* CT_2 monitors the current into the distribution system.

Secondary windings of each current transformer are connected via two relay coils; these windings are formed in the opposite direction. When current flows through the feeder, there is equal current in both coils; the induced EMF is balanced, so no current flows. If a fault develops in the feeder line, current CT_1 flows (but not CT_2), thereby creating an unbalanced condition.

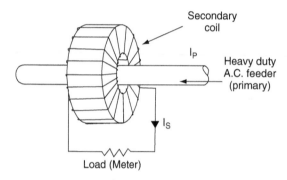

Figure 8.16 Current transformer

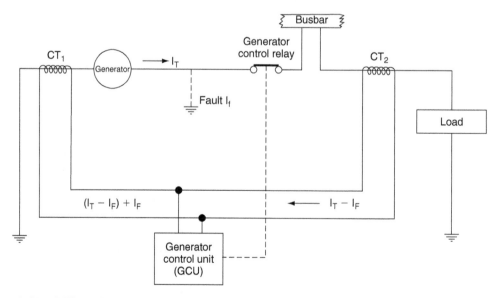

Figure 8.17 Differential current protection

Current flow in either of the coils opens the contacts and disconnects the feeder line at both ends.

8.9.4 Breakers/contactors

Breakers (sometimes referred to as contactors) are used in power generation systems for connecting feeder lines to busbars and for interconnecting various busbars. Unlike conventional circuit-breakers, these devices can be tripped on or off remotely. Referring to Fig. 8.19, they have several heavy-duty main contacts to switch power and a number of auxiliary contacts for the control of other circuits, e.g. warning lights, relays, etc.

The breaker is closed by an external control switch via contacts A; the coil remains energized via contacts B to ground. With the coil energized, the main and auxiliary contacts are closed and the spring is compressed. Contacts A latch the breaker closed, assisted by the permanent magnet. When a trip signal is applied (either by a fault condition or manual selection) current flows to ground in the opposite direction. The spring assists the reversed electromagnetic field and this breaks the permanent magnet latch. A Zener diode suppresses arcing of coil current across

the contacts. An electrical power breaker installation is shown in Fig. 8.20.

8.10 Load-shedding

Load-shedding can be defined as deciding on which systems to switch off and in what order this should be done to reduce the power consumption for the remaining power sources. (The worst-case scenario is when the aircraft is operating from the battery alone.) Non-essential loads are not required for the safe continuation of flight; these loads include the galleys, in-flight entertainment (IFE) system and main cabin lights as previously described. Essential loads are required for the safe continuation of flight. Emergency loads are required for the safe landing of the aircraft, e.g. radios, fuel control, landing gear and fire protection.

8.10.1 General aviation aircraft

In small general aviation (GA) aircraft, pilots need to be familiar with the electrical systems of their aircraft. Most light GA aircraft are only concerned about an alternator or generator failure. This type of aircraft

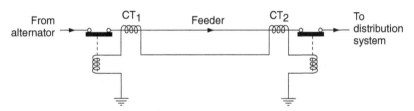

Figure 8.18   Phase protection (Merz Price circuit)

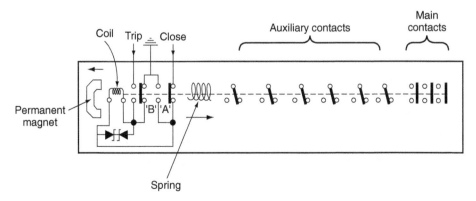

Figure 8.19   Electrical power breaker schematic

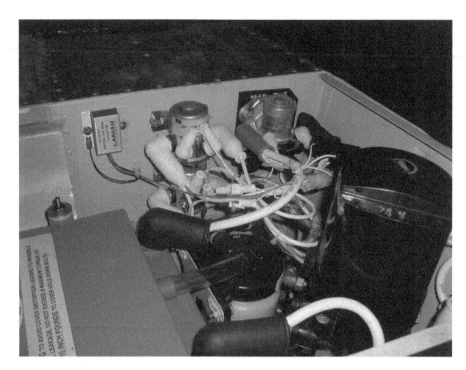

Figure 8.20   Electrical power breaker installation

Figure 8.21   Generator three-phase output and neutral wires

Figure 8.22 Typical power distribution

is normally fitted with an ammeter and/or a voltmeter. On some light GA aircraft an over-voltage light and ammeter can be used to determine the nature of electrical malfunctions. Some aircraft are also fitted with a light that comes on in the event of generator failure, where the battery alone is being used to supply the avionic bus. The aircraft battery has a finite capacity; the pilot has to make decisions on the use of the aircraft's power consumption in order to be able to make a safe landing.

8.10.2 Larger aircraft

More complex aircraft and/or twin-engined aircraft require that pilots have greater knowledge about the electrical systems. Automatic electrical load-shedding is provided on modern aircraft where a control unit detects the loss of electrical power, i.e. the loss of a single or both generators. Automatic load-shedding disconnects specified electrical equipment to maintain sufficient power and systems functionality for the safe continuation of flight.

Split bus or parallel systems need to have a centralized means of controlling power distribution. If a generator or bus fault develops, the relevant bus tie breakers (BTBs) or generator circuit-breakers (GCBs) must be configured to protect the remaining generators, whilst still delivering power to essential loads. Load-shedding is controlled automatically by a solid-state bus power control unit (BPCU). This unit receives inputs from the generator control unit (GCU), ground power control unit (GPCU) and BTBs, together with control transformers that sense real system current through the main power leads. The BPCU is programmed to detect abnormal conditions and open/close pre-determined BTBs.

Galley power represents a high percentage of the total electrical system's load since it comprises equipment such as beverage makers and ovens. These systems have a major influence on the electrical system's design.

8.10.3 Load-sharing of AC circuits

It is necessary to discriminate between real and reactive loads in AC load-sharing circuits. Current transformers in the feeder lines sense each load; these feeder lines are heavy-gauge wires connected to the output of the generator (Fig. 8.21). The feeder lines are connected through the various bus hierarchies as described earlier before finally being distributed to the individual loads via smaller-gauge wires and terminal blocks (Fig. 8.22).

Resistive loads consume **real power**; unbalances are corrected by adjusting the generator shaft input power (from the IDG or CSD). Inductive or capacitive loads consume **reactive power**; unbalances are corrected by adjusting the generator output via the field current control of the regulator. **Apparent power** is consumed by the entire circuit. Reactive power is the vector sum of inductive and capacitive currents and voltages, see Fig. 8.23.

- True power is measured in watts (W)
- Apparent power is measured in volt-amperes (VA)
- Reactive power is measured in volt-amperes-reactive (VAR).

Power factor (PF) is the ratio of true and apparent power (W/VA); this is a measure of circuit efficiency.

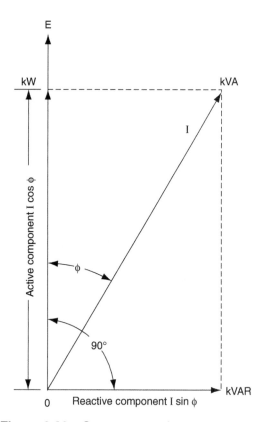

Figure 8.23 Components of currents due to phase difference

A power factor of one (unity) is for a purely resistive circuit; a power factor of zero is for a purely reactive circuit. Aircraft generators and inverters are rated in kVA, with a stated power factor over a given range.

8.11 Multiple choice questions

1. The reverse current relay is needed on any DC generation system to prevent the:
   (a) battery from feeding excess current back through to the generator's armature
   (b) generator from feeding excess current back through to the battery
   (c) battery from feeding excess current to the starter motor.

2. The current limiter device:
   (a) increases the output from a DC generator under fault conditions
   (b) reduces the output from a DC generator under fault conditions
   (c) reduces the output from a DC generator under normal conditions.

3. Battery charge/discharge current is monitored via which type of instrument?
   (a) voltmeter
   (b) contactor
   (c) ammeter.

4. Referring to Fig. 8.24 different size pins are used on external DC power connectors to:
   (a) prevent a reverse polarity voltage being applied
   (b) prevent excessive power being applied
   (c) prevent power being applied when the battery is discharged.

5. The main distribution bus is sometimes called the:
   (a) essential bus
   (b) emergency bus
   (c) non-essential bus.

6. The split bus system is sometimes called a:
   (a) non-parallel system
   (b) parallel system
   (c) standby and essential power system.

Figure 8.24   See Question 4

7. Inductive or capacitive loads consume:
   (a) real power
   (b) DC power
   (c) reactive power.

8. Essential DC power could be supplied from a:
   (a) transformer rectifier unit (TRU) powered from the essential DC bus
   (b) transformer rectifier unit (TRU) powered from the essential AC bus
   (c) inverter powered from the essential AC bus.

9. Referring to Fig. 8.25 apparent power is measured in:
   (a) volt-amperes (VA)
   (b) watts (W)
   (c) volt-amperes-reactive (VAR).

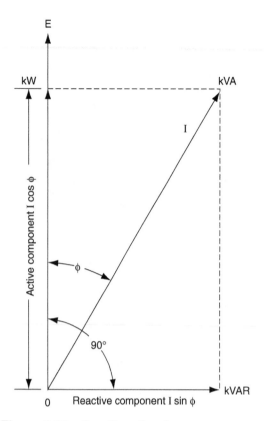

Figure 8.25   See Question 9

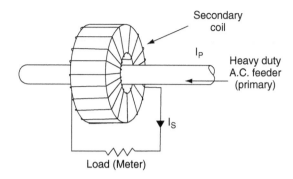

Figure 8.27   See Question 12

10. The third pin on external DC power connectors is used to:
    (a) charge the battery
    (b) energize the reverse current relay
    (c) energize the ground power relay.

11. Referring to Fig. 8.26, this device is
    (a) an external power relay
    (b) a current limiter
    (c) a reverse current relay

12. Referring to Fig. 8.27, this is a schematic diagram for a
    (a) contactor
    (b) current transformer
    (c) auto-transformer

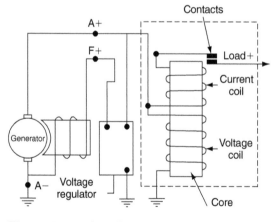

Figure 8.26   See Question 11

Chapter 9 Controls and transducers

There are many systems on aircraft that need to be controlled and/or monitored, either manually by the crew, or automatically. A switch provides the simplest form of control and monitoring. For example, the crew need to know if any doors are not closed as part of their pre-flight check; on larger aircraft the position of control surfaces is displayed in the cockpit. Using these two examples, doors can be either open or closed; control surfaces can move through an infinite number of positions (within their normal limits of travel). Many other aircraft parameters need to be measured, e.g. temperature and pressure. Measurements are made by a variety of **transducers**; these are devices used to convert the desired parameter, e.g. pressure, temperature, displacement, etc. into electrical energy. This chapter describes generic controls and transducer devices used on aircraft.

9.1 Switches

The simplest form of switch is the on/off device used to isolate circuits. Other switch types are used to direct the current into pre-determined parts of a circuit. Switches are characterized by:

- number of poles
- number of switched positions
- type of switched contacts (permanent or momentary).

Switches are sometimes guarded or wire-locked with fuse wire to prevent inadvertent operation. Some switch designs have to be pulled out of a detent position before the position can be changed. They are designed to be operated in a number of ways, e.g. toggle, push/pull, rocker or rotary selectors. These switches are designed with multiple contacts; they can be arranged as permanent or momentary contacts.

The **toggle switch** is a very basic device; Fig. 9.1 illustrates its internal schematic and external features. Operating the lever/arm opens and closes switch contacts. Operating levers on toggle switches are sometimes ganged so that more than one circuit is operated.

The simplest switch has two contact surfaces that provide a link between circuits; these links are referred to as **poles**. Switch contacts can be normally open or closed and this is normally marked on the switch (NO/NC). The number of circuits that can be linked by a single switch operation is called the **throw**. The simplest form of switch would be single pole, single throw (SPST). Schematics of switch configurations commonly found in aircraft are illustrated in Fig. 9.2.

Some switches are designed with instinctive tactile features so that the risk of selecting the wrong system is minimized, e.g. the flap up/down switch-operating lever would be shaped in the form of an aerofoil; the undercarriage selection switch-operating lever would be shaped in the form of a wheel (see Chapter 14 for examples).

Push/pull-operated switches incorporate a spring to hold the contacts open or closed; the switch contacts are therefore push-to-make or push-to-break. Rocker switches (Fig. 9.2b) combine the action of toggle and push/pull devices. Rotary switches are formed by discs mounted onto a shaft; the contacts are opened and closed by the control knob.

9.1.1 Combined switch/light devices

Modern aircraft panels utilize a combined switch and light display; the display is engraved with a legend or caption indicating system status. These can be used in a variety of ways e.g. to show system on/off. The switch portion of the device can be momentary or continuous; small level signals are sent to a computer or high-impedance device. Internal backlighting is from two lights per legend for redundancy; these are projected via coloured filters. The two captions provide such information as press to test (P/TEST). Examples of combined switch/light devices are given in Fig. 9.3.

Test your understanding 9.1

Explain the switch terms 'throw' and 'pole'.

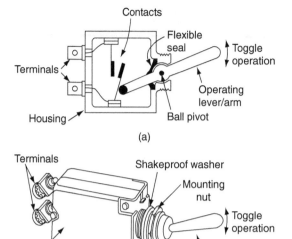

(a)

(b)

Figure 9.1 Toggle switch: (a) internal schematic, (b) external features

9.1.2 Micro-switches

These are used to sense if a device has moved or has reached its limit of travel, e.g. flap drive or undercarriage mechanisms. Figure. 9.4 illustrates the internal schematic and electrical contacts of a typical micro-switch product. The contacts open and close with a very small movement of the plunger. The distance travelled by the armature between make/break is measured in thousandths of an inch, hence the name '**micro**'. A snap action is achieved with a contact mechanism that has a pre-tensioned spring.

Micro-switches are attached to the structure and the wiring is connected into a control circuit. An example of micro-switch application is to sense when the aircraft is on the ground; this is achieved by mounting a micro-switch on the **oleo leg**. When the aircraft is on the ground, the oleo leg compresses and the switch is operated. Micro-switches are used to sense the mechanical displacement of a variety of devices, including:

- control surfaces
- undercarriage
- pressure capsules
- bi-metallic temperature sensors
- mechanical timers.

Proximity switches perform the same function as micro-switches; they sense the presence of an object

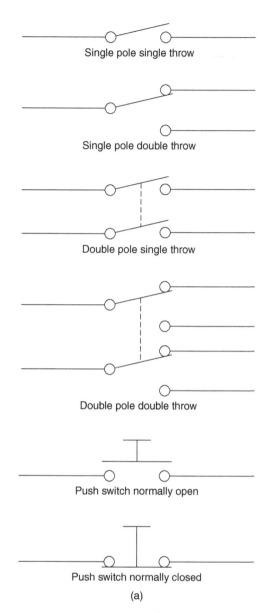

(a)

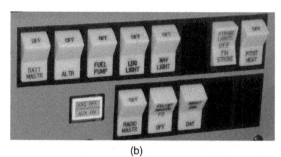

(b)

Figure 9.2 (a) Switch configurations, (b) rocker switches

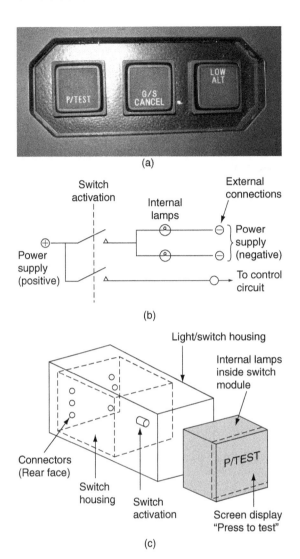

Figure 9.3 Combined switch/light: (a) typical installation, (b) electrical schematic, (c) external features

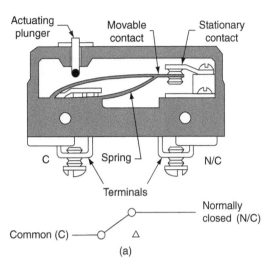

Figure 9.4 Micro-switch schematic: (a) internal schematic, (b) external features

by the interruption of a magnetic circuit. There are two types of proximity switch: reed and solid state.

The **reed switch** device comprises two hermetically sealed sections as illustrated in Fig. 9.5. One section (the actuator) contains a magnet; the other section (the sensor) contains a reed armature with rhodium-plated contacts. The usual arrangement is for the sensor unit to be fixed to the aircraft structure; the actuator is attached to the item being monitored, e.g. a door. When the gap between the actuator and the sensor reaches a pre-determined distance, the reed contacts close thereby completing the circuit. They open again when the actuator and sensor are moved apart.

The solid state proximity switch is based on an **inductance loop** and steel target as illustrated in Fig. 9.6. This inductance loop is the input stage of an electronic switch unit incorporated as part of the actuator. As the target moves closer to the coil, the inductance of the coil changes. An electronic circuit determines when the inductance has reached a pre-determined level. The obvious advantage of this type of switch is that there are no switch contacts, hence higher reliability.

9.1.3 Proximity switch electronic unit

Some aircraft are installed with a proximity switch electronic unit (PSEU). This unit receives the position of various items, e.g. flaps, gear, doors, etc and communicates this information to other systems including:

- take-off and landing configuration warnings
- landing gear position indicating and warning

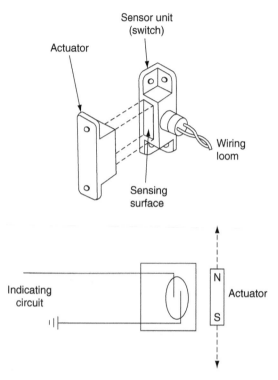

Figure 9.5 Proximity switch (reed type) schematic

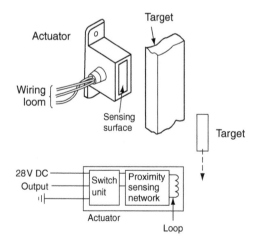

Figure 9.6 Proximity switch (solid state) schematic

- air/ground relays
- airstairs and door warnings.

The PSEU is integrated with the master caution system, and used to indicate if a problem exists that has to be corrected before flight. (The above systems are discussed in subsequent chapters.)

Key point

Typical applications for micro-switches and proximity switches include sensing the positions of flaps, gear, doors, etc. On a large transport aircraft there can be up to 100 such devices.

Key maintenance point

The maintenance engineer will often need to adjust micro-switch or proximity switch positions to ensure that they are correctly positioned; this will compensate for any displacement of the switch.

Test your understanding 9.2

What is the difference between proximity and micro-switches?

9.2 Relays and contactors

These electromechanical devices interrupt or complete a circuit when activated from a remote source, see Fig. 9.7. **Changeover** relays consist of a coil, moving contact (armature) and external connections. When the coil has current flowing through it the electromagnetic effect pulls in the contact **armature**. The armature is pivoted and is held in position by the spring force; with no current flowing, the armature returns to its original position by the spring force.

Contactors operate in the same way; the difference between them is their physical construction and application. Relays are generally used for low current applications; contactors (also known as breakers) are used for switching higher currents, e.g. for connecting battery power to the aircraft. The features of a contactor include the main power contacts and auxiliary contacts used for indication and control of other devices, e.g. lights and relays in power distribution systems (contactors/breakers are described in this context in Chapter 9).

Reed relays are used in control circuit applications; they are generally found within components, e.g. mounted onto printed circuit boards. Figure. 9.8 illustrates the principles of a reed relay. **Slugged relays** have delayed operating times and are needed in specialized applications. The delay in opening/closing the contacts is achieved by a second coil wound around the main coil; the turns are arranged such that the build up of magnetic flux in the main coil is opposed by the build up of magnetic flux in the secondary coil.

9.2.1 Relay configurations

The simplest relay has two contact surfaces that provide a link between circuits; these links are referred to as **poles**. Relay contacts can be normally open or closed (NO/NC) and this is normally marked on the body of

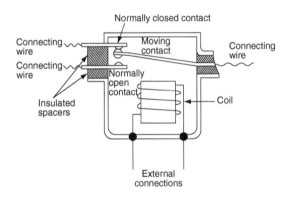

Figure 9.7 Changeover relay schematic

the relay. The number of circuits that can be linked by a single relay operation is called the **throw**. The simplest form of relay would be single pole, single throw (SPST). Relay configurations commonly found in aircraft are illustrated in Fig. 9.9. Typical relay installations on a general aviation aircraft are shown in Fig. 9.10.

Key maintenance point

The pull-in voltage of a relay is the specified voltage needed to attract the armature and close or open the switch contacts.

Key maintenance point

The drop-out voltage of a relay is the specified point at which the armature returns to its relaxed position when the magnetic force is overcome by the spring force.

9.2.2 Polarized relays

These are used in control circuits with very low voltages or currents; the relay is extremely sensitive and can respond to levels in the order of mA/mV. This low level is often not suitable for the conventional spring-loaded armature device since (at very low pull-in/drop-out voltages) the contacts would chatter, leading to spark erosion. Polarized relays use magnetic forces to attract and repel the armature instead of a spring force. The armature is a permanent magnet, pivoted between two pole faces formed by the frame of high

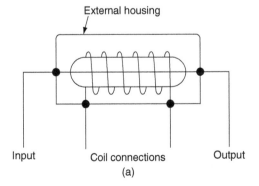

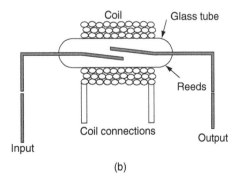

Figure 9.8 Reed relay schematic (a) external schematic, (b) internal schematic

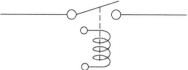
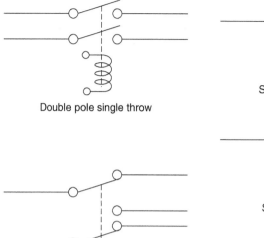

Double pole single throw

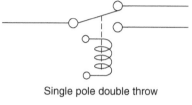

Single pole single throw

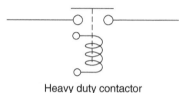

Single pole double throw

Double pole double throw

Heavy duty contactor

Figure 9.9   Relay configurations

Figure 9.10   Relay installation (GA aircraft)

permeability material. When current flows, the poles change and the frame becomes an electromagnet; this exceeds the force exerted by the permanent magnet and the armature changes position. In this position the N-S poles form a strong attractive force and the armature is retained in position. If the supply is interrupted the electromagnetic force is reduced to less than the permanent magnet and the armature returns to its original position.

Key maintenance point

Contacts in relays and contactors are made from silver alloy. The contacts need to have sufficient current flowing/switching to provide '**wetting**' of the contacts. In-service problems and failure associated with these contacts include:

- contact erosion
- welded contacts
- contact corrosion.

9.3 Variable resistors

Variable resistors are mounted on a linear slider or rotary shaft to provide a user-adjustable resistance; typical applications include the control of lighting, audio volume or generator regulator trimming. They are sometimes combined with micro-switches to provide an on/off control function. Variable resistors are produced as either:

- potentiometers
- preset resistors
- rheostats.

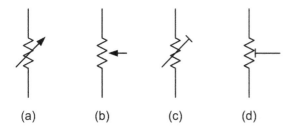

Figure 9.11 Variable resistors: potentiometers, preset or rheostats

Figure 9.11 provides some examples of symbols used for variable resistors. When the intention is for the pilot or maintenance engineer to adjust the circuit resistance for control purposes, e.g. audio volume or lighting intensity, the variable resistor device is used. If the circuit resistance is only intended to be adjusted in the workshop, a preset device is used.

A **potentiometer** (often called a 'pot' for short) is a type of variable resistor that is normally used as a **voltage divider**; this is a circuit used to supply a portion of the power supply voltage from a resistive contact. The potentiometer is typically a three-terminal resistor with a sliding centre contact (the wiper). If all three terminals are used, it can be used in the voltage divider application. If only two terminals are used (one end of the resistor and the wiper), it acts as a variable resistor.

Key maintenance point

Corrosion or wearing of the sliding contact can lead to audio volume or lighting intensity problems, especially if the centre contact (the wiper) is kept in one position for long periods.

Key point

A **rheostat** performs the same function as the potentiometer, but is physically much larger, being designed to handle much higher voltage and/or current. Rheostats are constructed with a resistive wire formed as a toroidal coil, with the centre contact/wiper moving over the surface of the windings.

Many aircraft parameters need to be measured, including: speed, altitude, temperature and pressure. Measurements are made by a variety of **transducers**; these are devices used to convert the desired parameter into an electrical signal that can be used in a control system and/or display.

9.4.1 Solenoids

The solenoid is a type of transducer that converts electrical energy into **linear displacement**. Typical applications include the actuation of pneumatic or hydraulic valves. They are electromechanical devices, consisting of an inductive coil wound around a steel or iron **armature**. The coil is formed such that the armature can be moved in and out of the solenoid's body. The armature is used to provide the mechanical force required to the item being moved.

9.4.2 LVDT

The **linear variable differential transformer** (LVDT) is a transducer used for measuring small linear displacements. The LVDT has three transformer windings located within a housing; the centre coil is the primary winding; the two outer coils are the secondary windings, see Fig. 9.12. A ferromagnetic core, connected to the item being measured, moves along the axis of the housing. The 26 V AC 400 Hz supply is connected to the primary windings, causing a voltage to be induced in each of the secondary windings proportional to its mutual inductance with the primary.

When the core is displaced from its neutral position, the mutual inductances change, causing the voltages induced in the secondary windings to change. The coils are wound such that the output voltage is the difference between the two secondary output voltages. When the core is in its neutral position (equidistant between the two secondaries) equal and opposite voltages are induced in the two secondary coils, so the output voltage is zero.

When the core is displaced, the voltage in one secondary coil increases as the other decreases; the resulting output voltage starts to increase in phase with the primary voltage. When the core moves in the opposite direction, the output voltage increases, but is out of phase with the primary. The output voltage is therefore proportional to displacement of the core; the

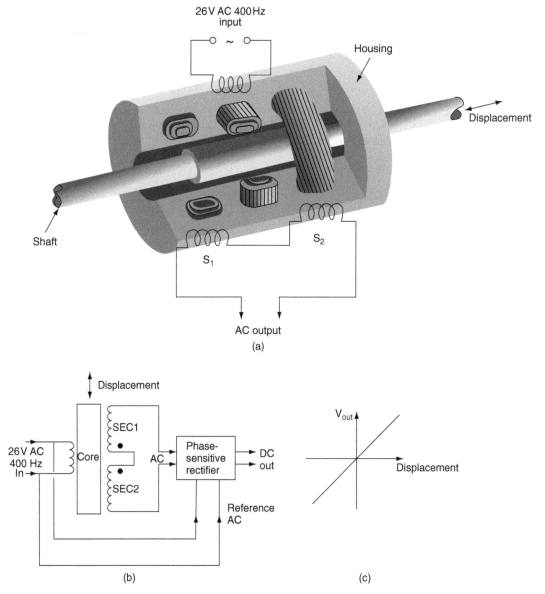

Figure 9.12 Linear variable differential transformer (LVDT) (a) mechanical schematic (b) electrical schematic (c) output voltage vs displacement

output voltage phase is used to determine the direction of movement. This AC output is then fed into a phase sensitive rectifier; the phase difference is then indicated by the +/− polarity of the DC output.

9.4.3 EI sensor

Another device used for measuring very small movements in control systems is the EI sensor. This device

gets its name from the shapes of the fixed E and pivoted I laminated cores, see Fig. 9.13(a). The centre limb of the E core is wound with the primary turns of a transformer; the outer limbs of the E core are wound with the secondary turns, see Fig. 9.13(b). These are wound in series, and in opposition to each other. The I bar is connected mechanically to the device being measured for displacement, e.g. a pressure capsule; it is pivoted and wound with a coil that is connected to a reference power supply.

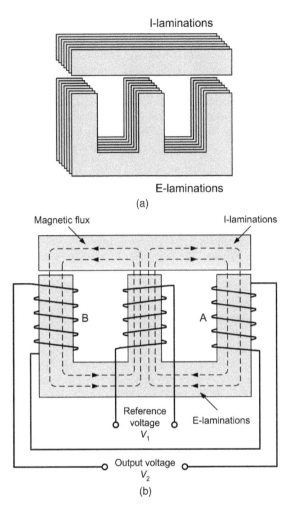

(a)

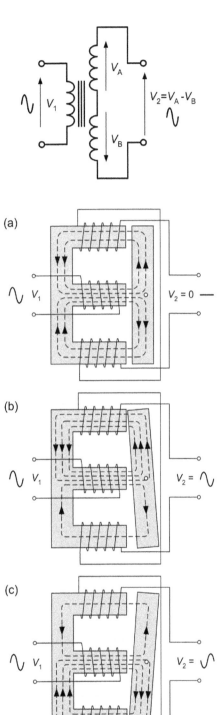

(a)

Figure 9.13 EI sensor – general arrangement: (a) core arrangement, (b) core windings

With the I bar in a neutral position:

- the air gaps at each end are equal
- the magnetic flux is equal
- equal and opposite voltages are generated in the outer limbs
- the output from the sensor is zero.

Referring to the schematic in Fig. 9.14, when the I bar moves, the air gaps become unequal, causing the reluctance to change. Flux in one limb (with the smaller gap) increases, flux in other limb decreases. The induced voltages change; the output voltage is amplified for use in a control system. This measures the phase angle and magnitude of the EI sensor.

Figure 9.14 EI sensor operation

9.5 Fluid pressure transducers

There are two basic methods used to measure fluid pressure: the Bourdon tube or a capsule. Typical fluid pressures being measured on the aircraft include hydraulic oil and fuel.

9.5.1 Bourdon tube

The **Bourdon tube** was invented by Eugène Bourdon (1808–84), a French watchmaker and engineer. The pressure-sensing element is a tube with either a flat or elliptical section; it is formed as a spiral or curve, see Fig. 9.15. One end of the tube is sealed and connected to a pointer mechanism, the open end is connected to the fluid system via a pipe.

As the applied pressure from the fluid system increases, the tube will tend to straighten out, while a reduced pressure will cause the tube to return to its original shape. This movement is transferred via the gear mechanism to move a pointer. The pointer moves across a scale thereby providing a **direct reading** of pressure. Materials used for the tube are selected for the pressure range being measured; these include phosphor bronze (0–1000 psi) and beryllium copper (0–10,000 psi). The Bourdon tube principle can also be used to remotely measure pressure, see Fig. 9.16.

9.5.2 Pressure capsule

The other method used for measuring fluid pressure is with a **capsule**, see Fig. 9.17. As pressure is applied by the fluid, the capsule expands. This moves an iron bobbin within the envelope of two inductor coils.

These coils are part of an **AC ratiometer system** as illustrated in Fig. 9.18. The linear shaded pole motors are single-phase induction motors. An unbalance of currents in the inductor coil circuits is caused by displacement of the cores within the coil. Current then flows in one of two directions into the motor circuits. The rotors of the motor are the aluminium discs attached to a shaft. This arrangement produces a low starting torque to produce a responsive indicating system.

An alternative arrangement is to connect the iron bobbin to a micro-switch to provide indications of low/high pressure. The advantage of this capsule system is that pressure pipes do not need to be run all the way up to the cockpit indicators.

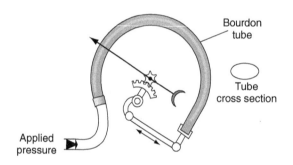

Figure 9.15　Bourdon tube principles

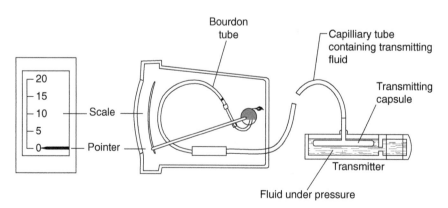

Figure 9.16　Bourdon tube/remote pressure sensing

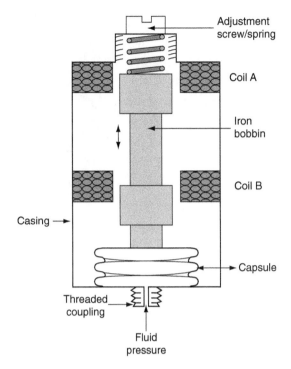

Figure 9.17   Pressure transducer

9.6 Temperature transducers

Aircraft use a variety of temperature transducers, or sensing devices, ranging from the simple bi-metallic strip or thermostat through to thermistors and thermocouples. The type of sensor used depends on the application and takes into account:

- temperature range
- accuracy requirements
- type of output required
- cost.

9.6.1 Bi-metallic strip

A bi-metallic strip converts temperature changes into mechanical displacement as an on/off measurement. The device consists of two strips of different metals with different coefficients of thermal expansion. The strips or elements are joined together by rivets, by brazing or by welding (see Fig. 9.19).

Differential expansion causes the element to bend one way when heated, and in the opposite direction when cooled below its nominal temperature. The metal

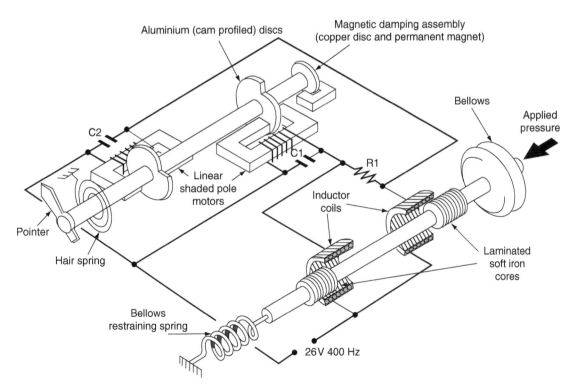

Figure 9.18   AC ratiometer

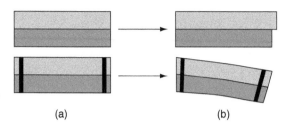

Figure 9.19 Bi-metallic strip principles: (a) before heating, (b) after heating.

with the higher coefficient of expansion is on the outer side of the curve whilst the element is heated and on the inner side when cooled. The element can be formed as a switch contact, called a **thermostat**, to open or close a circuit when temperature limits are exceeded. Alternatively the element is formed in a spiral shape so that temperature changes cause a shaft to rotate; the typical application for this is an outside air temperature indicator, see Fig. 9.20 (a). The sensing portion of the indicator projects through windscreen (Fig. 9.20(b)) into the ambient air.

9.6.2 Thermistors

A thermistor is a type of resistor (*therm*al res*istor*) used to measure temperature; the principle is based on the known relationship between resistance and temperature. The thermistor product used on aircraft is the resistance temperature device (RTD), also known as a resistance thermometer detector, or **temperature bulb.** Figure. 9.21 illustrates some features of an RTD. A length of very small diameter wire is wound on a bobbin to form resistance windings. The bobbin is contained within a stainless steel body to protect the windings against contaminants. Each end of the windings is soldered onto pins that are formed into an electrical connector; this is welded to the flange, resulting in a hermetically sealed unit.

The principle of this temperature-measuring device is based on measuring the change of resistance of a metal element and interpreting this as temperature. Metal elements have a **positive temperature coefficient**; certain metals, e.g. nickel and platinum, have a very stable and linear relationship between resistance and temperature, see Fig. 9.22. Nickel is less accurate than platinum, and is non-linear below 200°C, but has lower cost.

RTDs are typically used within a **DC ratiometer circuit**, see Fig. 9.23. A permanent magnet is attached

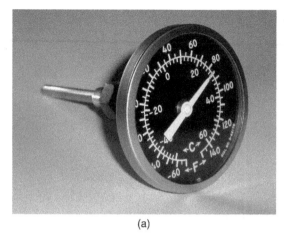

(a)

(b)

Figure 9.20 Bi-metallic strip application: (a) outside air temperature indicator, (b) outside air temperature (sensor projecting through windscreen)

to the pointer of the measuring instrument. When the RTD resistance is equal to the fixed resistor value, the current in each leg is equal and the pointer takes up the centre position. With a change in temperature, the RTD

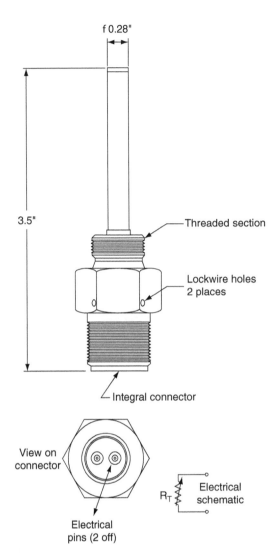

Figure 9.21 Resistance temperature device (RTD) features

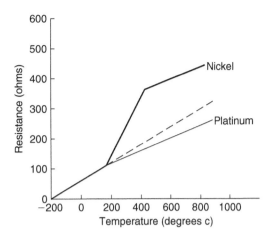

Figure 9.22 RTD temperature and resistance chart

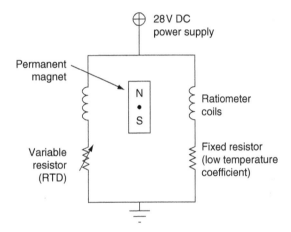

Figure 9.23 DC ratiometer

resistance value changes and the current through each leg is not equal, thereby causing the pointer to take up a new position. The position of the pointer is in proportion to the **ratio** of currents in each leg of the network.

Alternatively they are used in a Wheatstone bridge circuit, named after Sir Charles Wheatstone (1802–75), a British scientist. The circuit was actually invented by Samuel Hunter Christie (1784–1865), a British scientist and mathematician; Sir Charles Wheatstone further developed the circuit. It was used to measure an unknown value of resistance by balancing two legs of a bridge circuit, one leg of which includes the unknown value.

This circuit developed by Wheatstone is based on Ohm's and Kirchhoff's laws. It is typically used as an accurate method to overcome the effect of power supply variations. Referring to Fig. 9.24, resistors R_1, R_2, and R_3 have fixed values; the RTD is the variable resistor in the bridge network. Current I_T splits from the power supply splits into I_1 and I_2. Depending on temperature, if the RTD value equals R_1 then $I_1 = I_2$ and $V_A = V_B$ so I_3 is zero and the meter pointer has a zero deflection. A change in value of RTD (as a result of temperature) means that V_A is no longer equal to V_B. Current I_3 increases and the meter pointer is

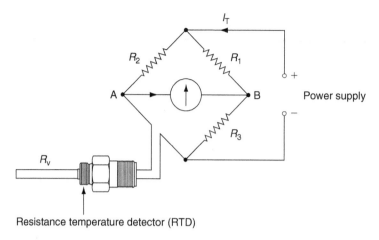

Figure 9.24 Wheatstone bridge

deflected, thereby indicating a change in temperature. For a given temperature, if the power supply voltage changes, then I_T will change in proportion, however I_1 and I_2 will be the same irrespective of the change in voltage. Thus, any variation in power supply does not affect the temperature reading. (R_1 and R_2 are used to prevent a short circuit if the RTD value is zero ohms.)

9.6.3 Thermocouples

The thermocouple principle is based on a **thermoelectric** effect; this is a generic expression used for temperature-dependent electrical properties of matter. Thermocouples use the potential difference that results from the difference in temperature between two junctions of dissimilar metals. This thermoelectric potential difference is called the **Seebeck effect**, named after the German physicist Thomas Seebeck (1770–1831). Two metal conductors made out of different materials are welded at each end to form **junctions**, see Fig. 9.25(a). The thermocouple junction is housed within a stainless steel tube to provide mechanical support; the wires are terminated as a junction in the thermocouple head.

When a temperature gradient between each junction is created, electrons will diffuse from the higher temperature junction to the colder junction. This creates an electric field within the conductors, which causes a current to flow, and an electromotive force (e.m.f.) to be generated. In a simple circuit, Fig. 9.25(b), the **hot junction** is exposed to an unknown temperature (T_2). The **cold junction** is contained within the indicator

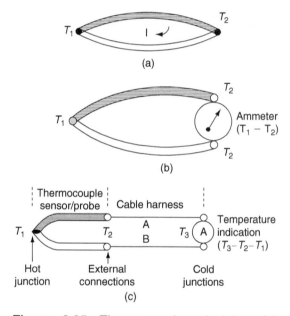

Figure 9.25 Thermocouple principles: (a) the Seebeck effect, (b) simple thermocouple circuit, (c) practical thermocouple circuit

(measuring current); this is at a known temperature (T_1). The practical thermocouple circuit is shown in Fig. 9.25(c). The hot junction is located in the thermocouple head, or probe. Connecting cables back to the meter are either built into the probe as flying leads, or they are attached onto external connectors.

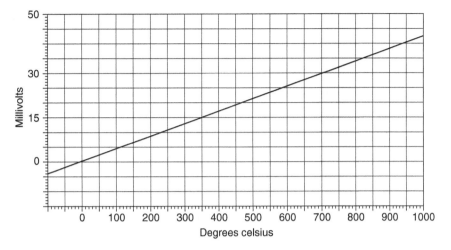

Figure 9.26 Thermocouple output (voltage/temperature)

Key maintenance point

It is important that the materials used in the harness are matched to the thermocouple type, otherwise additional junctions would be formed by dissimilar materials causing unwanted junctions and additional temperature gradients.

The relationship between the temperature gradient and EMF is linear; a typical thermocouple will generate approximately 4 mV per 100°C of temperature difference, see Fig. 9.26.

There are many different combinations of materials used for the two thermocouple wires; the choice of materials depends mainly on the operational range of temperatures to be measured. Typical materials and upper temperatures are:

- nickel-chrome/nickel-aluminium alloys* (1000°C)
- iron/constantan (400°C)
- copper/constantan (300°C)

Another term associated with thermoelectric principles is the Peltier effect, named after Jean Peltier (1785–1845). This is the opposite of the Seebeck effect; by applying a potential difference between two junctions, the temperature of the materials can be controlled.

---

*These are often referred to as chromel and alumel; tradenames of the Hoskins Manufacturing Company. Further details of thermocouple applications are given in Chapter 10 (engine systems).

Test your understanding 9.3

Explain the basic principles of bi-metallic switches, RTDs and thermocouples.

Key point

In an ideal thermocouple circuit, the cold junction should be maintained at constant temperature to ensure that the temperature difference is being accurately measured.

9.7 Strain transducers

Strain is defined as the deformation of a material caused by the action of **stress**. Defined as force per unit area, stress is a measure of the intensity of internal forces acting within a body. **Strain** is determined by a change that takes place between two material states, from the initial state to the final state. The physical displacement of two points in this material between these two states is used to express the numerical value of strain. A **strain gauge** is a device used to measure the deformation of a material when forces act upon it; the principles of a metallic foil strain gauge are illustrated in Fig. 9.27. Edward E. Simmons, an electrical engineer from the USA, developed the principle of the metallic foil type in 1938.

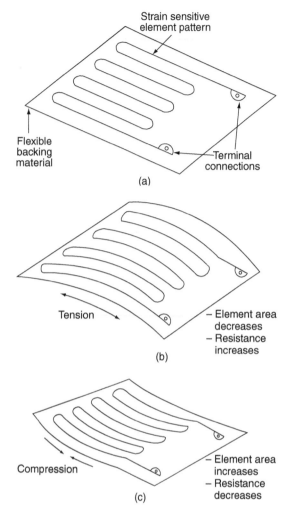

Figure 9.27 Strain gauge principles (a) no strain, (b) tension, (c) compression

The **piezo-resistive** strain gauge is based on the material's change of electrical resistance when subjected to mechanical stress. This effect was first discovered in 1856 by Lord Kelvin (1824–1907), a British mathematician, physicist and engineer. The principle was subsequently developed with silicon and germanium semiconductors during the 1950s.

Strain gauges have a number of applications on aircraft, e.g. the measurement of torque in an engine drive shaft. The metallic foil gauge is attached to the surface of the shaft, either by a suitable adhesive (metallic foil strain gauge) or it can be fabricated as an integral section of the tube surface. As the shaft is twisted by engine power, the strain gauge is deformed, causing its electrical resistance to change. Applications of this principle are discussed further in Chapter 10 (engine systems).

9.8 Rotary position transducers

There are many examples on an aircraft that require the measure of angular position; for example the position of shafts that drive control surfaces. By measuring the angle of the shaft mechanism that drives a control surface, an indication can be provided to the flight crew, or used in a computer. The type of power supply available (DC or AC) determines the main types of **synchro** systems used on aircraft.

9.8.1 DC synchro

The **DC synchro** consists of a transmitter and receiver as illustrated in Fig. 9.28. The transmitter is wire-wound circular resistance with three pick offs. The input shaft (of the device being monitored) is attached to the two contacts that slide across the resistance windings. These contacts are insulated from each other and connected to the power supply. The receiver contains a permanent magnet attached to a shaft; the magnet is positioned within three stator windings. DC synchros are known by various trade names including **Selsyn** and **Desyn**.

With the two contacts in the positions shown in Fig. 9.28, the power supply current enters the resistor at point A, current flows into the top coil (A) of the receiver and splits into the lower coils (B and C). This current sets up three magnetic fields around each of the three coils; the permanent magnet in the receiver takes up a position that aligns with the resulting magnetic field. A pointer attached to the receiver shaft moves to a position inside the indicator. When the input shaft of the transmitter is rotated, the contacts move to a new position. This alters the balance of currents through the resistance windings resulting in a change of currents in the receiver coils. The resulting magnetic flux rotates and the permanent magnet aligns to the new direction, moving the pointer to a different position.

Key maintenance point

The DC synchro suffers from contact wear on the resistance windings; this can lead to spurious operation.

9.8.2 AC torque synchro

The AC **torque synchro** also consists of a transmitter and receiver. These are similar in form; stator and rotor windings are illustrated in Fig. 9.29. Interconnections are shown in Fig. 9.30. The rotors consist of an iron core, single winding, slip-rings and brushes. Stators

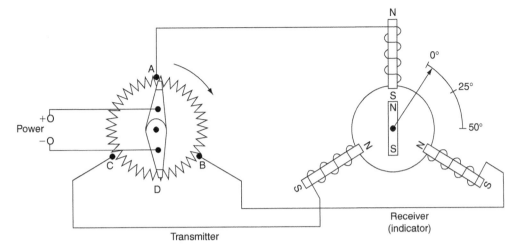

Figure 9.28 DC synchro (electrical schematic)

consist of three sets of windings each at 120 degrees to each other. The rotors of the transmitter and receiver are supplied from the same power supply (normally 26V AC); this alternating supply flows through both the rotor coils via **slip-rings**. Current flowing through the transmitter's rotor creates an alternating magnetic field; this induces currents into each of its stator windings.

Key point

Torque synchros have the advantage of no contact wear on resistance windings.

Referring to Fig. 9.31, the current from each of the stators flows out of the transmitter and into the receiver where it creates magnetic fields in each of the stator windings. The receiver's rotor will be aligned in angular position with the transmitter's rotor. Fig. 9.31(a) shows the system in correspondence. When the transmitter rotor is moved, this unbalances the induced voltages in the receiver until the receiver's rotor is aligned with the resulting magnetic field.

The system is then out of correspondence as shown in Fig. 9.31(b). When the rotor of the receiver aligns itself with the field again, Fig. 9.31(c), the system is back in correspondence.

Key point

Torque synchros are sometimes referred to by the trade name of **Autosyn**.

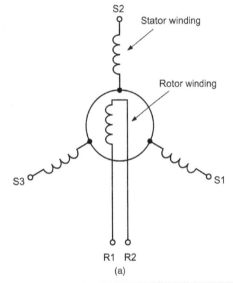

Figure 9.29 AC torque synchro system: (a) transmitter/receiver schematic, (b) transmitter/receiver hardware

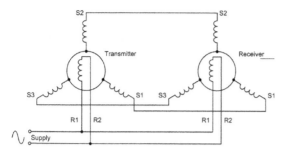

Figure 9.30 AC torque synchro system schematic

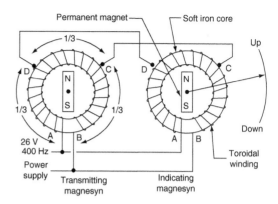

Figure 9.32 AC Magnesyn synchro

9.8.3 Magnesyn

An alternative AC synchro is known by the trade name of **Magnesyn**. The transmitter and receiver are similar in construction; see Fig. 9.32. The rotors are permanent magnets, the stators are soft iron cores with toroidal windings. The transmitter rotor's permanent magnetic field is superimposed onto the stator field (generated from the power supply).

When the input shaft is rotated to a given position, the resulting magnetic field rotates and this causes the receiver's rotor shaft to take up the same position as the input shaft.

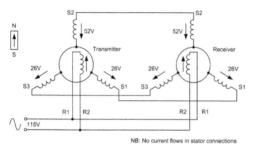

(a) System in correspondence

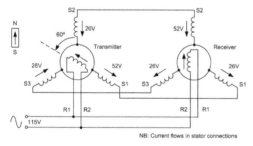

(b) System out of correspondence

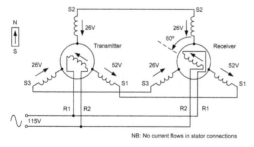

(c) System again in correspondence

Figure 9.31 AC torque synchro system operation

9.9 Accelerometers

Accelerometers can be single or three axis devices; a typical single axis device is packaged in a 25 × 25 mm casing weighing 45 grams (see Figure 9.33). This contains a pendulum (proof-mass) that senses acceleration over the range +/− 40 g; relative displacement between the pendulum and casing is sensed by a high gain capacitance pickoff and a pair of coils. A closed loop servomechanism feedback signal (proportional to acceleration) is then amplified and demodulated. This feedback signal (analogue current or digital pulses) is applied to the coils to restrain the pendulum at the null position. The feedback required to maintain the null position is proportional to the sensed acceleration; this becomes the accelerometer's output signal. Because of the high gain of the servomechanism electronics used, pendulum displacements are limited to micro radians. An integral temperature sensor provides thermal compensation.

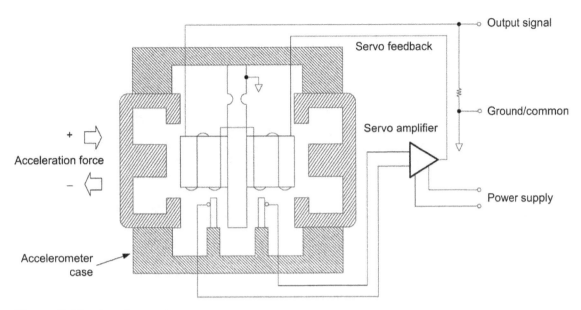

Figure 9.33 Accelerometer

9.10 Solid state technology

Integrated solid-state technology replaces mechanical pressure sensing capsules with micro-electromechanical sensors (MEMS). MEMS is a technology that can be defined as miniaturized electro-mechanical elements that are made using the techniques of micro-fabrication. The physical dimensions of MEMS devices vary from one micron to several millimetres. The types of MEMS devices can vary from relatively simple structures having no moving elements, to extremely complex electromechanical systems with multiple moving elements, all under the control of integrated microelectronics. There are at least some elements having some sort of mechanical functionality whether or not these elements can actually move. MEMS technology has enabled miniaturization and subsequent integration of microelectronics. On a MEMS chip, movement of the silicon crystal structure causes a change in electrical current that can be measured and processed.

9.11 Multiple choice questions

1. The number of circuits that can be linked by a single switch operation is called the:
   (a) pole
   (b) throw
   (c) NO/NC contacts.

2. The bi-metallic temperature sensor consists of:
   (a) two strips of metal with different coefficients of thermal expansion
   (b) two strips of metal with the same coefficients of thermal expansion
   (c) two wires welded into a junction.

3. When a foil strain gauge is deformed, this causes:
   (a) an electromotive force (EMF) to be generated
   (b) its electrical resistance to change
   (c) different coefficients of thermal expansion.

4. Metal elements used in RTDs have a temperature coefficient that is:
   (a) negative, temperature increases cause an increase in resistance
   (b) positive, temperature decreases cause an increase in resistance
   (c) positive, temperature increases cause an increase in resistance.

5. The rotors of a torque synchro transmitter and receiver are supplied from:
   (a) the same power supply (normally 26 V AC)
   (b) different power supplies (normally 26 V AC)
   (c) the same power supply (normally 26 V DC).

6. The thermocouple principle is based on the Seebeck effect; when heat is applied:
   (a) a change of resistance is measured

(b) this causes the element to bend

(c) an electromotive force (e.m.f.) is generated.

7. A rheostat performs the same function as a:
   (a) resistance temperature detector
   (b) potentiometer
   (c) Wheatstone bridge.

8. The solenoid is a type of transducer that converts:
   (a) electrical energy into linear motion
   (b) linear motion into electrical energy
   (c) electrical energy into thermal energy.

9. Proximity switches perform the same function as:
   (a) micro-switches
   (b) relays
   (c) toggle switches.

10. The linear variable differential transformer (LVDT) is used for measuring:
    (a) small rotary displacements
    (b) variable resistance
    (c) small linear displacements.

11. Referring to Fig. 9.34, these control devices are:
    (a) toggle switches
    (b) micro switches
    (c) rocker switches.

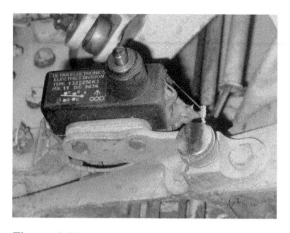

Figure 9.35

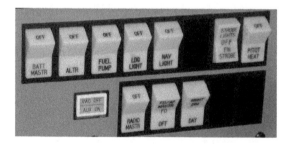

Figure 9.34

12. Referring to Fig. 9.35, this device is a:
    (a) toggle switch
    (b) micro switch
    (c) changeover relay.

13. Referring to Fig. 9.36, this circuit is called a:
    (a) wheatstone bridge
    (b) thermocouple bridge
    (c) ratiometer.

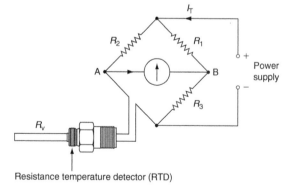

Figure 9.36

14. Referring to Fig. 9.37, this schematic is for:
    (a) an AC magnesyn synchro
    (b) an AC torque synchro
    (c) a DC synchro.

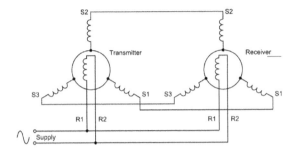

Figure 9.37

15. Referring to Fig. 9.38, this chart shows the output of a:
    (a) RTD
    (b) thermocouple
    (c) strain gauge.

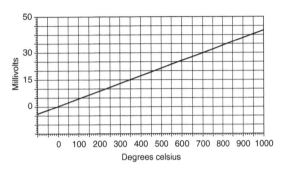

Figure 9.38

Engine systems

The aircraft engine is installed with many systems requiring electrical power. The predominant requirement (in terms of current consumption) is for the starting system. General aviation aircraft use electrical starter motors for both piston and gas turbine engines; larger transport aircraft use an air-start system (controlled electrically) derived from ground support equipment or by air cross-fed from another engine. Electrical starting systems on piston and gas turbine engines are very different. The trend towards the all-electric aircraft will see more aircraft types using electrical starting methods. The engine also requires electrical power for the ignition system. Once again, the needs of piston and gas turbine engines are quite different. Although starting and ignition systems are described in this chapter as separate systems, they are both required on a co-ordinated basis, i.e. a means to rotate the engine and ignite the air/fuel mixture.

Electrical and electronic requirements for engines also include the variety of indicating systems required to operate and manage the engine. These indicating systems include (but are not limited to) the measurement and indication of: rotational speed, thrust, torque, temperature, fuel flow and oil pressure. Indications can be provided by individual indicators or by electronic displays. This chapter describes engine starting, ignition and indicating systems for both piston and gas turbine engines.

10.1 Starting and ignition

10.1.1 Piston engines

The original way of starting piston (internal combustion) engines was by 'swinging' the propeller; this involves using the propeller as a lever to turn the engine shaft. The method in widespread use on most engines now is a starter motor powered by the aircraft battery. Basic electrical starting systems comprise a series or compound wound motor with engaging mechanism.

The simplest method of physically connecting the motor to the reciprocating engine is via a pinion on the motor that engages with a gear ring attached to the crankshaft; this mechanism disconnects after the engine has started. This pinion and gear ring provides a gear ratio in the order of 100:1 to turn the engine at sufficient speed to overcome compression and bearing friction.

Referring to Fig. 10.1, the battery master switch is selected on; this energizes the battery relay and power is applied to the busbar and starter relay. When the starter switch is closed, this energizes the starter relay and applies power to the starter motor. As soon as the engine fires and starts, the starter switch is released and power is removed from the motor; this opens the starter relay contacts.

10.1.2 Twin-engine (piston) starting system

The twin-piston engine aircraft is started in a similar way to the single engine. Referring to Fig. 10.2, power is applied from the battery to the battery relays. When the battery master switch is turned on, power is available at both starter relays. Power is also made available at the busbar and the starter circuit-breaker is fed to the dual starter switch. Each engine is started by its own set of contacts. When an engine has been started the switch is released and the spring-loaded contacts return it to the centre-off position.

10.1.2.1 Magneto ignition (high-tension type)

Ignition energy for piston engines is generated from a **magneto**; this provides pulses of electrical power via a **distributor** to spark plugs in each of the engine cylinders. The magneto operates on the principle of electromagnet induction (Fig. 10.3); it is a combined four-pole permanent magnet generator and autotransformer and can be used where there is no aircraft battery. The engine drives the input shaft of the magneto

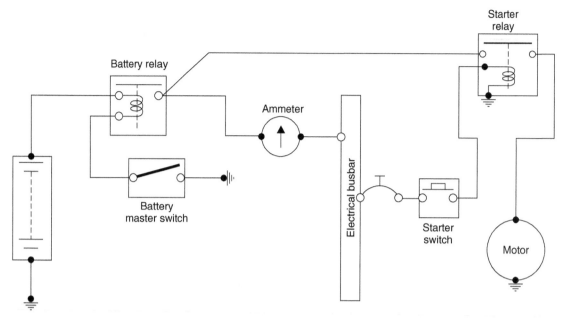

Figure 10.1 Electrical starting system

rotor via a gearbox; the relative movement of transformer windings and the poles of a permanent magnet can be arranged in one of three ways:

1. The transformer coils are on the shaft and the magnet is fixed to the housing (rotating armature type)
2. The permanent magnet is rotated by the shaft within stationary coils of the transformer (rotating magnet type)
3. A soft iron inductor is rotated between the permanent magnet and transformer windings (polar inductor type).

For every revolution of the shaft, a cam opens the contact breaker, interrupting the primary current; this causes the electromagnetic field in the primary coil winding to collapse. As the field collapses there is a voltage generated across the **primary coil**. A capacitor connected across the contacts discharges when the breaker contacts are closed and charges when they open. When the capacitor discharges, a high current flows through the primary coil, inducing high secondary voltages. The capacitor also prevents arcing across the breaker contacts, and determines the voltage across the primary coil thereby controlling the rate at which the electrical energy dissipates through the primary coil. The magneto's output is directed to the spark plugs via a distributor. The distributor

shaft is connected via gears to the magneto shaft; this ensures that energy is applied to the spark plugs with the correct timing.

In the aircraft engine, each cylinder normally has two spark plugs, each driven from a separate magneto. This arrangement provides redundancy in the event of failure of one of the magnetos. Two sparks per cylinder also provides a more complete and efficient burn of the fuel/air mixture. The magneto's simplicity and self-contained design provides reliability as well as light weight. An on/off switch controls the system; in the off position, the primary winding is connected to ground, and this prevents current from being induced in the primary windings.

10.1.2.2 Magneto ignition (low-tension type)

A larger piston engine has more cylinders and is designed to operate at higher altitudes. Decreased atmospheric pressure means that the high-tension magneto system is prone to electrical insulation breakdown in the distribution cables. The low-tension magneto system is based on the **polar inductor** method. The output from the magneto is a low voltage; this is increased to a high voltage by secondary transformers located near the plugs. This reduces the length of high-tension cable and reduces the risk of insulation breakdown in the distribution cables.

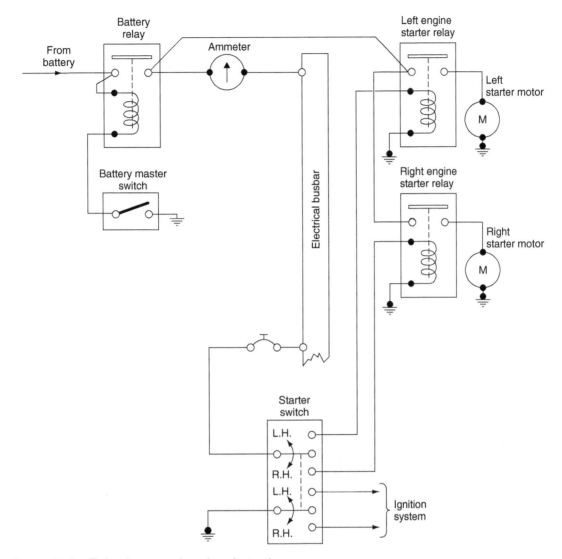

Figure 10.2   Twin-piston engine aircraft starting

Brushes and commutators form the low-tension magneto distributor. **Ignition cables** carry the high energy from the magneto to the plugs. The single core of stranded conductors is insulated by a substantial thickness of material.

10.1.2.3 Spark plugs

These conduct the high-energy output from the magneto across an air gap, see Fig. 10.4. The fuel/air mixture across the gap is an insulator; as the voltage from the magneto increases, it alters the structure of the fuel/air mixture between the electrodes. Once the voltage exceeds the dielectric strength of the fuel/air mixture, it becomes ionized. The ionized fuel/air mixture then becomes a conductor and allows electrons to flow across the gap. When the electrons flow across the gap, this raises the local temperature to approximately 60,000 K. The electrical energy is discharged as heat and light across the air gap, thereby appearing as a spark with an audible 'clicking' sound. This energy ignites the fuel mixture in the cylinder.

The plug is fitted into the cylinder head's combustion chamber and is therefore exposed to high pressures and temperatures; plugs have to operate with

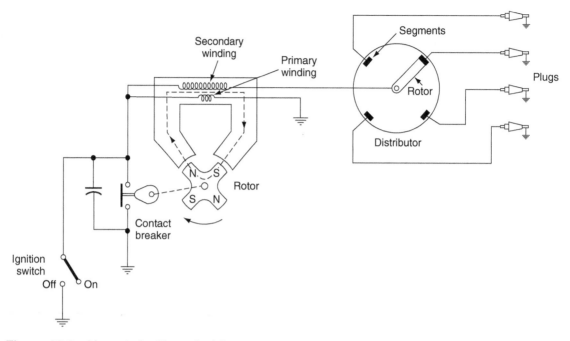

Figure 10.3   Magneto ignition principles

Figure 10.4   Spark plugs (new and used)

Figure 10.5   Spark plug installation

minimal deterioration over long periods of time in this harsh environment. The outer shell of the plug is made from high tensile steel with close tolerance threads to locate the plug in the cylinder head; a copper crush-washer completes the seal against high gas pressures. The outer shell is connected electrically to the cylinder head through its body. The centre electrode carries the high tension energy to the spark gap; this is formed from a material that is resistant to the repetitive arcing, typically nickel, platinum or iridium. An insulator separates the inner and outer

sections of the plug, typical materials include mica, ceramic, aluminium oxide ceramic. Examples of spark plug installations, together with ignition cables, are shown in Fig. 10.5.

Key point

Two spark plugs per cylinder provides redundancy in the event of magneto failure and an efficient burn of the fuel/air mixture.

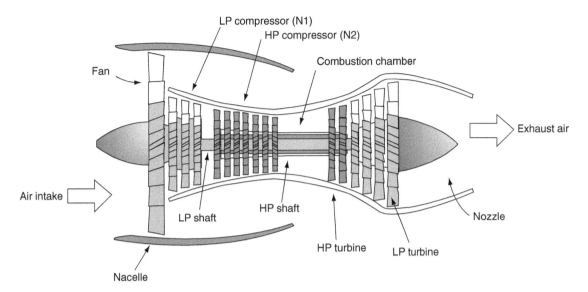

Figure 10.6   Gas turbine engine features

10.1.3 Turbine engine starting

Starting a turbine engine requires a higher-duty motor compared with piston engines; the motor has to overcome higher inertia and needs to achieve higher cranking speeds. The general features of a gas turbine engine are illustrated in Fig. 10.6. Air is compressed through a multi-stage system before entering the combustion chamber where it is mixed with fuel and ignited. The expanding exhaust gases are then directed through a turbine to produce thrust. The turbine also turns shafts to drive the compressor stages.

The starter motor has to overcome the inertia of the compressor; a large volume of air has to be drawn through the engine and then accelerated and compressed until the engine's **self-sustaining speed** is reached. Self-sustaining speed is when sufficient energy is being developed by the engine to provide continuous operation without the starting device. When the turbine engine is driving a propeller (commonly called a **turboprop**) the starter has to overcome the additional inertia of the propeller (helicopter engines are normally connected to the rotor via a clutch).

A typical turbine engine starter circuit is illustrated in Fig. 10.7. Current from the busbar is connected to the coil of a starter relay via a push-button switch. This makes a circuit to a timer switch and starter motor via a limiting resistor. (The resistor limits excessive currents that would otherwise occur in overcoming the initial torque of the motor.) Once

the motor speed has reached its nominal operating speed, its starting torque reduces and the timer switch operates contacts for the shorting relay. With the relay energized, current from the busbar is switched directly into the motor, i.e. bypassing the resistor. Ignition is switched on when self-sustaining speed is reached; power to the motor is removed.

Key point

The initial current through the turbine engine starter motor is in the order of 1000–1500 A, hence the need for a limiting resistor and timing circuit.

Direct-cranking electrical starting systems are similar to those used on piston engines. A typical 28 V DC four-pole motor produces 15–20 lb/ft of torque. The output shaft is clutched into a gear mechanism via the accessory gearbox. Typical duty cycles for motors in small- to medium-size engine requires speeds of 3800 r.p.m. for up to 90 seconds with current peaking at 1000 A. The motor must transmit sufficient torque to the engine's rotating assembly to provide smooth acceleration from rest up to the self-sustaining speed; at this point, the motor is disengaged.

Some aircraft are installed with a combined **starter-generator.** This involves a permanent coupling of the starter-generator shaft to the engine

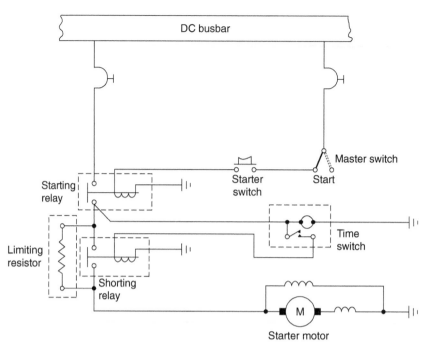

Figure 10.7   Turbine engine starter circuit

via a gearbox drive. The dual-purpose machine is compound-wound and the field is connected via a changeover relay. During engine start it acts like a conventional motor until the engine is up to speed. The changeover relay then automatically connects the field windings to a voltage regulator and it becomes a conventional generator. The starter-generator system has reduced weight and component parts compared with having a separate starter motor and generator, thereby reducing overall operating costs.

10.1.3.1 High-energy ignition unit (HEIU)

High-energy ignition is required for starting gas turbine engines; a dual system is normally installed for the main engines. The system comprises two HEIUs and two igniter plugs per engine. A typical HEIU installed on an engine is shown in Fig. 10.8. Turbine ignition systems are switched off after the engine has reached self-sustaining speed; the system is used as a precaution during certain flight conditions e.g. icing, rain or snow. **In-flight start** uses the wind-milling effect of the engine within the specified flight envelope of air speed and altitude. High voltages are required for the igniter plugs to accommodate the variations in atomized fuel over the range of atmospheric

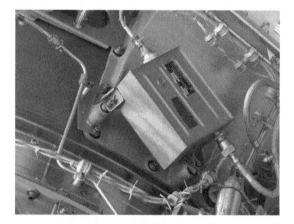

Figure 10.8   Ignition unit installation

conditions. Electrical energy is stored in the HEIU and then dissipated across the igniter plug.

Key maintenance point

The very high voltage output from an HEIU is potentially lethal. The HEIU can remain charged for several minutes; always refer to the maintenance manual for operating procedures.

10.1.3.2 Igniter plugs

Referring to Fig. 10.9, the end of the igniter plug is formed with the outer casing insulated from the centre electrode. The end of the igniter is coated with a semi-conductive ceramic material. The output from the HEIU heats the surface of the ceramic material and lowers its resistance. This creates a high-intensity flashover from the centre electrode to the outer

casing. (Note this is a surface discharge, not a spark across an air gap.)

10.1.4 Auxiliary power unit (APU) start and ignition

The APU on large transport aircraft is started electrically; air is then bled into the air distribution system for the main engines. The APU starter motor is engaged and disengaged automatically as part of the starting system. Referring to Fig. 10.10, a centrifugal switch connected to the APU shaft controls the start sequence via three sets of contacts that operate at 35%, 95% and 110% of maximum speed. An oil pressure switch ensures that the system cannot start until the lubrication oil pump builds up sufficient pressure.

With the system circuit-breaker closed, and the start switch selected, the start and control relays are supplied via the closed 110% switch contacts; the starter motor is now connected to the electrical supply via the start relay. The starting sequence is confirmed via a green annunciator light on the control panel; the

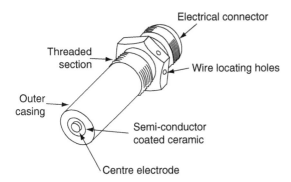

Figure 10.9   Igniter plug features

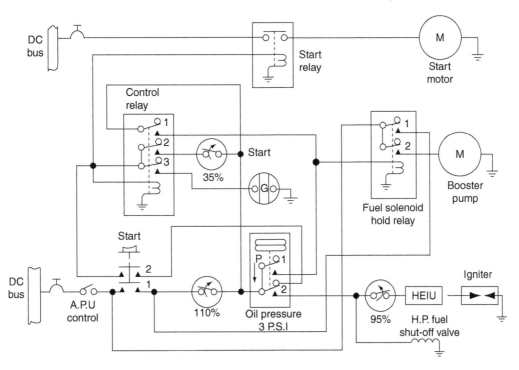

Figure 10.10   Auxiliary power unit (APU) start and ignition

light is connected to the power supply via contacts on the control relay. The fuel solenoid holding relay (FSHR) is operated via the control relay, and the 35% switch contacts; the low-pressure (LP) fuel pump motor is connected to the power supply via the FSHR. The start switch is now released; the starter circuits are maintained via the FSHR, 100/35% switch contacts and control relay. When the oil lube pump builds up sufficient pressure, this closes a switch that provides a retaining supply for the FSHR. The high-pressure fuel shut-off valve is then energized open (allowing fuel to be delivered under pressure into the APU) and the ignition system is supplied via the 95% contacts. The APU continues to accelerate; at 35% engine speed, the start and control relays open; the starter motor is disengaged, and the 'start' light is switched off. At 95% of maximum speed, the contacts open and the ignition is switched off. The APU start sequence is now completed, and the engine runs constantly at 100% r.p.m. (There is no throttle control on the APU.)

The APU can be shut down either manually (by selecting the APU to off) or automatically. Loss of oil pressure at any time will automatically shut the APU down. An overspeed condition (sensed by the 110% switch opening its contacts) will de-energize the fuel solenoid holding relay, close the high-pressure fuel shut-off valve and remove power from the low-pressure booster pump.

10.1.5 Main engine start

The main engines are normally started via air-driven motors; there are three sources of air for starting the main engines:

- APU
- ground air supply cart
- another engine.

A typical air distribution system is illustrated in Fig. 10.11. Valves, controlled either manually or automatically, are operated by motors. With the APU started and running at normal speed, a switch on the start control panel opens the APU bleed valve. Air is directed through the isolation valve and engine bleed valve to the engine start valve.

When using the ground air supply cart, an external connection is made and air is directed through the engine bleed valve to the start valve, or through the isolation valve and engine bleed valve to the start valve. When using another engine (that is already

running), air is supplied from its bleed valve, through the isolation and bleed valves of the engine to be started and through to the start valve.

For illustration purposes, a twin-engined aircraft starting and ignition system is described; refer to Fig. 10.12. There is a combined start and ignition control panel located in the overhead panel; in the illustration, this is fitted with a rotary switch for each engine. The operation and functions of this switch are identical for each engine. The switch has to be pushed in before any selections can be made; this is to prevent accidental movement of the switch. Selecting ground (GRD) connects 28 V DC to energize the start switch holding coil.

The circuit is completed through the cut-out contacts in the engine starter valve. The start switch is now held in the GRD position and the ground start sequence is initiated. The 28 V DC supply also energizes the start valve solenoid and this opens the valve, supplying air to drive a small turbine in the starter motor. The turbine connects through an accessory gearbox onto the engine's HP compressor shaft.

At approximately 16% of maximum rotational speed, the start lever is moved from the cut-off position to 'idle'. This applies 28 V DC through a second pair of contacts of the start switch and ignition switch to supply the HEIU. Each igniter plug discharges at a high level, typically 20 joules of energy, at 60–90 discharges per minute. (This can be heard outside the engine as an audible 'clicking' sound.) At a predetermined cut-out speed, the centrifugal switch in the starter motor opens: the start switch is de-energized and returns (under spring force) to the off position. The 28 V DC power supply is removed from the HEIU and the start valve motor drives to its closed position. The engine continues to accelerate to the **ground idle** speed; this is slightly above self-sustaining speed and occurs when the engine has stabilized. For a twin-shaft axial flow engine, ground idle is typically 60% of the high-pressure (HP) compressor speed.

Low-energy ignition (typically 4 joules of energy, at 30 discharges per minute) is used in certain phases of flight including take-off, turbulence and landing. Furthermore, if the aircraft is flying through clouds, rain or snow, continuous low-energy ignition is selected on the control panel. This closes a contact on the rotary switch and applies power to a second HEIU input.

In the event of an engine **flameout** during flight, the crew will attempt an in-flight start of the engine; this requires a modified procedure to that of the ground start. The engine will be wind-milling due to the forward speed of the aircraft. The starter valve

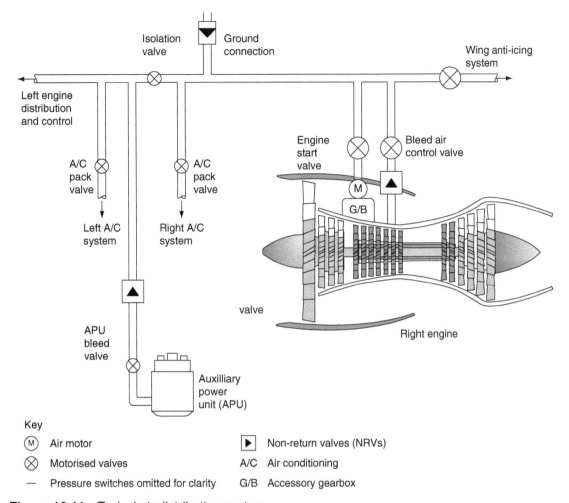

Figure 10.11 Typical air distribution system

and motor are not selected as with the ground start. Low ignition (LOW IGN) and flight (FLT) are manually selected on the control panel until the engine reaches **flight idle** speed. In-flight restarts can only be attempted within certain airspeed and altitude limits.

Key maintenance point

The starting sequence for a gas turbine engine is to: (i) develop sufficient airflow to compress the air, (ii) turn on the ignition, and (iii) open the fuel valves. This sequence is critical since there must be sufficient airflow through the engine to support combustion before the fuel/air mixture is ignited.

Key maintenance point

Facing the aircraft into the wind augments gas turbine engine starting; this assists with engine acceleration, particularly for turbo-prop engines. The propellers are normally designed with a fine-blade angle for starting and ground running.

Gas turbine engines sometimes suffer from a starting problem that results in fuel entering the combustion chamber, but no ignition; this is sometimes referred to as a **wet start**. Engine indications would be the engine turning at the correct starter speed, with indications of fuel flow, but no increase in exhaust gas temperature

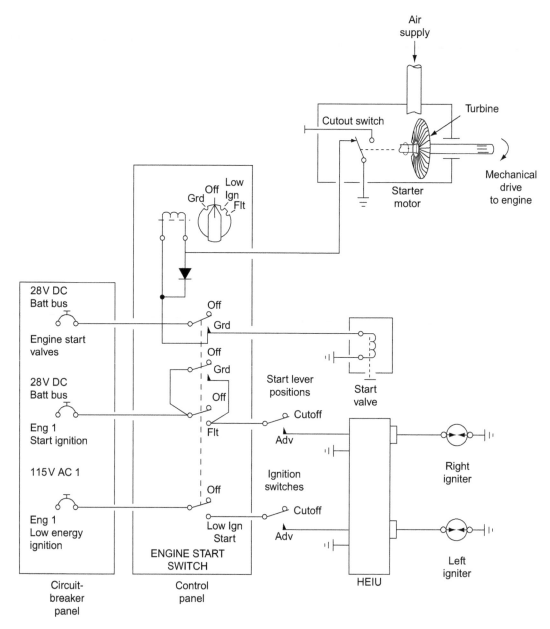

Figure 10.12 Turbine starting and ignition system

(EGT). Observers outside the aircraft could see atomized fuel or vapour from the engine exhaust.

Test your understanding 10.1

Explain the term 'self sustaining speed'.

The cause of a wet start is most likely to be a defective HEIU and/or igniter plug. The net result is no ignition in the combustion chamber and the accumulation of fuel. If compressor outlet air gets hot enough, it could ignite the fuel, causing a rapid expansion of the fuel/air mix (effectively an explosion) that could lead to damage of the turbine section and eject flames from the engine exhaust. The procedure is to shut off the fuel supply to the engine and continue turning over the engine with the starter motor to clear out (or blow out) the fuel. Some starter panels have a selectable **blow out** position to achieve this procedure.

Test your understanding 10.2

What is the difference between a flameout and a wet start?

10.2 Indicating systems overview

Engine indications can be broadly divided into primary and secondary systems. Some indication systems are unique to gas turbine, turboprop or piston engines, some are common to all types. Primary indicators include:

- speed
- temperature
- thrust
- fuel flow.

  Secondary indicators include (but are not limited to):

- oil temperature
- oil quantity
- oil pressure
- vibration.

Measurements are made by a variety of **transducers**; these are devices used to convert the desired parameter, e.g. pressure, temperature, displacement, etc. into electrical energy. The locations of engine instruments is normally between the two pilot's panels, see Fig. 10.13.

10.3 Primary indicating systems

10.3.1 Engine speed

This is a primary engine indication used on both piston and gas turbine engines. It is one of two methods used to indicate thrust on gas turbine engines (the other being EPR). In gas turbine engines, the usual practice is to display a **percentage** of maximum revolutions per minute (r.p.m.) rather than actual r.p.m. Typical gas turbine engine speeds are in the order of

Figure 10.13   Typical engine instruments

8000–12,000 r.p.m. The mathematical notation for rotational speed is N; gas turbine engines can have up to three shafts (or spools), these are referred to as low, intermediate and high-pressure shafts (**LP, IP and HP**). Alternatively, they are referred to numerically; the individual shaft speeds are therefore N_1, N_2 and N_3. Engine speed is monitored by the crew at all times; particularly during start and take-off to make sure that engine limits are not exceeded. The two principal types of engine speed transducer are the tachometer and variable reluctance device.

10.3.1.1 Tachometer system

The **tachometer** indicating system is a small three-phase AC generator connected via a mechanical link to engine accessory gearbox. A tachometer system is found on most general aviation aircraft. Referring to fundamental principles, the tachometer's output increases with increased engine speed; the output is rectified and connected to a moving coil meter. The output from the generator is supplied to a three-phase AC synchronous motor in the indicator; see Fig. 10.14.

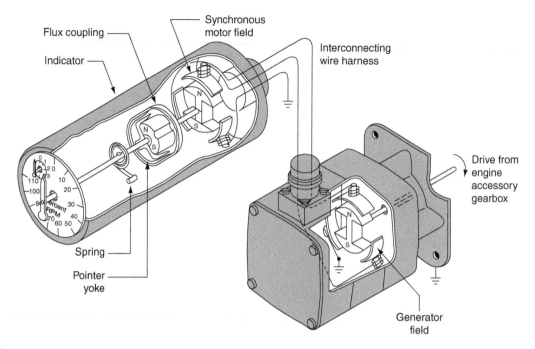

Figure 10.14   Tachometer system schematic

The AC generator tachometer consists of a **permanent magnet** that is rotated inside stator windings wound in a three-phase star configuration. Three stator outputs are connected to the stator windings of the motor. As the engine turns, the permanent magnet induces currents in the stator windings. The three-phase output induces a rotating field in the motor stator windings.

Referring to Fig. 10.15, a permanent magnet is attached to the rotor shaft of the synchronous motor; this is connected to the pointer of the indicator. As the stator field rotates, the permanent magnet keeps itself aligned with the field. A second permanent magnet is attached to the indicator rotor; this is located within a **drag cup**, see Fig. 10.16. As this second magnet rotates, it induces **eddy currents** in the drag cup. These currents produce their own magnetic fields in opposition to the rotating magnet.

When the rotating magnet increases in speed, the drag (or torque) on the drag cup increases. A hairspring attached to the shaft opposes this torque; the net result is the pointer moves across the scale in proportion to the speed of the engine shaft.

10.3.1.2 Variable reluctance

The N1 **variable reluctance** speed transducer comprises a coil wound onto a permanent magnet core

Figure 10.15   Tachometer system installation

as shown in Fig. 10.17. This type of sensor is found on most gas turbine engines. When the blade tip passes the sensor, the magnetic field is disturbed. This induces a voltage into the coil; as the shaft speed increases the fan tips pass the sensor at an increased rate. The output from the coil is in the form of voltage 'spikes'. These are counted by a processor and used to determine engine speed. Some engines are fitted with a low permeability material inserted into the blade tip to provide a distinctive pulse each time that a specific blade passes the sensor.

The N2 speed transducer is located within the accessory gearbox, see Fig. 10.18. A gearbox-driven **target wheel** contains a permanent magnet inserted on the periphery of the wheel. The coil/core sensor field is disturbed each time that the target passes the sensor. The output is supplied into a processor that drives the indicator.

Test your understanding 10.3

Explain the terms N1, N2 and N3. How do these relate to the HP, IP and LP shafts?

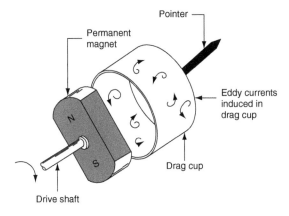

Figure 10.16   Drag cup features

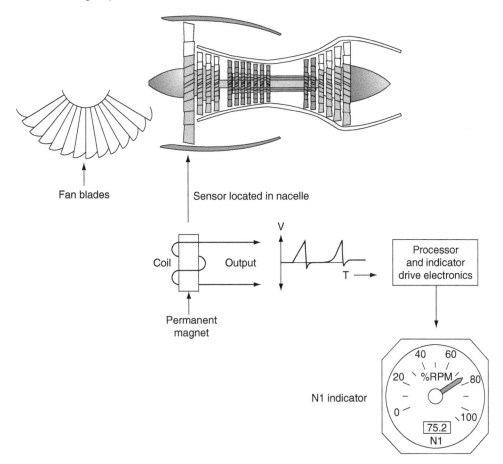

Figure 10.17   N1 engine speed indication

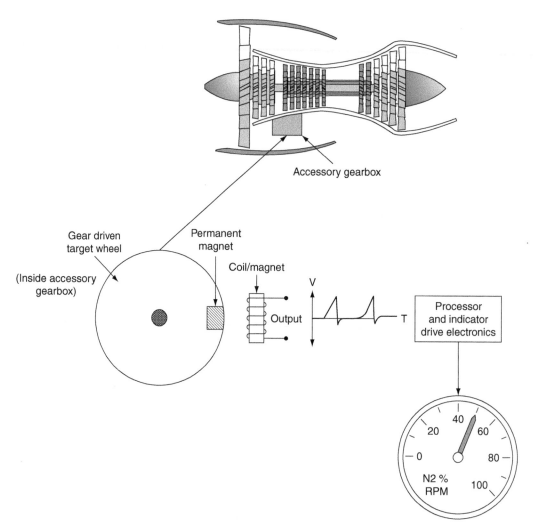

Figure 10.18   N2 engine speed indication

10.3.2 Engine temperature

Exhaust gas temperature is a primary gas turbine engine indication. Engine temperature is closely monitored at all times, particularly during start and take-off, to make sure that engine limits are not exceeded. It is sometimes referred to as:

- turbine inlet temperature (TIT)
- inter-turbine temperature (ITT)
- turbine outlet temperature (TOT)
- exhaust gas temperature (EGT)
- turbine gas temperature (TGT)
- jet pipe temperature (JPT).

The type of measurement depends on where the probe is located and according to individual engine manufacturer's terminology; Fig. 10.19 illustrates some of these locations. The turbine section runs at very high temperatures, typically 1000°C; the transducer used in this application is the **thermocouple** (sometimes referred to as a 'temperature probe').

A typical engine temperature measurement system is illustrated in Fig. 10.20. On larger engines, it is possible to have a range of temperatures in the exhaust zone due to the turbulence of gases.

Some installations feature a thermocouple that has two or even three hot junctions within the same

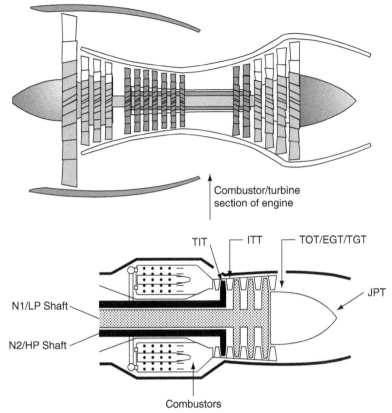

TIT – Turbine Inlet Temperature
ITT – Inter Turbine Temperature
TOT – Turbine Outlet Temperature
EGT – Exhaust Gas Temperature
TGT – Turbine Gas Temperature
JPT – Jet Pipe Temperature

Figure 10.19   Engine temperature measurement

outer tube. This arrangement provides an average of temperature within the zone to provide an average temperature at different immersion distances in the engine, see Fig. 10.21.

Key point

Thermocouple junctions alone do not generate electromotive force (e.m.f.). The potential difference that develops at the heated junction is a function of both the hot and cold junction temperatures.

Small gas turbine engines are normally fitted with several thermocouples to provide an average temperature and some redundancy in case of failure. Larger engines can be fitted with up to 21 thermocouples; these are connected in parallel to provide an average reading of the gas temperature in the exhaust zone. The interconnecting cables between the thermocouple(s) and indicator have to be the same material throughout the system otherwise addition junctions will be formed, thereby generating unwanted voltages. A typical thermocouple cable installation is shown in Fig. 10.22.

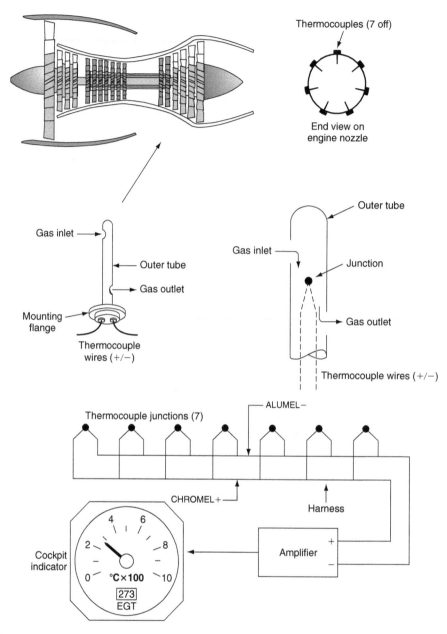

Figure 10.20　Engine temperature system

Key maintenance point

Thermocouple cables are colour-coded to reduce the likelihood of different materials being cross-connected, or mixed in the same installation (note that these codes vary in some countries; always refer to the maintenance manual):

- nickel-chrome (white)
- nickel-aluminium (green)
- iron (black)
- constantan (yellow)
- copper (red).

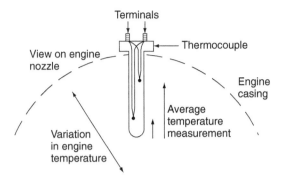

Figure 10.21 Average temperature measurement

Figure 10.22 Thermocouple/cable installation

On piston engines, the EGT indication has a different application; it is used to adjust the fuel/air mixture for economical running of the engine. The thermocouple is located in the exhaust system, reducing the amount of fuel into the cylinders (called **leaning** the mixture); this results in a more efficient combustion and the exhaust temperature increases. When this temperature reaches a peak, maximum efficiency has been achieved. Using one thermocouple per cylinder averages out individual differences and provides enhanced fuel efficiencies.

Piston engines also use thermocouples to monitor the **cylinder head temperature** (CHT). This indication is used to monitor the temperature of an air-cooled engine. The system can be arranged with either one CHT thermocouple per cylinder or just one CHT thermocouple per engine. In the latter case, the thermocouple is installed in the hottest running cylinder; on a horizontally opposed engine this would be the rear cylinder. The thermocouple can be formed to fit under the spark plug or directly into the head itself. The advantage of monitoring each cylinder is that trend indications can be derived.

Test your understanding 10.4

Why are thermocouple cables colour-coded?

10.3.3 Engine pressure ratio (EPR)

This is a primary gas turbine engine indication system and is one of two methods used to indicate **thrust** (the alternative method is described under engine speed). The principle of an EPR indication system is to measure the exhaust and inlet pressures and derive a ratio between them. EPR probes are located at the inlet and exhaust of the engine, see Fig. 10.23. The inlet pressure probe is a single device, located in the nacelle; there can be several exhaust probes connected via a manifold.

Pressures from the inlet and exhaust are sent via small diameter pipes to the EPR transducer; this comprises pressure sensors (capsules). These are coupled to a mechanism that determines the displacement of the capsules as a ratio. Outputs from the ratio sensor are sent via a **linear variable differential transformer (LVDT)** to the indicator. The EPR signal from the transducer is transmitted as a voltage to the indicator. The typical range of EPR indications is between one and two.

Test your understanding 10.5

What EPR reading would be displayed when an engine is not running?

10.3.4 Fuel flow

This is a primary engine indication for which there are various types of transducer; these are located in the fuel delivery pipeline. A typical transducer type is based on the **metering vane** principle; see Fig. 10.24. The metering vane is attached to a shaft; this

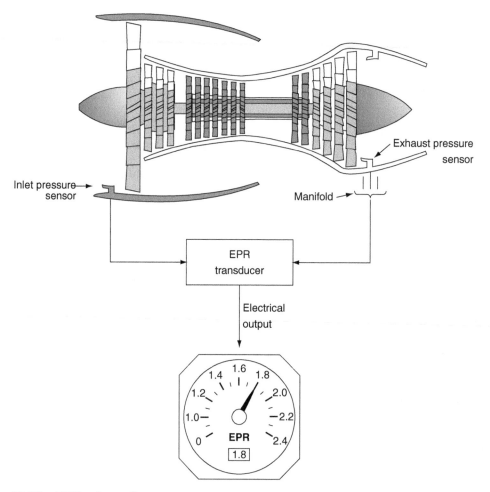

Figure 10.23   EPR schematic

rotates when the fuel supply passes through the body of the transducer. The circular chamber has sufficient clearance to allow fuel to pass through with minimal restriction.

Rotation of the vane is opposed by a spring; the angular position of the vane is measured and indicated by a synchro system, see Fig. 10.25. The metering vane system gives an indication of **volume flow**, e.g. gallons or litres of fuel flow per hour.

An alternative fuel flow transducer is the **motor-driven impeller** principle; see Fig. 10.26. An impeller is driven at a constant speed by an AC synchronous motor. The fuel that flows through the impeller is imparted with a rotational force; this causes the tur-

bine blades to rotate against the force of a spring. The speed of turbine rotation is therefore proportional to fuel flow. Denser fuel imparts more energy into the turbine; the transducer is therefore measuring mass flow. The turbine is connected to the indicator via a synchro system; typical units of **mass flow** are given in pounds or kilograms of fuel per hour.

10.3.5 Torque

The power delivered by the engine to the propeller shaft can be derived from the relationship:

$$\text{Power} = \text{torque} \times \text{speed}$$

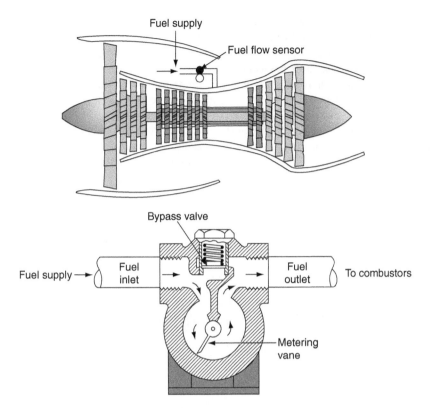

Figure 10.24   Fuel flow-metering vane principles

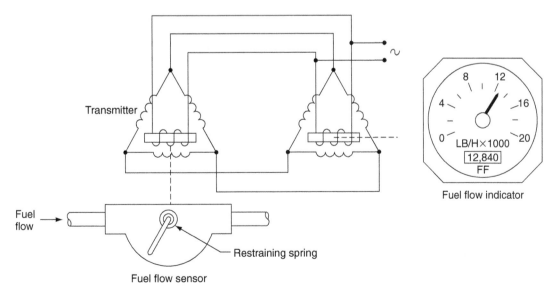

Figure 10.25   Fuel flow indicating system

Power is derived from measuring both torque and speed; the indication is normally used for turboprops and helicopter rotors. Indicators are calibrated in a percentage of maximum torque or **shaft horsepower** (SHP). A typical torque indicator used on a twin-engined helicopter is shown in Fig. 10.27.

Indications are given for the output from both engines and main rotor (M/R) torque. This is the most effective way of indicating the power being produced by the engine.

There are several methods used to measure torque. The torque shaft is formed with toothed or **phonic wheels**, see Fig. 10.28. As the applied torque changes the phase difference between the signals obtained from the two speed sensors increases. The torque applied to the shaft results in a phase difference between the outputs of the two sensors, see Fig. 10.29.

Torque can also be measured by embedding **strain gauges** into the shaft to measure the deformation (strain) of the shaft; these can either be **metallic foil**

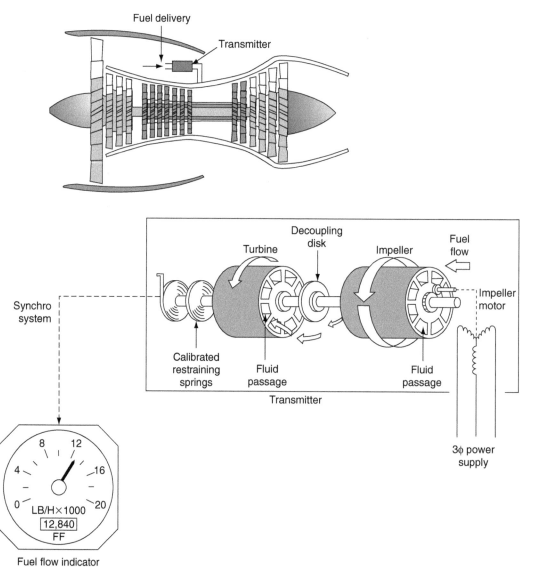

Figure 10.26 Fuel flow – impeller principle

or semiconductor **piezo-resistors** (as described in Chapter 9).

Test your understanding 10.6

What is the difference between torque and thrust?

10.4 Secondary indicating systems

10.4.1 Oil/fuel temperature

There is a need for accurate measurement of fluid temperatures on and around the engine, e.g. fuel, engine lubrication (lube) oil and hydraulic oil. The typical temperature range of these fluids is between −40 and +150°C. When lube oil operates at high temperatures, its viscosity reduces and its lubrication

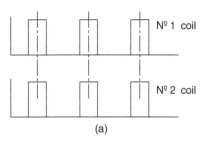

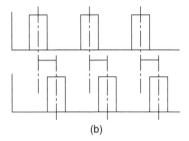

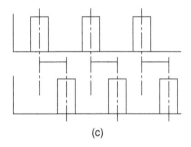

Figure 10.29 Shaft torque sensor phase differences (a) no load conditions, (b) engine lightly loaded, (c) engine heavily loaded.

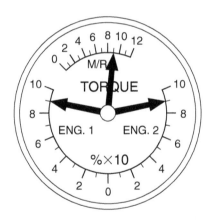

Figure 10.27 Typical helicopter torque indicator (twin engine)

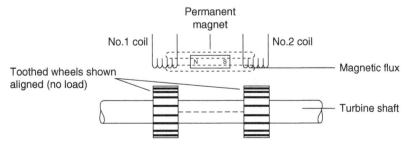

Figure 10.28 Torque transducer principles

performance decreases. This leads to engine wear and eventual failure of bearings or other engine components. When lube oil operates at low temperatures, its **viscosity** increases, which has an effect on starting. When fuel is exposed to high temperatures, it can vaporize, causing fuel delivery problems and a potential risk of explosion; at low temperatures, ice formation can occur leading to blocked filters. (Fuel systems are further described in Chapter 11.)

The flight compartment indicator is a moving coil meter connected in series with the resistance winding of a resistance temperature device (RTD); this item is described in more detail in Chapter 9. As the temperature changes, the resistance changes and this causes the pointer to respond accordingly. The meter is acting as an ammeter, but since there is a linear relationship between temperature and resistance, and since we know (from Ohm's law) that the current in a circuit will vary in accordance with resistance, the moving coil meter can be calibrated as a temperature indicator.

This is all very well provided that the power supply voltage does not change. To illustrate this problem, assume a given temperature and hence a specific resistance of the RTD. The effect of a varying power supply voltage will be to vary the current through the RTD and meter; this would change the reading even though temperature has not changed; an undesirable situation. The method employed to overcome this is a circuit called the **Wheatstone bridge**, as described in Chapter 9. This circuit has a number of applications and is ideally suited for the RTD.

10.4.2 Vibration

Unbalanced shafts on a gas turbine engine can cause damage, particularly at high rotational speeds. These conditions can be anticipated by measuring the engine's vibration. Sensors used to detect vibration are based on **piezoelectric** crystals; a small electrical charge is created when the crystal is vibrated. Each engine has a sensor(s) located at strategic locations; the output is fed into a processor and displayed on an indicator, see Fig. 10.30.

The output from the processor can also be used to illuminate a warning light when pre-determined limits are exceeded. A test switch (incorporated into the light) is used to energize a relay (or equivalent electronic switch); this inserts a known frequency into the processing circuit to generate a warning. The vibration warning circuit also activates the master caution system.

10.4.3 Fluid pressure

Typical fluid pressures being measured on the engine include lubrication oil and fuel. There are two basic methods used to measure fluid pressure: the **Bourdon tube** or a capsule. The Bourdon tube comprises a tube formed into a curved or spiral shape. As the applied pressure from the fluid system increases, the tube will tend to straighten out, while a reduced pressure will cause the tube to return to its original shape. This movement is transferred via the gear mechanism to move a pointer. The pointer moves across a scale, thereby providing a **direct reading** of pressure. The other method used for measuring fluid pressure is with a **capsule**. As pressure is applied by the fluid, the capsule expands. This moves an iron bobbin within the envelope of two inductor coils. (The principles of Bourdon tubes, pressure capsules and measurement systems are described in Chapter 9.)

10.4.4 Propeller synchronization

This is a mechanism that automatically synchronizes all propellers of a multiengine aircraft to ensure that they rotate at the same speed. The main reason for synchronizing the propeller speed is for the comfort of crew and passengers; propellers that are turning at slightly different speeds create a 'beat' through **heterodyning**. This is the creation of new frequencies through mixing two different frequencies; one at the difference of the two mixed frequencies, and the other at their sum. Heterodyning creates beats that can become very tiresome to persons inside the aircraft over long periods of time. Some aircraft are installed with a visual indicator of propeller synchronization. Pilots can use this indicator to decide whether or not to engage automatic synchronization, or to assist with manually synchronization of propeller speeds by adjusting the throttles.

10.5 Electronic indicating systems

There are two formats of electronic indicating systems in widespread use on larger passenger aircraft: EICAS and ECAM. The **Engine Indication and Crew Alerting System** (EICAS) is a Boeing-developed system that provides all engine instrumentation and crew annunciations in an integrated format. The system used on Airbus aircraft is the **Electronic**

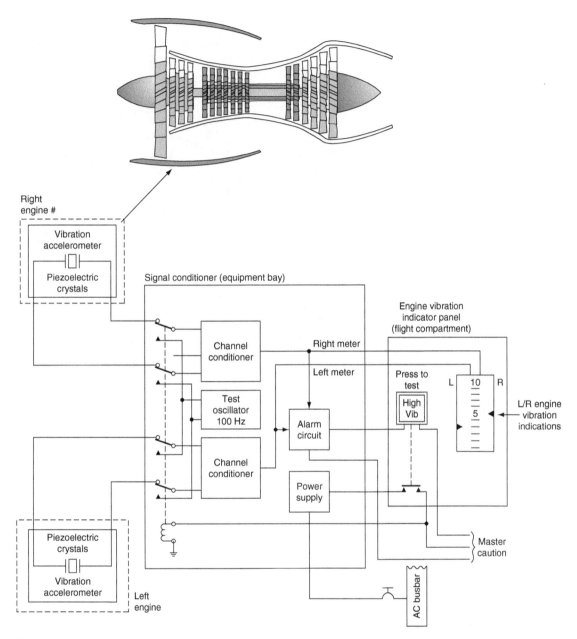

Figure 10.30   Vibration sensor system

Centralized Aircraft Monitoring (ECAM) system. The two systems operate on different philosophies; however, their basic functions are to monitor aircraft systems and display relevant information to the pilots. Both systems produce warnings, cautions and advisory messages that need to be evaluated by the crew. In certain cases, the system provides the procedures

required to address the problem. Each colour display unit uses either an active matrix liquid crystal display (AMLCD) or a cathode ray tube (CRT).

EICAS is arranged with two displays in the centre instrument panel (upper and lower), see Fig. 10.31. ECAM uses either a left and right display, see Fig. 10.32, or an upper and lower arrangement. The

Figure 10.31 EICAS

information displayed by EICAS/ECAM includes engine thrust, engine r.p.m., fuel flow, oil temperature and pressure. Graphical depiction of aircraft systems can also be displayed; this includes electrical, hydraulic, de-icing, environmental and control surface positions.

10.5.1 EICAS

System features include:

- primary engine parameters displayed on a full-time basis
- secondary engine parameters displayed as required
- automatic monitoring of systems
- fault codes, fault history, self test (lower display)
- two computers (includes interfaces, symbol generator, memory)
- select panel, control switches, annunciation, standby display.

This integrated system improves reliability through elimination of traditional engine gauges and simplifies

the flight deck through fewer stand-alone indicators. EICAS also reduces crew workload by employing a graphical presentation that can be rapidly assimilated. A typical EICAS comprises two large high-resolution colour displays together with associated control panels, two or three EICAS data concentrator units and a lamp driver unit. The primary EICAS display presents primary engine indication instruments and relevant crew alerts. It has a fixed format providing engine data including:

- engine thrust (EPR)
- engine rotational speed (N_1)
- exhaust gas temperature (EGT).

If N1 or EGT limits are exceeded, their respective pointers and digital readouts change from white to yellow, and eventually to red. Exceedance information is stored in non-volatile memory (NVM) for access by maintenance engineers when troubleshooting the systems.

Messages are displayed as part of the Crew Alerting System (CAS); these are colour-coded to indicate their importance. Typical messages include

Figure 10.32   ECAM

low pressure in the fuel or hydraulic oil systems, doors not closed, etc. System warnings are automatically colour-coded and prioritized; they are accompanied by an audible warning depending on the message priority:

- **Warning** messages are red, accompanied by an audio alert (prompt action is required by the crew).
- **Caution** messages are yellow accompanied by an audio alert (timely action is required by the crew).
- **Advisory** messages are yellow, no audio alert (time available attention is required by the crew).

The secondary EICAS display indicates a wide variety of options to the crew and serves as a backup to the primary display. Pages are selectable using the EICAS control panel and includes many types of system information, for example:

- hydraulic systems pressure
- flying controls position
- cargo bay temperature
- brake temperatures
- tyre pressures.

The information on the aircraft system data buses is routed to both EICAS displays and both multifunction displays. A data concentrator unit (DCU) receives data in various formats from a variety of sensors, including the high- and low-speed ARINC 429 bus, from analogue and discrete inputs from the engines and other aircraft systems. In the event of a display system failure, the other system displays essential information in a compacted format. Primary engine indications are as before, secondary indications are digital readings only. In the event that both systems fail, a standby LED display is used for EPR, N1 and EGT.

10.5.2 ECAM

This is a flight-phase-related architecture; displays are automatically selected for:

- pre-flight
- take-off
- climb
- cruise
- descent
- approach.

To illustrate ECAM features, during the pre-flight phase, checklists are displayed on the left-hand screen; this includes information such as brake temperatures, APU status and advisory messages. During the pre-flight checks, the right-hand display provides a graphic representation of the aircraft, e.g. if doors are open or closed. The system automatically changes to the relevant pages needed for the next flight phase. System warnings are prioritized, ranging from level one to level three; the warning hierarchy is similar to EICAS.

Both EICAS and ECAM reduce flight deck clutter by integrating the many electro-mechanical instruments that previously monitored engine and aircraft systems. Reliability is increased, and the pilot workload is reduced. Further information on ECAM is given in Chapter 14.

10.6 Multiple choice questions

1. HEIUs can remain charged for several:
   (a) seconds
   (b) minutes
   (c) hours.

2. Ground idle speed occurs when the engine has:
   (a) stabilized (slightly above self-sustaining speed)
   (b) stabilized (slightly below self-sustaining speed)
   (c) just been started.

3. Power from an engine is derived from measuring:
   (a) torque and speed
   (b) temperature and speed
   (c) engine pressure ratio (EPR).

4. When lube oil operates at high temperatures, its viscosity:
   (a) reduces and its lubrication performance decreases
   (b) reduces and its lubrication performance increases
   (c) increases and its lubrication performance increases.

5. The starting sequence for a gas turbine engine is to:
   (a) turn on the ignition, develop sufficient airflow to compress the air, and then open the fuel valves
   (b) develop sufficient airflow to compress the air, open the fuel valves and then turn on the ignition
   (c) develop sufficient airflow to compress the air, turn on the ignition and then open the fuel valves.

6. EICAS warning messages are:
   (a) red, accompanied by an audio alert (prompt action is required by the crew)
   (b) yellow, accompanied by an audio alert (timely action is required by the crew)
   (c) yellow, no audio alert (time available attention is required by the crew).

7. Engine pressure ratio (EPR) is used to measure a gas turbine engine's:
   (a) torque
   (b) thrust
   (c) temperature.

8. Low-, intermediate- and high-pressure shafts are also referred to as:
   (a) N1, N2 and N3
   (b) N2, N3 and N1
   (c) N3, N1 and N2.

9. Typical units of fuel mass flow are given in:
   (a) pounds or kilograms per hour
   (b) gallons per hour
   (c) litres per hour.

10. A gas turbine engine's self-sustaining speed is when sufficient energy is being developed by the engine to provide continuous operation:
    (a) with the starting device still engaged
    (b) with the ignition system in operation
    (c) without the starting device and ignition.

11. Referring to Fig. 10.33, this transducer measures:
    (a) speed
    (b) torque
    (c) linear displacement.

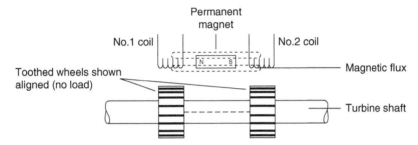

Figure 10.33   See Question 11

12. In turboprop engines, power is measured from:
    (a) engine pressure ratio
    (b) torque
    (c) torque × speed.

Chapter 11

Fuel management

The management of fuel is essential for the safe and economic operation of aircraft. The scope of fuel management depends on the size and type of aircraft; fuel is delivered to the engines using a variety of methods. The system comprises fuel quantity indication, distribution, refuelling, defuelling and fuel jettison. On a typical passenger aircraft, the fuel is contained within the sealed wing box structure. The fuel tanks are divided into main tanks, reserve tanks and centre wing tanks. Fuel tanks on general aviation (GA) aircraft are rubberized bags (bladder tanks) contained within the structure of the aircraft; smaller GA aircraft use metal fuel tanks attached to the wings and/or fuselage.

In the first instance, we need to measure the quantity of fuel on board the aircraft. Various technologies and methods are used to measure fuel quantity: this depends mainly on the type and size of aircraft. Technologies range from sight gauges through to electronic sensors. On larger aircraft, fuel is fed to the engines by electrically driven pumps. On smaller aircraft, an engine-driven pump is used with electrical pumps used as back-up devices. Solenoid or motorized valves are used to isolate the fuel supply to engines under abnormal conditions. On larger aircraft, the fuel can be transferred between tanks; this is controlled manually by the crew, automatically by a fuel control computer. This chapter provides an overview of fuel management on a range of aircraft types.

Key Point

Fuel safety training is a requirement of EASA Part 66, EASA Part-M and Part-145 regulations, which requires that personnel involved in Continued Airworthiness Management and Maintenance of Aircraft Fuel Systems, are given suitable training appropriate to their job function. The following is for training purposes only; always refer to approved data.

11.1 Storage overview

Rigid tanks are usually found in smaller general aviation aircraft. They are installed within the fuselage and/or wings, and are designed to be removable for inspection, replacement, or repair. They do not form an integral part of the aircraft structure.

Bladder tanks are reinforced rubberized bags installed within specific areas of aircraft structure. The bladder can be inserted/removed via the fuel filler inlet or a dedicated access panel; the bladder is then secured to the airframe by clips inside the compartment.

Integral fuel tanks are located within the structure on larger aircraft; these are sealed to accommodate fuel storage. These tanks form part of the aircraft structure; they cannot be removed for service or inspection. Access panels are installed to allow internal inspection, repair, and servicing. On large passenger aircraft, there are four main tanks, two in each wing. The area between the forward and aft spars is divided into tank sections by solid ribs. The wing skin (upper and lower surfaces) completes the tank. Wing ribs act as baffles to prevent the fuel from 'sloshing' around the tank. The inter-spar area of the wing centre section is also used to store fuel. This centre wing tank is similar in construction to the main tanks.

11.2 Fuel quantity measurement and indication

Various technologies and methods are used to measure and display fuel quantity: this depends mainly on the type and size of aircraft. The fuel quantity methods described here could equally apply to other fluids, e.g. oil, hydraulic fluid or water. Some of the methods used are not actually part of an electrical system; however, they are described here to provide a complete review of the methods employed across a range of aircraft types. The methods used for measuring fuel quantity can be summarized as:

- sight glass
- float gauge

- resistance gauge
- under-wing measurement
- capacitance units.

11.2.1 Sight glass

The **sight glass** method illustrated in Fig. 11.1 is based on a simple glass or plastic tube located on the outside of the tank, and visible to the pilot. Fluid level in the tube is the same as the level in the tank; graduations on the tube provide an indication of tank contents. The advantage of this method is that it has no moving parts and is suitable for fuel or oil. This method is suitable for small aircraft where the pilot can see the sight glass; alternatively it is suited for ground serving applications.

Key point

The space above the fuel in a storage tank is called the ullage.

11.2.2 Float gauge

The **float gauge** uses a rod projecting through a hole in the tank cap; see Fig. 11.2. A float is attached to the base of the rod and this rises and falls with the fuel level. The pilot checks the amount of rod protruding through the cap and this provides a direct reading of fuel quantity. One disadvantage of this method is that it is not very stable during aircraft manoeuvres.

The majority of small general aviation (GA) aircraft use a float gauge system similar to that used in motor vehicles as shown in Fig. 11.3; this is based on a float connected to a variable resistor adjacent to the tank. The variable resistor is connected into a **DC ratiometer** circuit where two opposing magnetic fields are created in each of the coils. The pointer is formed with a permanent magnet and is aligned with the resulting magnetic field created by the coils; the pointer moves in accordance with the ratio of currents in the coil.

Key point

The DC ratiometer pointer position is not adversely affected by variations in power supply voltage.

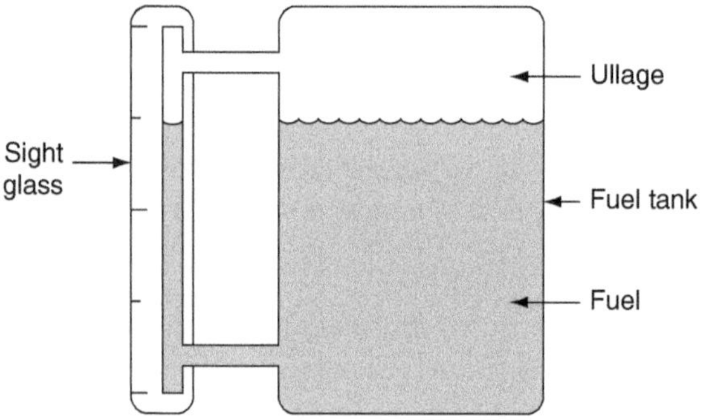

Figure 11.1   Fuel quantity sight glass

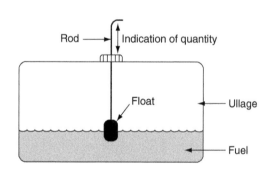

Figure 11.2   Fuel quantity float gauge

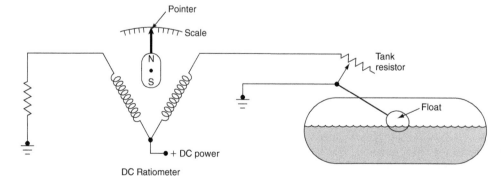

Figure 11.3   Fuel quantity float valve

11.2.3 Under-wing measurement

Under-wing measurement of fuel quantity is used during ground servicing only. The drip stick method uses a hollow tube pushed into the tank. During flight, the tube is stowed into a latched position, which is flush with the aircraft wing or fuselage surface. To take a reading, the tube is released from its stowed position and slowly withdrawn from the tank; when fuel starts to drip from the tank, a reading is taken from graduations on the tube.

Key maintenance point

Spilt fuel presents a fire hazard; take all necessary precautions to clean up any spillage and dispose of any rags soaked in fuel.

A safer alternative to the drip stick method utilizes a transparent plastic rod. This method uses the principle of **light refraction**, and is based on fuel and air having different refractive indexes. When the tip of the rod is moved above/below the surface of the fuel, the light intensity emerging from the viewing end of the rod changes. A reading is then taken from graduations on the rod.

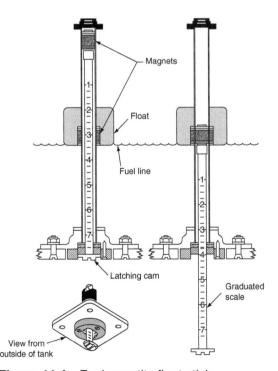

Figure 11.4   Fuel quantity float stick

Another version of the under-wing fuel gauge uses a **floatstick** that comprises a rod, float and magnets located inside the tank as shown in Fig. 11.4. The floatstick is stowed when not in use and released via a quarter-turn cam mechanism; it slides out of the tank until the two magnets align and is then retained in this position. The floatstick is moved in and out of the tank until the attraction of the magnets can be sensed. The fuel quantity reading is then taken from a reference point on the surface of the wing. When the reading has been taken, the rod is pushed back and locked into the stowed position.

Floatsticks can be used when the electronic fuel quantity system (see capacitive fuel quantity system) is unserviceable. All floatsticks are read for each tank and the measurements recorded, the quantity is then calculated from tables in the aircraft documentation. A typical medium-sized aircraft has six floatsticks in each wing tank and four in the centre tank.

Key point

Under-wing fuel quantity measurements are used during ground servicing only.

11.2.4 Capacitive fuel quantity system

The majority of turbine-powered aircraft fuel quantity-indicating systems (**FQISs**) use the **capacitive fuel quantity system**. On a typical twin-engine passenger aircraft the fuel tanks are contained within the aircraft structure as shown in Fig. 11.5. Fuel tank sensors are located throughout each tank and are monitored by an electronic control system. Accurate readings can be obtained by capacitive fuel quantity systems in large and irregular-shaped tanks.

11.2.4.1 Principle of operation

Fuel tank units are formed by concentric aluminium tubes; the inner and outer tubes are the capacitor plates, see Fig. 11.6. The primary advantages of this technology are no moving parts and fuel quantity is measured in **mass** rather than volume. (The mass of fuel determines the amount of energy available.) The **dielectric** is either fuel or air depending on the quantity of fuel in tank. Air has a dielectric of 1.0006 (practically unity); fuel has a dielectric of approximately

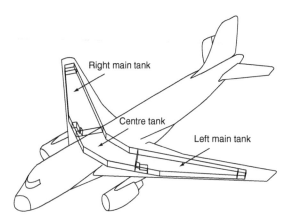

Figure 11.5 Integral fuel tanks

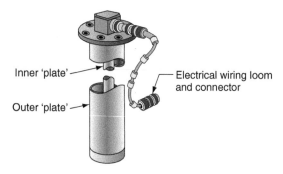

Figure 11.6 Capacitance fuel quantity system – sensor

two; this provides a good relationship for the measurement of a variable quantity.

From basic fundamentals, we know that capacitance is proportional to:

- plate area
- air gap
- dielectric strength.

In the capacitive tank unit, the first two parameters are fixed; the capacitance varies in accordance with the dielectric, i.e. the amount of fuel in the tank as illustrated in Fig. 11.7. With a high quantity of fuel in the tank, the capacitance is high; capacitance varies in direct proportion to the amount of fuel in the tank.

The tank's capacitance unit is connected into an **impedance bridge** circuit, see Fig. 11.8. Variation in capacitance (C) of the fuel tank unit causes a change in reactance (X_C); this is given by the formula:

$$X_c = \frac{1}{2}\pi F C$$

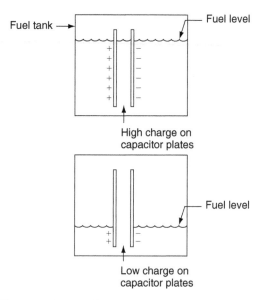

Figure 11.7 Capacitance fuel quantity system – principles

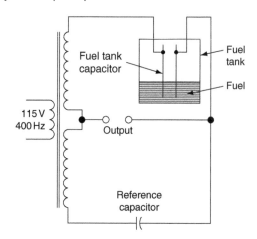

Figure 11.8 Capacitance fuel quantity system – impedance bridge circuit

where X_C is the reactance, Π is a constant, F is the frequency (400 Hz from the aircraft power supply) and C is the capacitance. The impedance (Z) of a capacitive network can be calculated from the formula:

$$Z = \sqrt{(R^2 + X_c^{\,2})}$$

The current in a capacitive network can now be calculated from the formula:

$$I = \frac{V}{Z}$$

When the fuel is consumed during a flight, and the fuel level decreases, the capacitance decreases and the reactance increases; less current flows in the tank unit (compared with the reference capacitor). This unbalance causes a potential difference at points X–Y, proportional to fuel quantity; this signal is amplified and used to drive an indicator. Capacitance trimmers are used for calibration of the fuel indicators (or gauges).

Key point

Air and fuel have dielectrics of approximately unity and two respectively.

11.2.4.2 Density compensation

The volume of fuel in a tank varies with temperature; as the temperature changes, the mass of fuel remains the same, but the volume changes. The dielectric is therefore affected by fuel **density**; this density will change with temperature. Increased density is a result of reduced temperature that will cause increased capacitance. Changes in fuel density are measured by a **compensation unit**. This is an additional tank unit located in the bottom of the fuel tank, therefore it is always immersed in fuel. The compensating unit is

connected into the impedance bridge such that changes in fuel density cause the bridge to become unbalanced and this compensates for the change in fuel level.

Key point

The volume of fuel in a tank varies with temperature; as the temperature changes, the mass of fuel remains the same, but the volume changes.

11.2.4.3 Multiple tanks

Multiple tank units are often employed in larger aircraft as illustrated in Fig. 11.9. On a typical medium-sized passenger aircraft there are twelve capacitive tank units in each main fuel tank, and nine in the centre fuel tank. The construction of all the tank units is the same, except for their length; this depends on the depth of the tank at that particular location. Each tank unit consists of two concentric aluminium tubes and a terminal block. The aluminium tubes are anodized and polyurethane coated to protect against corrosion. The air gap between the tubes is relatively wide to avoid electrical short-circuiting, caused by contamination in the fuel or fungus coating on the tubes.

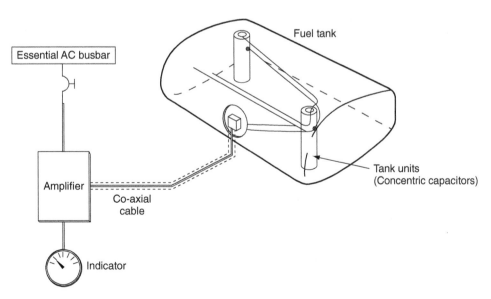

Figure 11.9   Capacitance fuel quantity system – multiple tank units

Key maintenance point

The fuel tank units are factory calibrated and cannot be adjusted on the aircraft.

A tank unit wiring harness attaches to the terminal block of the tank unit. The terminals on the tank wiring harness and the terminal posts have different dimensions to prevent cross-connection during installation. Tank unit end caps are insulated to ensure that the unit does not short to the ground; they also control stray capacitance that forms between the tank unit and ground plane (**capacitance fringing**). Two brackets made of nonferrous glass-filled nylon attach each tank unit to the fuel tank structure.

Intrinsic safety is a technique used for safe operation of electrical and electronic equipment in explosive atmospheres. It is essential that the available electrical and thermal energy in the fuel quantity system is always low enough that ignition of the hazardous atmosphere cannot occur.

Key maintenance point

When working with fuel systems, always use intrinsically safe lamps or torches.

11.3 Fuel feed and distribution

GA aircraft are normally fitted with an **engine-driven pump** (EDP), with electrical **boost pumps** fitted to prime the system during starting. The electrical boost pumps also provide fuel pressure should the EDP fail. A simple fuel pump system comprises an electrically driven boost pump motor controlled by an on/off switch; see Fig. 11.10.

This system is enhanced by a two-stage throttle control system, as illustrated in Fig. 11.11. When the boost pump selector switch is set at the 'low' setting, electrical power is switched through the resistor and the motor runs at a low speed. With the engine running, the selector is moved to the 'high' setting; this provides power through the normally closed (NC) contacts of the throttle micro-switch. When the throttle is set below one-third open, the resistor remains in series, and the motor continues to run at the low speed. When the throttle is advanced, the throttle micro-

switch changes over via the normally open (NO) contacts to bypass the resistor and apply full power to the motor.

The typical fuel feed arrangement on a medium- to large-sized passenger aircraft comprises two booster pumps for each main tank; the motor is located on the tank bulkhead, with the pump located inside the tank. The **fuel distribution** system requires electrical power and is controlled by a panel in the flight compartment as illustrated in Fig. 11.12. Fuel shut-off valves are connected to the battery bus, and controlled by the engine start lever (see engine systems, Chapter 10) and fire handle (see fire protection, Chapter 16). The fuel system normally has the means of transferring fuel between tanks (see fuel transfer); this is controlled by a selector switch that operates a cross-feed valve. Each boost pump is driven by a 115 V AC three-phase motor selected on/off on the control panel via a relay.

The delivery output from each pump feeds into the system via a non-return valve (NRV). Under normal operating conditions, each pump feeds its own engine via a motor driven **low-pressure (LP) cock**. If a centre tank is fitted as part of a three-tank installation, this can feed either engine by a **fuel transfer** system as shown in Fig. 11.13. On some aircraft, tank pumps are located in a dry area of the wing root; on other aircraft the pumps are actually inside the fuel tank. In the latter case, the crew have to maintain certain minimum fuel levels. Control switches for all pumps, cocks and valves together with warning indications are located on the overhead panel of the flight engineer's station. LP cocks are automatically closed if the fire handle is activated (see fire protection, Chapter 16).

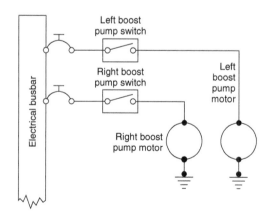

Figure 11.10   Simple fuel pump system

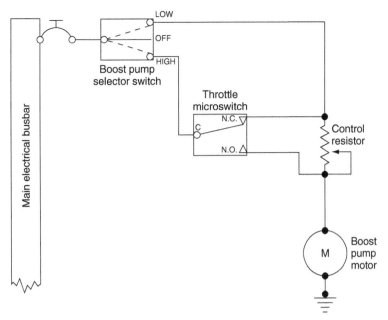

Figure 11.11 Two-stage fuel pump system

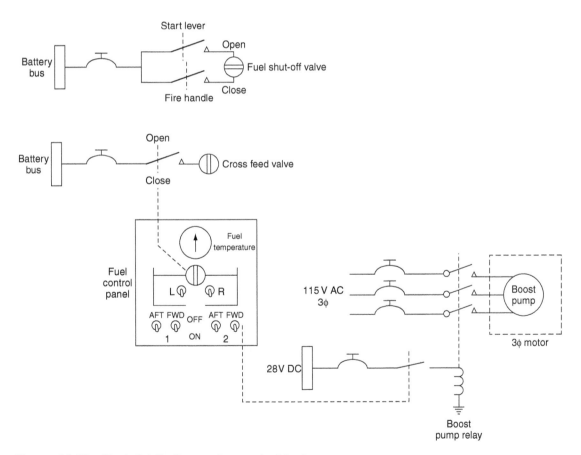

Figure 11.12 Fuel distribution system – electrical power

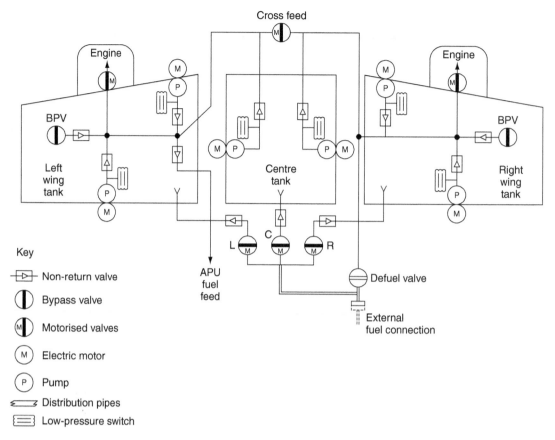

Figure 11.13   Fuel distribution – three-tank system

Key maintenance point

Leaving a fuel pump switched on with low fuel quantity is an explosion risk since the pump is likely to overheat.

Key maintenance point

If a pump is left running dry for extended periods, typically 10 minutes, it will have insufficient fuel for priming which will render it inoperative when the tank is subsequently refuelled.

11.4 Fuel transfer

This system is used to selectively transfer fuel between tanks; electrically driven fuel pumps and motorized valves are controlled either manually by the crew or by an automatic control system. A simple example of this is illustrated in Fig. 11.13, where fuel can be cross-fed between the left and right tanks. On larger aircraft, the complexity of this fuel transfer system increases. The system comprises a number of motorized valves. Engine valves are activated by the start lever or fire handles. The left, centre and right refuel/defuel valves are operated from an under-wing panel. Bypass valves (BPVs) are operated if an electrical pump's fuel filter is blocked. (The BPV senses differential pressure across the filter.) Controls and indications are located on an overhead panel or flight engineer's station. An electrically operated cross-feed valve is normally closed unless fuel is being transferred.

It is essential that **fuel temperature** is monitored, either manually or automatically. The maximum fuel temperature is typically +49°C; the minimum fuel temperature is –45°C or freezing point +3°C, whichever is higher; the typical freezing point of Jet

A1 fuel is −47°C. Fuel temperature is measured by an RTD. If the fuel temperature is approaching the lower limits, some fuel could be transferred between tanks; alternatively the aircraft would have to descend into warmer air or accelerate to increase the kinetic heating.

11.5 Refuelling and defuelling

A refuelling control panel and pressure connections are normally located in one or both of the wing areas allowing the fuel to be supplied directly into the main fuel system. A **bonding lead** is always connected between the fuel bowser and aircraft to minimize the risk of static discharge. The fuel tank supply line is connected to the aircraft and pressure applied. Selective control of the system's motorized valves allow specific tanks to be filled as required. **Defuelling** is often required before maintenance, or if the aircraft is to be weighed. The fuel is transferred from the aircraft into a suitable container, typically a fuel bowser. This is achieved via a defuelling valve and applying suction through valves.

11.6 Fuel jettison

When an aircraft takes off fully loaded with passengers and fuel, and then needs to make an emergency landing, it will almost certainly be over its maximum landing weight. Fuel has to be disposed of to reduce the aircraft weight to prepare for the emergency landing. A large aircraft such as the Boeing 747 can be carrying over 100 tonnes of fuel; this is almost 50% of the aircraft's gross weight. One way of burning off fuel and reducing aircraft weight is to fly in a high drag configuration, e.g. 250 knots with the gear down (speed-brakes will further increase the drag). Aircraft can be certified for landings up to the maximum takeoff weight (MTOW) in an emergency; however, overweight landings would only be made if burning off fuel exposed the aircraft to additional hazards. Some aircraft are installed with a **fuel jettison**, or **fuel dumping** system, Fig. 11.14. This provides a means of pumping fuel overboard to rapidly decrease the aircraft's weight. Fuel can normally be jettisoned with landing gear and/or flaps extended. Two **jettison pumps** are installed in each main tank, fuel is pumped via a jettison manifold to nozzle valves located at each wing tip trailing edge.

Key maintenance point

Bonding of fuel system components is essential to ensure intrinsic safety.

11.7 Fuel tank venting

The venting system takes ram air from intakes on the underside of the wing for two specific purposes. When the aircraft is flying, ram air is used to pressurize the fuel to prevent vaporization at lower atmospheric pressures. The ram air also pressurizes the fuel tanks to ensure positive pressure on the inlet ports of each pump. The expansion space above the fuel, called **ullage**, changes with aircraft attitude. Float-operated vent valves are located at key points in the tank to allow fuel to escape into the vent system. Venting tanks (sometimes called **surge tanks**) in the wing tips collect this overspill from the main tanks; the fuel is then pumped back into one of the main tanks. Float-operated vent valves prevent inadvertent transfer of fuel between the tanks.

11.8 Fuel tank inerting

Aircraft have been destroyed due to explosions in empty centre wing fuel tanks. The most likely cause is the centre tank fuel pumps being left running in high ambient temperatures with low fuel quantity. The first centre wing tank (CWT) explosion on a large passenger aircraft occurred on a B707 in 1959; there have been other events in more recent times. In May 1990, shortly after pushback, the centre wing tank (CWT) exploded on a Philippine Airlines Boeing 737, killing eight people. The CWT had not been filled since March 9, 1990. Air conditioning (A/C) packs had been running on the ground before pushback for approximately 30 to 45 minutes, ambient air was 35°C/95°F. On March 3, 2001, the centre wing tank explosion on Thai Airways International B737 was followed 18 minutes later by an explosion in the right wing tank. Residual fuel was in the CWT. Air conditioning packs had been running continuously since the aircraft's previous flight, including some 40 minutes on the ground. Ambient air temperature was in excess of 30°C. On July 17, 1996, the CWT exploded on a TWA B747-100 shortly after takeoff from JFK International Airport with fatal consequences. The CWT contained a low amount of residual fuel. The A/C packs had been

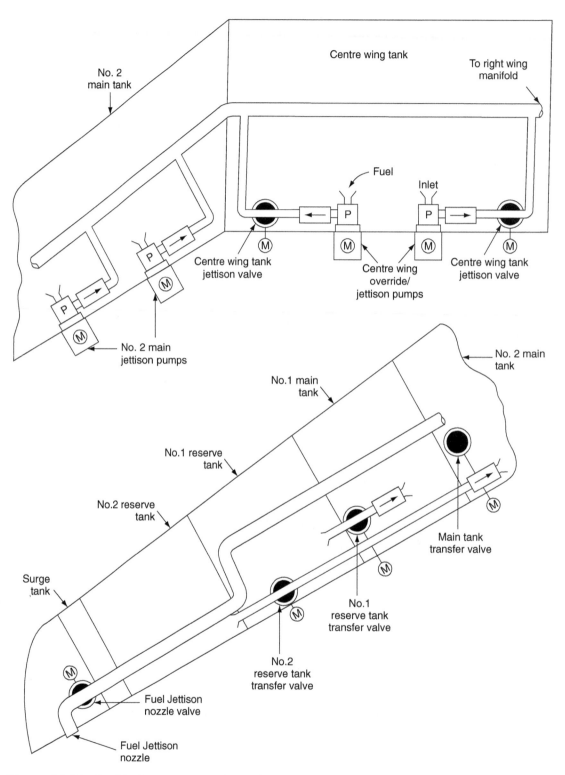

Figure 11.14   Fuel jettison system schematic

running on the ground for 2.5 hours before takeoff, ambient temperature was 28°C/82°F. These three fatal CWT fuel tank explosions all had certain elements in common:

- all three aircraft had only a small amount of fuel in the wing tank
- the air conditioning packs, located in non-vented bays directly under the CWT, had been running before the explosions
- the outside air temperatures were quite warm.

A small amount of fuel in the centre wing tank means that there is a large volume for fuel vapour to occupy. Air conditioning packs generate heat, which contributes to fuel vaporization in the non-insulated tanks. The vaporization is increased with high outside air temperatures.

Empty fuel tanks retain some **unusable fuel** which, in high ambient temperature conditions, will evaporate and create an explosive mixture when combined with oxygen in the ullage. Regulatory mandates have introduced improvements to the design and maintenance of fuel tanks to reduce the chances of such explosions. These improvements include the enhanced design of the FQIS and fuel pumps, inspection regimes for wiring in the tanks and insulating fuel tanks that are in close proximity to sources of high temperature. The longer-term solution that will contribute more towards fuel tank safety is **inerting**; possible solutions include:

- ground-based inerting
- on-board ground inerting
- on-board inert gas generation system (OBIGGS)
- liquid nitrogen from a ground source.

For ground-based inerting, the fuel tanks would receive an amount of nitrogen-enriched air before pushback. Inerting would last for the taxi, takeoff and climb phases of flight, when the fuel vapours would be warmest. On-board ground inerting achieves the same objective; however, the inerting equipment is an aircraft system.

Although OBIGGS offers tremendous benefits, it is expensive. Nitrogen generating systems (NGSs) have been developed which decrease the flammability risks of the tanks. The NGS is an onboard inert gas system; external (atmospheric) air is directed into an air separation module (ASM), this separates out the oxygen and nitrogen via a **molecular sieve**. After separation, the nitrogen-enriched air (NEA) is supplied into the centre wing tank and the oxygen-enriched air (OEA) is vented overboard. NEA decreases the oxygen content in the

ullage to prevent combustion. Trials have determined that an oxygen level of 12% is sufficient to prevent ignition; this is achievable with one ASM on a medium-sized aircraft and up to six ASMs on larger wide-bodied aircraft.

In the fourth potential method, the aircraft would receive a supply of **liquid nitrogen** from a ground source. The liquid nitrogen would be stored on the aircraft in a vacuum-sealed insulated container, and would be fed to the fuel tanks under low pressure. Oxygen sensors in the fuel tanks provide feedback to the system's computer control function; the supply of liquid nitrogen is regulated to keep the fuel tanks inerted for critical phases of flight. The liquid nitrogen could also be used to supplement fire suppression in other areas of the aircraft. A computer would monitor temperature in the fuel tanks to determine if the ullage is within a flammable temperature range. This system would know the tank temperature at all times, together with the oxygen content of the tanks.

There remains an on-going industry debate regarding the effectiveness and cost of inerting. There needs to be a practical means of inerting fuel tanks for in-service and new-production civil transports.

11.9 Multiple choice questions

1. The volume of fuel in a tank varies with temperature; as the temperature changes:
   (a) the mass and volume of fuel remains the same
   (b) the mass of fuel changes but the volume remains the same
   (c) the mass of fuel remains the same, but the volume changes.

2. In the DC ratiometer fuel quantity circuit:
   (a) two opposing magnetic fields are created in each of the coils
   (b) three opposing magnetic fields are created in each of the coils
   (c) two complementing magnetic fields are created in each of the coils.

3. Air and fuel have dielectrics of approximately:
   (a) unity and zero respectively
   (b) unity and two respectively
   (c) two and unity respectively.

4. Fuel tank ullage is the:
   (a) expansion space above the fuel, and changes with aircraft attitude

(b) unusable fuel, and changes with aircraft
attitude

(c) expansion space above the fuel, remaining
constant with aircraft attitude.

5. When fuel level decreases, the capacitance of the
fuel quantity sensor:
   (a) decreases and the reactance increases
   (b) increases and the reactance increases
   (c) increases and the reactance decreases.

6. Defuelling is often required:
   (a) following aircraft maintenance
   (b) before maintenance, or if the aircraft is to be
weighed
   (c) after the aircraft has been weighed.

7. Under-wing fuel quantity measurements are used
during:
   (a) level flight
   (b) all flight conditions
   (c) ground servicing only.

8. Intrinsic safety is a technique used for:
   (a) safe operation of electrical/electronic
equipment in explosive atmospheres
   (b) ensuring fuel temperature does not become
too low/high
   (c) reducing fuel quantity.

9. Fuel tank inerting is a system where:
   (a) nitrogen-enriched air (NEA) is vented
overboard
   (b) nitrogen-enriched air (NEA) is supplied into
the centre wing tank
   (c) oxygen-enriched air (OEA) is supplied into
the centre wing tank.

10. Increased fuel density is a result of:
   (a) increased temperature and increased
capacitance
   (b) reduced temperature and increased
capacitance
   (c) reduced temperature and reduced
capacitance.

Chapter 12 Lights

Lighting is installed on aircraft for a number of reasons including: safety, operational needs, servicing and for the convenience of passengers. The applications of aircraft lights can be broadly grouped into four areas: flight compartment (cockpit), passenger cabin, exterior and servicing (cargo and equipment bays). There are many types of lighting technologies used on aircraft. Lights are controlled by on/off switches, variable resistors or by automatic control circuits. This chapter will review each of these lighting technologies and the type of equipment used in specific aircraft applications.

12.1 Lighting technologies

Aircraft lighting is based on a number of technologies:

- incandescence
- light-emitting diode (LED)
- electro-luminescent
- fluorescence
- strobe.

Incandescence is the radiation of light from an electrical filament due to an increase in its temperature. The filament is a small length of wire, e.g. tungsten, which resists the flow of electrons when a voltage is applied, thereby heating the filament. Tungsten can be drawn into a very thin wire filament and has a very high melting point (3,659 K). The electron flow creates a voltage drop that heats the filament to a temperature where radiation is emitted in the visible spectrum.

Electro-luminescence is a combined optical and electrical phenomenon that causes visible light to be emitted. This can be achieved with electron flow through a semi-conductor material, or by a strong electric field applied across a phosphor material. Electro-luminescence is the effect of recombining electrons and holes in a **light-emitting diode (LED)** or phosphor material. The electrons are imparted with an external energy source and release their own energy as photons; this is radiated as light energy. In the LED semiconductor, electrons and holes are separated by a doping process to form the p-n junction. With the electro-luminescent phosphor display, electrons are imparted with energy by the impact of high-energy electrons that are accelerated by a strong electric field.

Fluorescent lamps are gas-discharge devices formed from a sealed tube of glass that is coated on the inside with phosphor; the glass tube contains mercury vapour mixed with an inert gas, e.g. argon or neon. The lamp uses a high voltage to energize the mercury vapour; this results in an **ionized gas** where the electrons are separated from the nucleus of their atoms creating **plasma**. The release of energy causes the phosphor coating to fluoresce, i.e. it produces visible light. Fluorescent lamps require a ballast resistor to regulate the flow of energy in the tube.

Strobe lights are formed from small diameter (typically 5 mm) sealed quartz or glass envelopes/tubes filled with xenon gas, see Fig. 12.1. Power from the aircraft bus is converted into a 400 V DC supply for the strobe. The tube is formed into the desired shape to suit the installation; Fig. 12.2 is a wing tip anti-collision light, normally located behind a clear plastic protective cover. **Xenon** is an inert (or noble) gas, chemically very stable, and has widespread use used in light-emitting devices, e.g. aircraft anti-collision lights. The emission of light is initiated by ionizing the xenon gas mixture by applying a high voltage across the electrodes.

Figure 12.3 illustrates a typical strobe light circuit; this comprises a low-voltage power supply and a 400 V DC power pack that provides a dual power supply. Shielded cables are required between the power pack and strobes to minimize the effect of electromagnetic interference (EMI). A high-energy pulse of current is triggered through the ionized gas on a cyclic basis. The ionization decreases the electrical resistance of the gas such that a pulse in the order of thousands of amperes is conducted through the gas.

When this current pulse travels through the tube, it imparts energy into the electrons surrounding

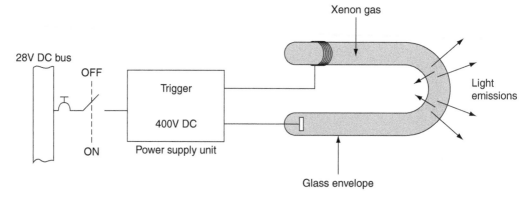

Figure 12.1   Strobe/xenon gas tube schematic

Figure 12.2   Strobe/xenon gas tube product

the xenon atoms causing them to move up to higher energy levels. The electrons immediately drop back into to lower energy levels, thereby producing photons. Strobes are a source of short-wavelength ultra-violet radiation together with intense emissions in the near infrared; the overall effect is a high-intensity white flash.

Key maintenance point

400 V DC used in strobe circuits is dangerous: take all necessary health and safety precautions when working near the system.

Key maintenance point

Do not handle strobe tubes with bare hands; moisture causes local hot-spots that can lead to premature failure.

Test your understanding 12.1

Briefly explain the difference between the following lighting technologies:

- incandescent
- light-emitting diode (LED)
- electro-luminescent
- fluorescent
- strobe.

12.2 Flight compartment lights

Lighting is needed for the illumination of instruments, switches and panels. **Dome lights** located on the ceiling provide non-directional distribution of light in the compartment; it typically contains an incandescent lamp and is powered from the battery or ground services bus. **Flood lighting** in the flight compartment from incandescent lamps and/or fluorescent tubes provides a general illumination of instruments, panels, pedestals, etc. Fluorescent tubes located beneath the glareshield provide overall illumination of the instrument panels. **Emergency lights** are installed in the flight compartment for escape purposes. The colour of flight compartment lights is normally white; this

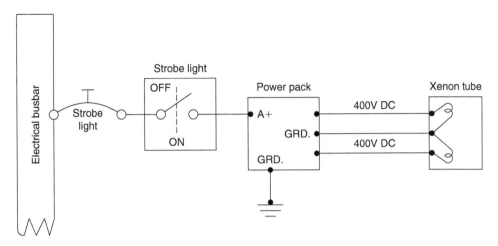

Figure 12.3 Strobe/xenon gas tube circuit

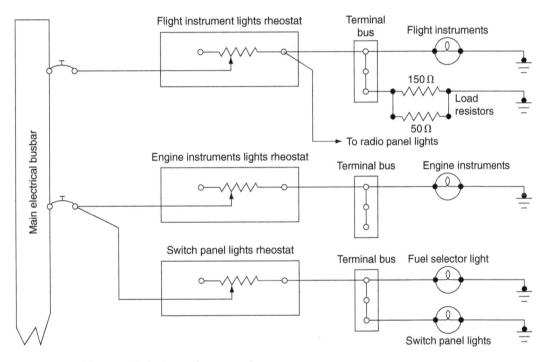

Figure 12.4 Rheostat light intensity control

reduces the power and heat, improves contrast on the instruments, and reduces eye fatigue.

12.2.1 Instruments

Internal instrument lighting is normally from incandescent lamps integrated within individual instruments; lighting must be shielded from causing any direct glare to the pilot and must be dimmable. External instrument lighting is provided by pillar (or bridge) lights positioned on the panels for individual instruments. The light intensity can be dimmed by a simple **rheostat** device as illustrated in Fig. 12.4; this

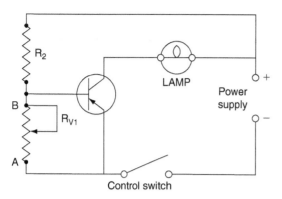

Figure 12.5 Transistor lighting control circuit

typical circuit is for flight instruments, engine instruments and switch panels.

The increased quantity of instruments requires electronic control in place of rheostat due to the higher loads. A **transistor circuit** provides electronic control as illustrated in Fig. 12.5; variable resistor R_{V1} varies the (relatively low) base current into the base of a PNP transistor; this controls the (relatively high) current through the collector and lamp. A typical transistor controlled lighting system is illustrated in Fig. 12.6. The relatively low base currents in the respective transistors can now control a variety of lighting circuits:

- radio navigation systems
- compass
- fuel panels
- engine indications.

Instrument panels are often constructed from Perspex; the surface is painted and then engraved with the identification of switches and controls; the panel is illuminated from the edges. The light is dispersed through the panel, but is only seen through the engravings. Alternatively, electro-luminescent panels are used; these are AC powered and energy efficient. Referring to Fig. 12.7, the phosphorus layer is laminated between front and rear clear plastic layers. The phosphorus material glows when AC power is applied; the front of the panel is painted to match the colour of other panels. Engraved lettering or symbols remain clear and transmit light from the glowing phosphorus layer.

12.2.2 Master warning

An increasing number of systems are being designed into aircraft; this leads to more warning lights and larger panels with an increased possibility of a warning light being missed by the crew. This has led to centralized 'attention getters', or **master warning** and **caution light** panels, a typical arrangement is shown in Fig. 12.8. The lights are located within pilot's immediate view, typically on the glareshield. Master caution and warning systems were developed to ease pilot workload, particularly on aircraft designed for operation without a flight engineer. In addition to the attention getter, it also directs the pilot towards the problem area concerned. The system annunciators are usually arranged such that the cautions are in the same orientation as the main system panels, e.g. the overhead panel.

A simplified master warning circuit is illustrated in Fig. 12.9. In this example, switch contacts 1 and 2 represent the individual warning signals from two different systems. In the event of a failure or malfunction, closure of either switch causes its corresponding warning light to illuminate together with the master warning (MW) light. The master warning and caution light is cancelled by pressing the light caption, whilst the crew react and investigate the reason for the warning. Typical panels could have up to 50 individual warning lights, any one of which also illuminates the master warning light. The individual lights could be located on an overhead or side panel.

Key point

A diode is required in the warning light circuit to ensure that only the relevant system light is illuminated.

When the master warning or caution light is illuminated, the pilot cross-refers to a centralized group of warning lights on the relevant panel, each connected to the warning devices of specific systems. The individual systems are identified by the system name and are located within the pilot's scan. Warning lights can be tested by a separate **test** switch, or by a centralized master dim and test switch. The night/day switch is used to reduce the intensity of warning lights during low ambient lighting conditions. Referring to Fig. 12.10(a), warning and caution lights affecting system operation and aircraft safety are defined by specific colours:

- **warning**, red, an unsafe condition exists
- **caution**, amber, an abnormal condition exists, but it is not unsafe

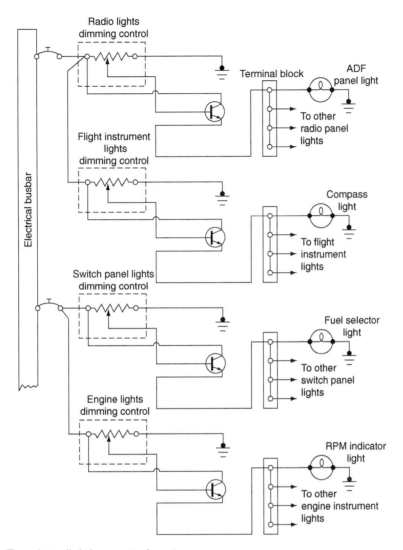

Figure 12.6   Transistor lighting control system

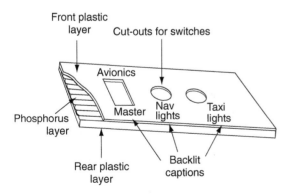

Figure 12.7   Phosphorus lighting panel

- **advisory**, green or blue, a safe condition exists, or for information e.g. gear down.

Some installations have a comprehensive master warning and caution light panel that occupies the entire upper instrument panel, see Fig. 12.10b.

12.2.3 Emerging technology

Digital technology is replacing traditional wire bundles from the flight compartment to the equipment bays. A digital processor on each warning and

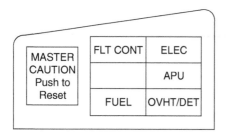
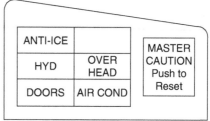

Figure 12.8 Master warning and caution lights

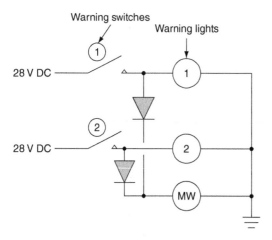

Figure 12.9 Simplified master warning circuit

caution module replaces lighted switches with the associated wire bundles. LEDs can now provide sufficient brightness to replace incandescent lighting with 30% power reduction compared with fluorescent lighting. The objectives are to accomplish weight, cost and reliability goals while giving flight crews what they are used to seeing in older flight compartments, including the tactile feel for push buttons and controls for lighting and dimming.

Key point

Lights used for certain warnings are not dimmable, e.g. fire and overheat. This is to ensure that the warnings are not missed under bright ambient lighting conditions.

12.3 Passenger cabin lights

Interior lighting installations for the passenger cabin vary depending on the size of aircraft; this ranges from a small quantity of roof-mounted incandescent lamps, through to integrated lighting concealed within the interior trim. These lights are controlled from the flight attendants' station. LED illumination is being specified on business and passenger aircraft that have pre-programmed settings for specific flight phases and time zones. The systems are automatically controlled to customize the mixing of colours and lighting levels; this is intended to help passengers combat the fatigue of long-distance travel.

Traditionally, each cabin light is controlled individually; this technology is being replaced by a central control unit connected to all the lights in the cabin. Those lights are linked via a data bus to their respective control units; each of the lights can be programmed for different scenarios.

Implemented through software, the lighting in the cabin is a mixed array of colours. LED technology offers higher reliability and reduced maintenance costs compared to incandescent and fluorescent lights. Cabin signs, e.g. 'return to seat' or 'no-smoking', are normally activated by the flight crew; on some aircraft the lights are armed by the crew and then activated automatically. Passenger reading lights are controlled from individual seat controls.

Additional entry floodlights are provided in the door areas. **Exit** lights are located adjacent to the emergency exits and are clearly visible, irrespective of whether the door is open or closed. **Floor path lighting** is used in emergency situations to provide visual identification of escape routes along cabin aisle floor. These systems have sufficient energy to enable passengers to identify aisle boundaries. The system guides the passengers to designated **emergency exits**

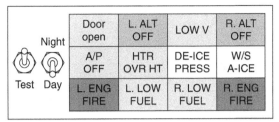

		Door open	L. ALT OFF	LOW V	R. ALT OFF
Night		A/P OFF	HTR OVR HT	DE-ICE PRESS	W/S A-ICE
Test	Day	L. ENG FIRE	L. LOW FUEL	R. LOW FUEL	R. ENG FIRE

Colour coding

R. ALT OFF	Amber
W/S A-ICE	Green
R. ENG FIRE	Red

(a)

(b)

Figure 12.10 Warning and caution panels: (a) typical arrangement, (b) helicopter installation

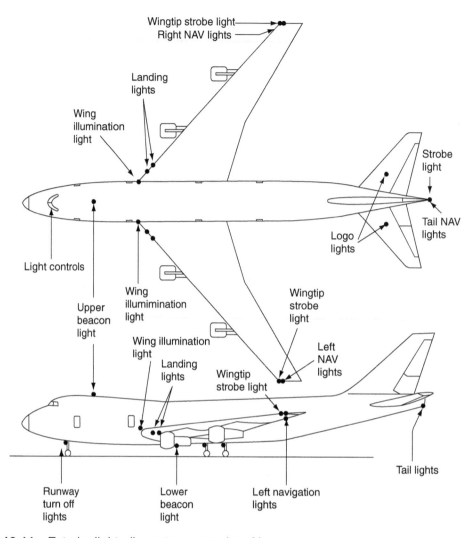

Figure 12.11   Exterior lights (large transport aircraft)

in accordance with the specific certification requirements of the aircraft.

12.4 Exterior lights

An overview of the exterior lighting arrangement on a large passenger aircraft is illustrated in Fig. 12.11. Exterior lighting is used for:

- logo illumination
- landing/taxiing
- wing illumination
- anti-collision/navigation.

12.4.1 Logo lights

Referring to Fig. 12.12, **logo lights** are used to illuminate the tail fin; this is primarily for promotional purposes, i.e. for the airline to highlight their logo during night operations at an airport. Apart from the advertising value at airports, they are often used for additional awareness in busy airspace. **Taxi lights** (or runway turn off lights) are sealed beam devices with 250 W filament lamps located on the nose, landing gear or wing roots. They are sometimes combined with the landing light and used when approaching or leaving the runway. Taxi lights improve visibility during ground operations; they are directed at higher

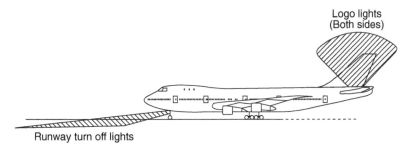

Figure 12.12   Exterior lights (turn off/logo)

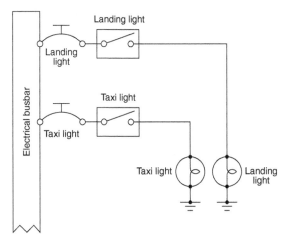

Figure 12.13   Fixed landing light circuit

angle than landing lights. Runway turnoff lights, in the wing roots, are normally only used at night on poorly lit runways. A typical fixed landing and taxi light circuit is shown in Fig. 12.13. On some aircraft types, taxi lights will switch off automatically with gear retraction. It is common practice to have taxi lights on whilst the aircraft is in motion as a warning to other aircraft and vehicles.

12.4.2 Landing lights

Landing lights are located on the wing tips, or on the front of the fuselage, usually at fixed angles to illuminate the runway. They are sealed beam devices with 600–1000 W filament lamps; a parabolic reflector concentrates light into a directional beam. The high current requirement is controlled via a relay. Some landing light installations have a retractable

assembly located on the underside of the wing. This has a reversible motor and gear mechanism to drive the light out against the airflow; a typical circuit is illustrated in Fig. 12.14. The alternative location for a landing light is in the wing leading edge; this has a transparent cover to provide aerodynamic fairing. Inboard and outboard landing lights (Fig. 12.15) provide extended illumination of the landing area.

12.4.3 Wing illumination

Ice inspection lights (Fig. 12.15) are often installed to check ice formation on wing leading edges and engine intakes. Typical lights are the sealed beam type with filament lamps of 50–250 watts. They are recessed into the fuselage or engine nacelle with a preset direction that illuminates a section of the wing that can be viewed from the flight compartment. (More details on ice detection are given in Chapter 15.)

12.4.4 Service lights

Service lights are provided throughout the aircraft as illustrated in Fig. 12.16. These lights are powered from the aircraft ground servicing bus. Examples include:

- cargo bays
- wheel wells
- equipment bays
- fuelling panels.

Wheel wells lights are normally only used during the turnaround at night during the pre-flight inspection; they can also be used to see the mechanical gear down-lock indications at night.

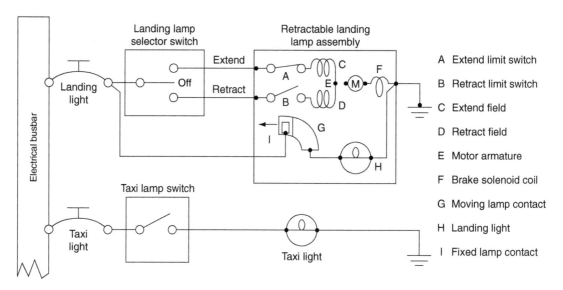

Figure 12.14   Retractable landing/taxi light circuit

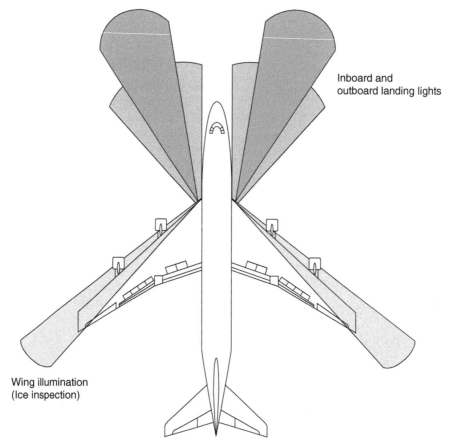

Figure 12.15   Exterior lights (landing/wing)

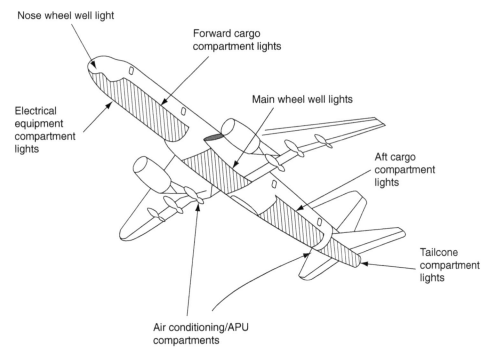

Figure 12.16 Exterior lights (servicing)

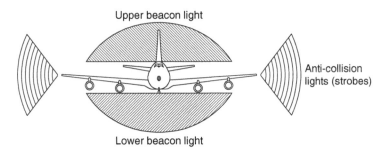

Figure 12.17 Exterior lights (beacons/strobes)

12.4.5 Navigation lights

The primary external lights required for navigation purposes are the **beacons** and **anti-collision** lights, see Fig. 12.17.

12.4.5.1 Navigation lights

The **navigation** (or position) lights are a legal requirement for night flying. Navigation lights are normally based on filament lamps, providing steady illumination. They are located at the extremes of the aircraft,

see Fig. 12.18, and provide an indication of the aircraft's direction and manoeuvres. Navigation lights are based on regulations that define the colour, location and beam divergence such that the aircraft is visible from any viewing angle; these colours and divergence angles are:

- **green**, starboard wing, divergence of 110 degrees
- **red**, port wing, divergence of 110 degrees
- **clear** (white), tail, divergence of ±70 degrees either side of aircraft centreline (140 degrees total).

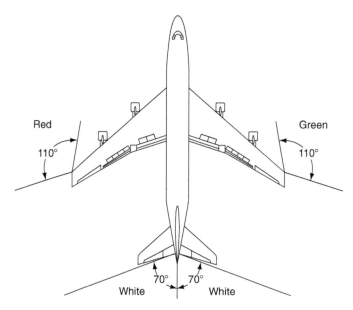

Figure 12.18   Navigation lights – angular coverage

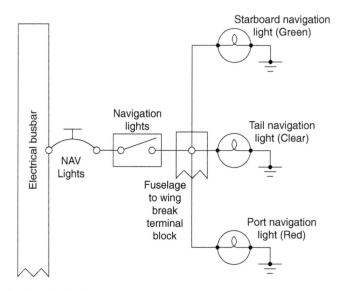

Figure 12.19   Navigation light circuit

The traditional location of the white light is on the tail cone or fin tip; some aircraft have the rear facing light on the trailing edge of each wing tip. The wing lamps are 20 W filament lamps, the tail lamp is 10 W. Coloured filters produce the specific colours; these filters must not shrink, fade or become opaque. A typical navigation light circuit is shown in Fig. 12.19. Note that the lights are controlled by single switch and protection device.

Some aircraft are installed with LED position lights; these are formed with a bank of LEDs, see Fig. 12.20. Helicopters have varying navigation light installations due to their specific geometry, see Fig. 12.21.

12.4.5.2 Anti-collision lights

Anti-collision lights often supplement navigation lights; these can be provided either by a strobe

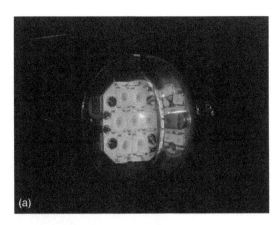 

Figure 12.20 Position lights: (a) LED type, (b) wing tip location

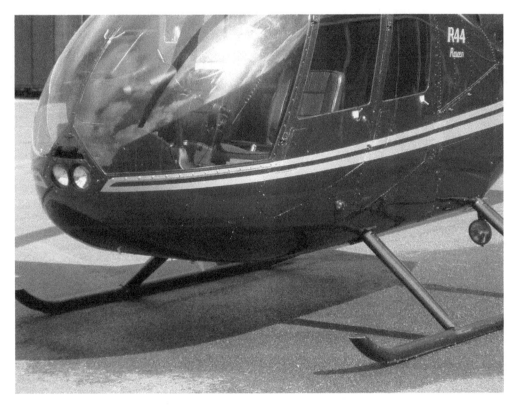

Figure 12.21 Helicopter lights: Twin landing lights on nose; Navigation light below door; Additional landing light on rear skid

light, rotating beacon or a combination of both, see Fig. 12.22. Anti-collision lights are also used as a warning that the engines are running or are about to be started. They are typically not switched off until it is considered safe for ground personnel to approach the aircraft. Strobe lights are typically located on the:

- vertical stabilizer
- wing tips

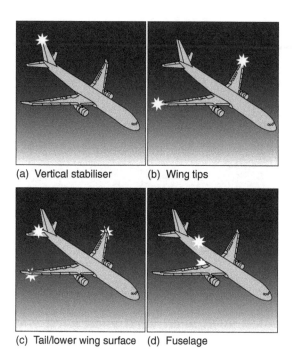

(a) Vertical stabiliser   (b) Wing tips

(c) Tail/lower wing surface   (d) Fuselage

Figure 12.22   Anti-collision strobe lights

Figure 12.23   Rotating beacon (top of fin)

- tail/lower wing surfaces
- fuselage.

These anti-collision lights are controlled by a single switch, with a single protection device. Anti-collision lights used in conjunction with the navigation lights enhance situational awareness for pilots in nearby aircraft, especially during night-time or in low-visibility conditions.

The **rotating beacon** comprises a filament lamp, reflector, motor and drive mechanism that gives the effect of a light through a red filter that flashes 40–50 times per second. A typical rotating beacon is illustrated in Fig. 12.23. They are located on tail fins and the upper and lower fuselage (or tail boom on a helicopter).

Strobe lights are wing tip and tail fin mounted to supplement navigation lights. The strobe light produces a high-intensity white flash of 1 mS duration at approximately 70 flashes per minute through a white or red filter; these provide light that can be seen from several miles.

Many external lights are based on **sealed beams**. A sealed beam combines an incandescent filament lamp and reflector into a single assembly. The reflector concentrates the light from the lamp into a

predetermined beam shape; the assembly is fitted with a clear glass front cover that is permanently sealed to the reflector and cannot be removed. The filament lamp is inserted through a hole in the rear of the reflector and retained by a locking mechanism.

12.5 Multiple choice questions

1. A high-intensity white flash is produced from a:
   (a) strobe light
   (b) fluorescent tube
   (c) landing light.

2. A clear (white) light with a divergence of ±70 degrees either side of aircraft centreline is the:
   (a) landing light
   (b) rear position light
   (c) logo light.

3. Incandescence is the radiation of light from:
   (a) a gas-discharge device
   (b) an electrical filament due to an increase in its temperature
   (c) a combined optical and electrical phenomenon.

4. Anti-collision lights can be provided by:
   (a) rotating beacon or strobe lights
   (b) fluorescent tubes
   (c) retractable assembly.

5. Green or blue lights in the instrument panel indicate:
   (a) a safe condition exists
   (b) an unsafe condition exists
   (c) an abnormal condition exists.

6. Flood lighting in the flight compartment is normally from:
   (a) strobe lights
   (b) incandescent lamps and/or fluorescent tubes
   (c) position lights.

7. Sealed quartz or glass tubes filled with xenon gas are called:
   (a) fluorescent tubes
   (b) LEDs
   (c) strobe lights.

8. The starboard wing tip navigation light has the following colour and divergence:
   (a) red and 110 degrees
   (b) green and 110 degrees
   (c) white and ±70 degrees.

9. Anti-collision lights are used in conjunction with the:
   (a) master warning lights
   (b) wing inspection lights
   (c) navigation lights.

10. Red warning lights in the instrument panel indicate the existence of:
    (a) unsafe conditions
    (b) abnormal conditions
    (c) safe conditions.

11. Referring to Fig. 12.24, this device is a:
    (a) strobe/xenon tube
    (b) fluorescent tube
    (c) LED.

12. Referring to Fig.12.25, the diode is required to:
    (a) provide dimming
    (b) ensure the correct light is illuminated
    (c) ensure both lights are illuminated.

13. Referring to Fig.12.26, this illustrates the angular coverage for:
    (a) wing illumination lights
    (b) rotating beacons
    (c) navigation (or position) lights.

Figure 12.24   See Question 11

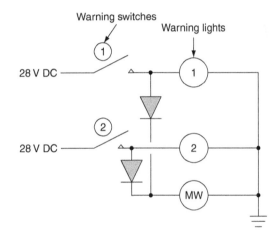

Figure 12.25   See Question 12

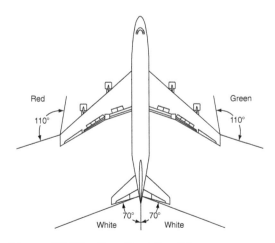

Figure 12.26   See Question 13

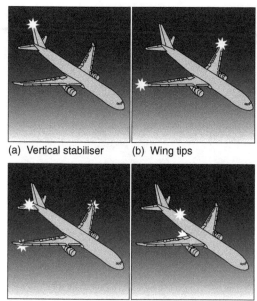

(a) Vertical stabiliser   (b) Wing tips

(c) Tail/lower wing surface (d) Fuselage

Figure 12.27   See Question 14

Figure 12.28   See Question 16

14. Referring to Fig. 12.27, these positions are for:
    (a) navigation lights
    (b) wing inspection lights
    (c) strobe lights.

15. Master caution and warning lights are located
    on the:
    (a) lower instrument panel
    (b) upper instrument panel
    (c) overhead panel.

16. Referring to Fig. 12.28, this light assembly
    comprises:
    (a) LEDs
    (b) strobes
    (c) incandescent elements.

Cabin systems

Passenger transport and business aircraft are fitted with a range of cabin systems and equipment for passenger safety, convenience and entertainment. Typical applications for this equipment include lighting, audio and visual systems. Cabin lighting is used for the safety and comfort of passengers; this was described in Chapter 12. Audio systems include the passenger address system used by the flight or cabin crew to give out safety announcements and other flight information. These announcements are made from hand-held microphones and are heard over loudspeakers in the cabin and passenger headsets. The same system can be used to play automatic sound tracks; this is often used for announcements in foreign languages, or to play background music during boarding and disembarkation. Audio systems are also available at the individual seat positions.

A range of galley equipment is installed on business and passenger aircraft. The nature of this equipment depends on the size and role of the aircraft. Air conditioning is provided in passenger aircraft for the comfort of passengers; pressurization is required for flying at high altitudes. Both these systems have electrical and electronic interfaces and control functions. Airstairs allow passengers, flight crew and ground personnel to board or depart the aircraft without the need for a mobile staircase or access to a terminal. This chapter describes the many types of systems and equipment used for passenger safety, convenience and entertainment.

13.1 Passenger address system

The passenger address (PA) system is primarily a safety system that provides passengers with voice announcements and chime signals from the flight or cabin crew; its secondary purpose is for the audio entertainment system. The crew make these voice announcements via the **interphone system** (see *Aircraft Communication and Navigation Systems*, a related book in the series). The PA amplifier provides a level of **sidetone** to the crew's handset or headsets during voice announcements. Sidetone is the technique of feeding back a small amount of sound from the mouthpiece and introducing this at low level back into the earpiece of the same handset, acting as feedback to confirm that the handset or headsets are functional.

The PA system audio outputs are transmitted through to speakers located in the cabin, see Fig. 13.1; typical locations include:

- passenger service units (PSUs)
- galleys
- washrooms
- cabin crew areas.

PA voice announcements are integrated with the passenger entertainment system so that safety announcements can be made over and above entertainment channels.

A typical PA system is controlled by a selector panel (Fig. 13.2) located at cabin crew stations; an amplifier makes chime sounds in response to discrete signals from the cabin interphone system; examples of these chimes include:

- a single low chime to indicate when the 'no smoking' and 'fasten seat belt' lights have been switched on
- a single high/low chime for calls between cabin crew stations
- three high/low chimes when the flight crew needs to gain the attention of the cabin crew
- a single high chime when a passenger needs to gain the attention of the cabin crew.

Test your understanding 13.1

What is the primary purpose of a cabin PA system?

PA systems are often integrated with video systems used for giving cabin safety information to

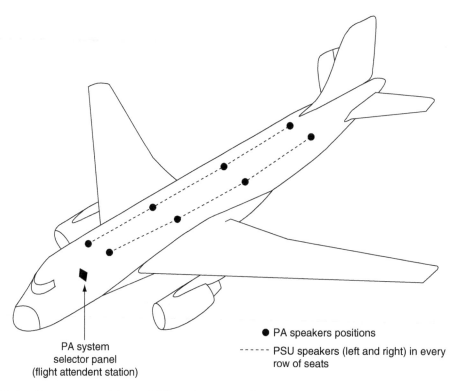

Figure 13.1 Passenger address (PA) system overview

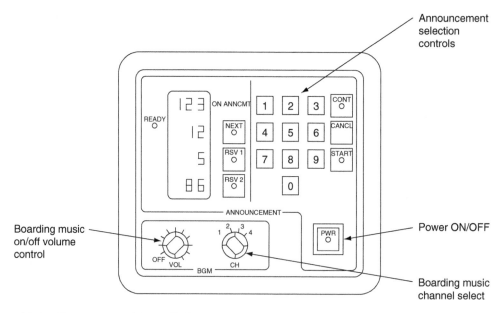

Figure 13.2 Passenger address (PA) selector panel

passengers. A complete **cabin management system** includes:

- crew announcements
- recorded announcements
- video system audio
- boarding music
- chimes.

13.2 Galley equipment

Galley equipment is a major consideration for the design of an aircraft's electrical power system since it can represent a high proportion of the total load requirements. The nature of the installed equipment depends on the size and role of aircraft, e.g. if it is operating short or long routes, the type of cabin classification (economy, business, first class, etc.). In the 1930s, the Douglas Aircraft DC-3 was the first aircraft to have a purpose-built galley area for the preparation of food and beverages. Typical galley equipment on a modern aircraft now includes:

- beverage makers
- ovens/cookers
- refrigeration units.

Typical power requirements for galley equipment on a typical narrow body aircraft are given in Table 13.1.

Power from the main distribution system is supplied to a dedicated galley bus which also contains its own protection circuits; galley power is the first non-essential load to be shed.

13.3 In-flight entertainment (IFE)

This was initially introduced onto aircraft to break the boredom of long flight times; IFE is now part of an airline's image and forms part of its competitive edge. There is now a vast range of cabin electronic equipment that is used for passenger safety, convenience and entertainment.

13.3.1 Overview

The first ever in-flight 'movie' was shown in 1925. These early systems were traditional projectors and screens. This was followed in the 1960s by centralized television screens placed at the front of each cabin. This concept was further developed into **video projectors** suspended on the aircraft ceiling and large screens on the bulkheads. Smaller business aircraft are installed with video screens fitted into the front and/or rear bulkheads; larger aircraft are installed with monitors integrated within the overhead panels or, more usually, into the back of the seat headrest.

Satellite-based communication is available with an individual transmitter/receiver and handset for business aircraft through to multiple access systems on larger passenger aircraft. The rapid development of broadband communications and satellite systems has also facilitated access to the Internet via laptop computers at an individual seat position. The use of mobile phones is not permitted by many regulatory authorities due to the potential interference with aircraft systems; technology is being developed to allow their use on aircraft.

Audio entertainment systems installed on aircraft range from a single source, e.g. a digital versatile disc (DVD) player on private aircraft through to multichannel systems with multiple channels on larger aircraft. Passengers are able to select from a wide range of music channels, news bulletins, current affairs, documentaries and comedy channels, etc. Audio channels are played through headphones that plug into an individual seat position. **Video entertainment** is traditionally provided via a large video screen at the

Table 13.1 Typical power requirements for galley equipment on a typical narrow body aircraft

Product	QPA*	Power supply	Rating	Protection
Beverage maker	10	200 V AC three-phase 400 Hz	2750 W	7A
Ovens	5	200 V AC three-phase 400 Hz	1000 W	5A
Refrigeration units	5	200 V AC three-phase 400 Hz	1000 W	5A

*QPA quantity per aircraft

front of each cabin zone, together with smaller monitors suspended from the ceiling situated every few rows. Audio is supplied via the same headphones used for audio entertainment. In addition to entertainment, these screens are also used to broadcast safety information to passengers.

Personal video screens (PVSs) for every passenger are becoming the norm, providing passengers with a selection of audio and visual channels. These screens are usually located in the seat backs or stowed in the armrests during take-off and landings. Some airlines also provide news and current affairs programmes, which are pre-recorded prior to the flight and made available to the aircraft before departure. PVS are operated via an in-flight management system which stores pre-recorded channels on a central server, and streams them to the individual seat positions during flight. Some airlines also provide video games to the individual seat positions. **Audio-video on demand** (AVOD) entertainment enables passengers to pause, rewind, fast-forward or stop a programme. AVOD also allows the passengers to choose between an assortment of audio-visual programmes stored in the aircraft computer system.

Passengers can obtain real-time flight information on a video channel at their seat position or on cabin video screens via **moving-map systems**. The typical display is an electronic screen, typically using **liquid crystal display (LCD)** technology that illustrates the aircraft current position and direction of travel; the system can also display information such as aircraft altitude, airspeed, distance to destination, distance from origination and local time. Moving-map system information is derived from the aircraft's navigation systems. (LCD technology is covered in more detail in *Aircraft Digital Electronic and Computer Systems*, a related book in the series.)

Test your understanding 13.2

What does audio-video on demand (AVOD) entertainment provide for passengers?

13.3.2 Typical product specifications

Companies that offer a complete range of IFE products include Flight Display Systems in the USA. Products include DVD/CD/MP3 players that provide high-resolution images with high-quality sound; high-definition (HD) LCD screens from 5 to 42 inches in size; moving-map systems incorporate an internal database that interfaces with the aircraft's navigation system to display:

- custom waypoints
- takeoff, landing and owner interest sites
- standard GPS position coordinates
- corporate logo
- personalized cabin greeting
- expanded city database
- regional flight information.

Another popular addition to the range of IFE accessories is the very small high-resolution colour camera that is typically mounted in the flight compartment headliner. This provides an NTSC or PAL video output to the cabin entertainment system, enabling passengers to see an outside view from the pilot's perspective on the cabin monitors. Typical IFE products are shown in Figure 13.3.

13.3.3 IFE system safety and regulation

One major consideration for designing and installing IFE systems is the system's safety. The primary consideration is the additional wiring required by the systems; insulation breakdown and arcing are potential sources of toxic fumes, overheating and fire. To address these potential issues, the IFE system is typically isolated from the aircraft's main electrical systems; many installations are fitted with a master switch in the flight compartment. Electrical equipment must not alter the safety or functionality of the aircraft as a result of a failure. IFE systems must be independent from the aircraft's main power supplies and digital systems. Protection of the aircraft power supplies and digital data links is required to mitigate against IFE failures and maintain the integrity of the aircraft's systems.

13.4 Satellite communications

Some aircraft are fitted with intranet-type **data communication** systems that provide full access to the Internet and email via satellite. A **passenger telephone system** is installed on some aircraft to allow telephone calls from passengers to the ground; these are used for outgoing calls on passenger aircraft, and two-way calls on private aircraft. Data communication is via the **Iridium** satellite communication system and allows passengers to connect to live **Internet**

(a)

(b)

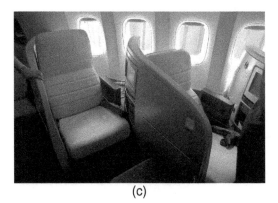

(c)

Figure 13.3 IFE products (a) Seat back display and controls (Andy Mabbett, 2013) (b) Economy class seats (Eric Salard, 2014) (c) Business class seats (Phillip Capper, 2011)

from the individual IFE units or their laptops. Next-generation IFE systems are being introduced to incorporate live data/TV reception and on-demand capabilities.

The Iridium system is a satellite-based, personal communications network providing global voice and data features. Iridium is a privately owned company based in the USA. The satellite communications system is illustrated in Fig. 13.4 and comprises three principal components:

- satellite network
- ground network (based on gateways)
- subscriber products (including phones and data modems).

The Iridium network allows voice and data messages to be routed anywhere in the world. The system operates between user and satellite in the L-band, 1616–1626.5 MHz. Voice and data messages are relayed from one satellite to another until they reach the satellite above the Iridium handset or terminal; the signal is then relayed back to a gateway. When an Iridium customer places a call from a handset or terminal, it connects to the nearest satellite, and is relayed among satellites around the globe to whatever satellite is above the appropriate gateway; this downlinks the call and transfers it to the global public voice network or Internet so that it reaches the recipient. Users can access the network via **aircraft earth stations** (AES) or Iridium subscriber units (ISU).

13.4.1 Satellite communication network

The **satellite communication network** comprises a constellation of 66 active satellites in a near-polar orbit at an altitude of 485 miles (780 km). The satellites fly in formation in six orbital planes, evenly spaced around the planet, each with 11 satellites equally spaced apart from each other in that orbital plane. A single satellite orbits the earth once every 100 minutes, travelling at a rate of 16,832 miles per hour; the time taken from horizon to horizon is approximately ten minutes. As a satellite moves out of view, the subscriber's call is seamlessly handed over to the next satellite coming into view.

Key point

The Iridium network allows voice and data messages to be routed anywhere in the world.

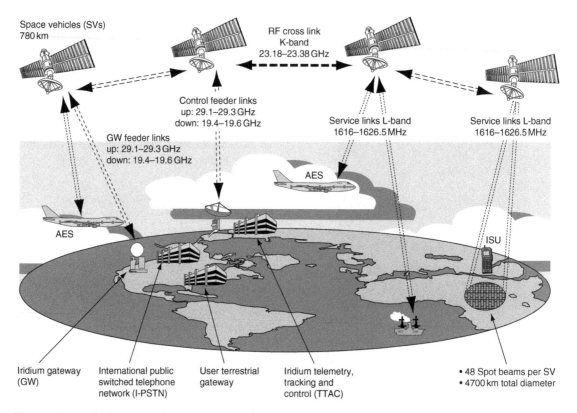

Figure 13.4   Iridium satellite communication system

Since Iridium is a **low earth orbit** (LEO) satellite system, voice delays are minimal. Communication systems using geostationary earth orbits (GEOs) have satellites located 22,300 miles above the equator. As a result, latency can be quite high, causing the users to have to wait for each other to finish. GEOs are also largely ineffective in more northern or southern latitudes. The curvature of the earth disrupts message transmission when attempted at the edge of a GEO satellite's footprint. (Global positioning system (GPS) satellites used for navigation are in 20,200 km orbits.) A comparison of LEO, GPS and GEO orbits is illustrated in Fig. 13.5.

Each Iridium satellite is cross-linked to four other satellites – two satellites in the same orbital plane and two in an adjacent plane. These links create a dynamic network in space – calls are routed directly between satellites without reference to the ground, creating a highly secure and reliable connection. Cross-links make the Iridium system particularly impervious to natural disasters – such as hurricanes, tsunamis and earthquakes – that can damage ground-based wireless towers.

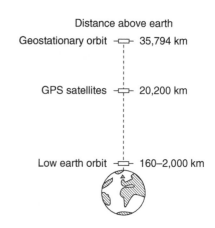

Figure 13.5   Comparison of satellite orbits

Key point

Satellite communication systems use a low earth orbit to minimize voice delays.

13.4.2 Iridium ground network

The Iridium **ground network** comprises the system control segment and telephony gateways used to connect into the terrestrial telephone system. With centralized management of the Iridium network, the system control segment supplies global operational support and control services for the satellite constellation, delivers satellite tracking data to the gateways, and controls the termination of Iridium messaging services. The system control segment consists of three primary components:

- four telemetry tracking and command/control (TTAC) stations
- the operational support network
- the satellite network operation center (SNOC).

Ku-band feeder links and cross-links throughout the satellite constellation supply the connections among the system control segment, the satellites and the gateways (ku-band is a section of the electromagnetic spectrum in the microwave range of frequencies between 12 and 18 GHz). **Telephony gateways** are the ground-based antennas and electronics that provide voice and data services, messaging, prepaid and postpaid billing services, as well as other customer services. The gateways are responsible for the support and management of mobile subscribers and the interconnection of the Iridium network to the terrestrial phone system. Gateways also provide management functions for their own network elements and links.

13.5 Multiplexing

Increasing the amount of IFE and communication system choices to each passenger position has the potential to rapidly increase the amount of wiring on the aircraft; this would make IFE very costly. One way of reducing the amount of wiring to a seat position is via **multiplexing**; this technology provides a means of selecting data from one of several sources. Equivalent circuits (depicted as switches for illustration purposes) of some common types of multiplexer are shown in Fig. 13.6.

The single two-way multiplexer (Fig. 13.6(a)) is equivalent to a simple single pole double throw (SPDT) changeover switch. The dual two-way multiplexer (Fig. 13.6(b)) performs the same function but two independent circuits are controlled from the same select signal. A single four-way multiplexer (Fig. 13.6(c)) requires two digital selector inputs (A and B) to place the switch in its four different states. These are coded in a simple truth table format with four

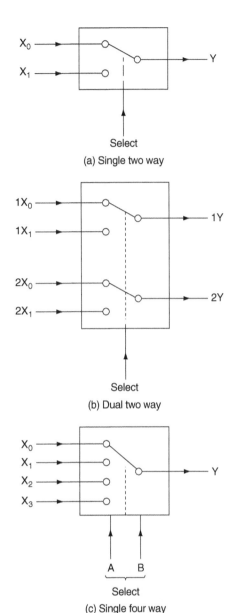

Figure 13.6 Multiplexer schematics

possible outcomes, or selectable switch positions. (The subject of multiplexing is covered in more detail in *Aircraft Digital Electronic and Computer Systems*, a companion volume.)

Test your understanding 13.3

What are the advantages of using multiplexing for IFE?

13.6 Fibre optics

Another technology adopted for IFE is based on **fibre optics**. This technology has been widely used as a transmission medium for ground-based data communications and in local area networks (LANs) for many years and they are now being introduced into passenger aircraft to satisfy the need for wideband networked avionic and cabin entertainment systems. Compared with copper wiring, optical fibres have a number of advantages:

- reduced weight, compact size
- immunity to electromagnet interference (EMI)
- exceptionally wide bandwidth
- significantly reduced noise and cross-talk
- relatively low values of attenuation within the medium
- high reliability coupled with long operational life
- electrical isolation and freedom from earth/ground loops.

These advantages mean that optical fibres are ideally suited as a replacement for conventional copper network cabling. The technology is relatively new in the civil aircraft industry and introduces new challenges for those involved with aircraft operation and maintenance. There are a number of disadvantages for optical fibres; these include:

- industry resistance to the introduction of new technology
- need for a high degree of precision when fitting cables and connectors
- concerns about the mechanical strength of fibres
- cable bends need to have a sufficiently large radius to minimize losses and damage.

The advantages are now outweighing the disadvantages and systems based on optical fibres are being installed on many aircraft types. A simple one-way (simplex) fibre optic data link is shown in Fig. 13.7. The optical transmitter consists of an infrared light-emitting diode (LED) or low-power semiconductor laser diode coupled directly to the optical fibre. The diode is supplied with pulses of current from a bus interface.

These pulses of current produce equivalent pulses of light that travel along the core of the fibre until they reach the optical receiver unit. The optical receiver unit consists of a photodiode or phototransistor that passes a relatively large current when illuminated and negligible current when not. The pulses of current at the transmitting end (logic inputs) are thus replicated at the receiving end (as logic outputs). **Cladding** is a layer of material with a lower refractive index than the core, and confines the optical signal inside the core.

13.6.1 Construction

The construction of a typical fibre optic cable is shown in Fig. 13.8; in this example, the cable consists of:

1. an outer jacket
2. aramid yarn (strength member)
3. separator tape (polyester)
4. two filler strands
5. five optical fibres.

Key maintenance point

Fibre optic cable bends need to have a sufficiently large radius to minimize signal losses and physical damage.

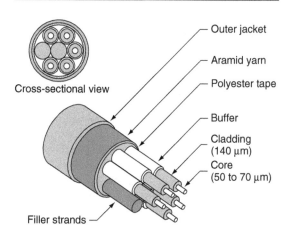

Cross-sectional view

- Outer jacket
- Aramid yarn
- Polyester tape
- Buffer
- Cladding (140 μm)
- Core (50 to 70 μm)
- Filler strands

Figure 13.8   (Simplex) fibre optic data link

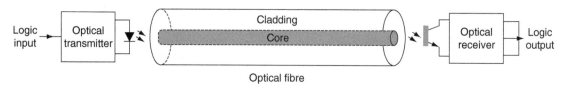

Optical fibre

Figure 13.7   Fibre optic cable connector

A protective **buffer** covers each fibre and protects it during manufacture, increases mechanical strength and diameter in order to make handling and assembly easier. The buffers are coded in order to identify the fibres using colours (blue, red, green, yellow and white). The filler strands are made from polyester and are approximately 0.035 inches in diameter. A polyester separator tape covers the group of five fibres and two filler strands. This tape is manufactured from low-friction polyester and it serves to make the cable more flexible. A layer of woven aramid (or Kevlar) yarn provides added mechanical strength and protection for the cable assembly. The outer thermoplastic jacket (usually purple in colour) is fitted to prevent moisture ingress and also to provide insulation. A typical fibre optic cable connector arrangement is shown in Fig. 13.9.

13.6.2 Connectors

Each connector has alignment keys on the plug and matching alignment grooves on the receptacle. These are used to accurately align the connector optical components; the guide pins in the plug fit into cavities in the receptacle when the plug and receptacle connect. In order to ensure that the connector is not over-tightened (which may cause damage to the fibres) the pins of the plug are designed to provide a buffer stop against the bottom of the cavities in the receptacle. Plugs and receptacles have ceramic contacts that are designed to make physical contact when properly connected.

The light signal passes through holes in the end of the ceramic contacts when they are in direct physical contact with each other. The coupling nut on the plug barrel has a yellow band whilst the receptacle barrel has a red and a yellow band. A correct connection is made when the red band on the receptacle is at least 50 per cent covered by the coupling nut. This position indicates an effective connection in which the optical fibres in the plug are aligned end-to-end with the

fibre in the receptacle. Three start threads on the plug and receptacle ensure a straight start when they join. The recessed receptacle components prevent damage from the plug if it strikes the receptacle at an angle. The plug and receptacle are automatically sealed in order to prevent the ingress of moisture, dust and other contamination. Colour-coded bands ensure that the plug and receptacle are fully mated. (The subject of optical fibres is covered in more detail in *Aircraft Digital Electronic and Computer Systems*, a companion volume.)

Key point

The optical fibre receiver unit consists of a photodiode or phototransistor that passes a relatively large current when illuminated and negligible current when not illuminated.

13.7 Air conditioning

Air conditioning is provided in passenger aircraft for the comfort of passengers. The cabin on larger aircraft is divided into passenger compartment locations, or temperature control zones, see Fig. 13.10. Air conditioning, and the **environmental control system** (ECS), normally derive high-pressure air from the compressor stage of each turbine engine. The temperature and pressure of this bleed air varies, depending upon rotational speed of the engine. A **pressure-regulating shutoff valve** (PRSOV) restricts the flow as necessary to maintain the desired pressure for the ECS.

13.7.1 Environmental control system

To increase efficiency of the ECS, air is normally bled from two or three positions on the engine. Aircraft types vary, but the principles of air conditioning systems comprise five salient features:

- air supply
- heating
- cooling
- temperature control
- distribution.

The air conditioning system is based on an air cycle machine (ACM) cooling device as illustrated in

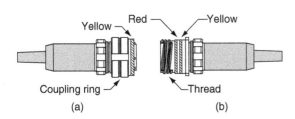

Figure 13.9 Fibre optic cable

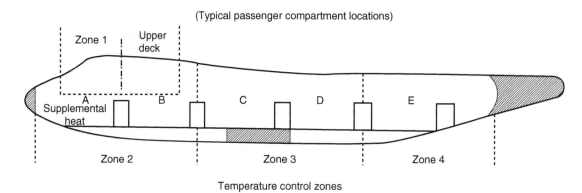

Figure 13.10   Air conditioning: passenger cabin zones

Fig. 13.11. The air conditioning packs (**A/C packs**) are located in varying places on aircraft, including:

- between the two wings in the lower fuselage
- in the rear fuselage tail section
- in the front of the aircraft beneath the flight deck.

Engine bleed air, with temperatures in the order of 150 and 200°C and a pressure of between 30–35 pounds per square inch (psi), is directed into a primary heat exchanger. External ram air (at ambient temperature and pressure) is the cooling medium for this air-to-air heat exchanger. The cooled bleed air then enters the centrifugal compressor of the air cycle machine (ACM). This compression heats the air (the maximum air temperature at this point is about 250°C) and it is directed into the secondary heat exchanger, which again uses ram air as the coolant. Pre-cooling through the primary heat exchanger increases the efficiency of the ACM by reducing the temperature of the air entering the compressor; less work is required to compress a given air mass (the energy required to compress a gas by a given ratio increases with higher temperature of the incoming air).

At this stage, the temperature of the heat exchanger air is higher than the outside air temperature; the temperature is sensed by an RTD and this is displayed on the air conditioning control panel and/or used as part of a control system input. The compressed, cooled air is then directed into the expansion turbine of the ACM; this extracts work from the air as it expands, cooling it to between −20°C and −30°C. (The ACM can cool the air to less than 0°C even when the aircraft is on the ground in a high ambient temperature.) The work extracted by the turbine drives a shaft to turn the

ACM's centrifugal compressor, together with a ram air inlet fan that draws in the external air during ground running. A motorized bypass valve controls the ratio of air being directed into the turbine. An electrically driven ram air fan within the system provides air flow across the heat exchangers when the aircraft is on the ground. To assist ram air recovery, some aircraft use modulating vanes on the ram air exhaust. Power for the air conditioning pack is obtained by reducing the pressure of the incoming bleed air relative to the cooled air output of the system.

Test your understanding 13.4

What type of sensor is used for regulation of cabin air temperature?

With the air now cooled, its water vapour condenses. To remove this, the moist air output of the expansion turbine is passed through a water separator; this uses centrifugal force to eject the water particles into a coalescing bag that absorbs the moisture. This condensate is sometimes fed back into the ram air entering the secondary heat exchanger to improve its performance. The cool dry air is now combined in a mixing chamber with a small amount of engine bleed air; the amount of trimming air mixed with cooled air is modulated to achieve the desired cabin air temperature before the air is ready for supply into the cabin. An RTD is installed to monitor the ACM outlet temperature; airflow into the cabin is monitored by a flow sensor.

The A/C pack outlet air for use in the cabin is mixed with filtered air from re-circulation fans, and

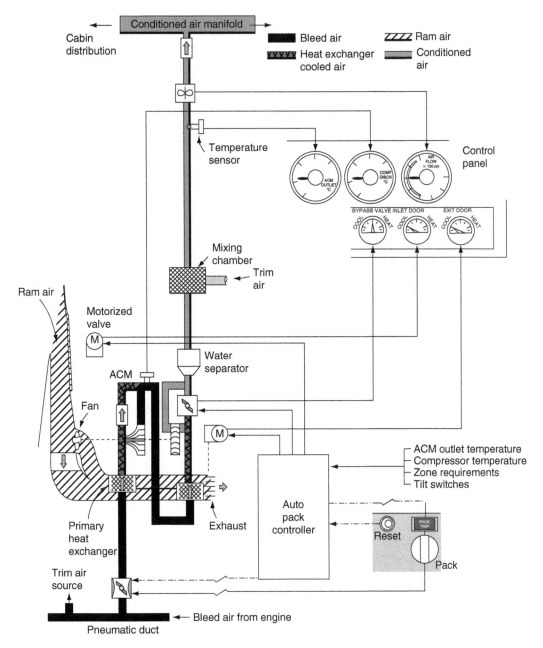

Figure 13.11   Air conditioning system

then fed into the mixing manifold. On modern passenger aircraft, the airflow into the cabin is approximately 50% bleed air and 50% filtered air. Control of the air conditioning system on a Boeing 757 aircraft is from an overhead control panel, see Fig. 13.12.

Key point

To assist ram air recovery, some aircraft use modulating vanes on the ram air exhaust.

Figure 13.12   Air conditioning system control (right/centre of overhead panel)

13.7.2 Ventilation

A **re-circulation fan** is used to re-circulate filtered cabin air back into the cabin to reduce bleed air requirements. Up to 50% of the cabin air can be re-circulated for passenger comfort. The fan will switch off if either A/C pack is in high flow, giving a net reduction in the ventilation rate. Optimum cooling and reduced bleed air demand (hence reduced fuel consumption) is achieved with a combination of air re-circulation and automatic operation of the A/C packs. The ventilation rate is increased on larger aircraft with additional re-circulation fans for comfort levels.

Some aircraft are installed with a **gasper fan**; this is an electric fan designed to increase pressure in the outlets above passenger seats. The gasper fan is typically used when there is a:

- low supply of air pressure
- high cold air demand
- high ambient temperature (on the ground on a hot day).

When the A/C packs are off, the gasper fan draws cabin air into the distribution ducts, into the main air manifold and mixing chamber where it is then blown (albeit not chilled) into the cabin.

13.7.3 Equipment cooling

Although not part of the cabin system for the comfort of passengers, some aircraft are installed with equipment cooling that takes air from the cabin system. Equipment cooling can be used for:

- flight compartment panels
- display units
- circuit-breaker panels
- electrical and electronic (E & E) bay.

Replacing the warm air generated by equipment with cool air from the cabin is achieved with dedicated electric fans. On the ground, the air is directed overboard through the flow control or exhaust valves. During flight, the warm air generated by equipment is sometimes used for heating the cargo holds by exhausting air around their walls.

13.8 Pressurization

In addition to temperature control, aircraft flying at high altitudes need to be pressurized since the amount of oxygen in the atmosphere decreases with altitude. The effects of insufficient oxygen on people can begin

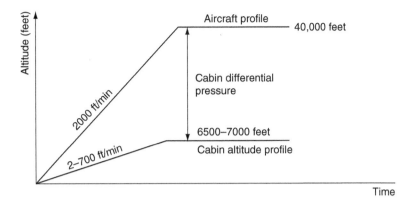

Figure 13.13   Cabin pressurization/altitude chart

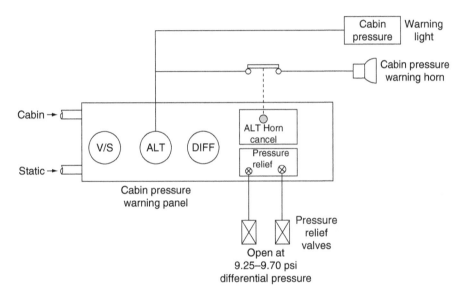

Figure 13.14   Cabin pressurization warning

as low as 1500 m (5000 ft) above sea level, although most passengers can tolerate altitudes of 2500 m (8000 ft) without ill effect. At this altitude, there is approximately 25% less oxygen compared to sea level. Flight operations above 3000 m (10,000 ft) generally require supplemental oxygen; this is achieved through pressurization on passenger aircraft.

The relationship between cabin pressure and atmospheric pressure is illustrated in Fig. 13.13. Cabin pressure is controlled automatically; this is managed by the crew via a control panel as illustrated in Fig. 13.12. Pressurization is monitored and

warnings provided in the event of a malfunction. Fig. 13.14 illustrates a typical warning system schematic; this illuminates a warning light on the master caution panel and sounds a horn in the event of reduced cabin altitude.

Engine bleed air is supplied into the cabin and then allowed to pass out of the fuselage via **outflow valves** (OFVs). Most aircraft have a single OFV located near the bottom aft end of the fuselage; some larger aircraft have two. By modulating the position of the outflow valve(s), the pressurization in the cabin can be maintained higher than atmospheric pressure. Modern commercial aircraft have a dual channel electronic controller for maintaining pressurization, with a manual back-up system. These systems maintain cabin air pressure at the equivalent to 2500 m (8000 ft) at high cruising altitudes.

A positive pressure relief valve opens in the event of excessive pressure in the cabin; this protects the aircraft structure from excessive loading. The maximum pressure differential between the cabin and atmospheric pressure is between 7.5 and 8 psi. This is the equivalent of approximately 10,000 feet; if the cabin were maintained at sea level pressure and then operated at cruise altitude, the differential would mean the structural life of the aircraft would be reduced.

Key point

Bleeding air from the engine comes at the expense of overall aircraft efficiency. Some aircraft such as the Boeing 787 are now using **electrically driven compressors** to provide pressurization.

A pressurization failure above 10,000 ft results in an emergency situation; the cabin pressure might be 10 psi, while the atmospheric pressure is only 2 psi. The pilot must make an emergency descent and activate oxygen masks for everyone aboard. In most passenger aircraft, oxygen masks are automatically deployed if the cabin pressure drops below an equivalent cabin altitude of 14,000 feet. Gradual decompression is dangerous to all on board since it may not be detected. Rapid decompression causes the lungs to decompress faster than the cabin. Explosive decompression is when cabin pressure reduces faster than the lungs can decompress (less than 0.5 seconds). Rapid decompression of commercial aircraft is extremely rare, but the results are life-threatening. Furthermore, the air temperature will rapidly fall due to the expansion of the cabin air, potentially resulting in frostbite.

Key point

Aircraft cabin pressure is commonly referred to as a 'cabin altitude', typically 8000 feet or less, therefore a lower cabin altitude relates to a higher pressure.

Key point

The cargo compartment is normally pressurized to the same level as the cabin; the temperature in this compartment may be controllable depending on the nature of cargo being carried.

13.9 Airstairs

These are located in either forward or aft cabin locations, and can be raised or lowered while the aircraft is on the ground. Airstairs allow passengers, flight crew and ground personnel to board or depart the aircraft without the need for a mobile staircase or access to a terminal.

On forward door positions, the stair folds and stows directly into the floor, beneath the door. This type of airstair is found on many short-range aircraft to provide operational flexibility, although the stairs are quite heavy; many airlines remove this system to reduce aircraft weight. Ventral airstairs are fitted on most tail-engined airliners, and are incorporated as ramps which lower from the rear fuselage.

They can be operated from either internal or external control panels; both panels typically have normal and standby systems. Normal operation requires both AC and DC power, standby only requires DC power. External standby system power typically comes directly from the battery bus and so does not require the battery switch to be on. Airstairs should not be operated more than is necessary to prevent the motors from overloading; their typical duty cycle is three consecutive cycles of normal system operation within a 20 minute period.

13.10 Multiple choice questions

1. Multiplexing is a technique used for:
   (a) increasing the amount of IFE wiring to a seat position
   (b) reducing the amount of IFE wiring to a seat position
   (c) increased immunity to electromagnet interference (EMI).

2. Fibre optic cable bends need to have a sufficiently large radius to:
   (a) minimize losses and damage
   (b) maximize immunity to electromagnet interference (EMI)

(c) accurately align the connector optical components.

3. ACM outlet temperature is measured using:
(a) a Bourdon tube
(b) an RTD
(c) a thermocouple.

4. The Iridium network allows voice and data messages to be routed:
(a) anywhere in the world
(b) between the flight crew and cabin crew
(c) via a fibre optic network.

5. The passenger address (PA) system is primarily a safety system that provides passengers with:
(a) in-flight entertainment
(b) reduced amount of IFE wiring to a seat position
(c) voice announcements and chime signals.

6. To assist ram air recovery, some aircraft use:
(a) modulating vanes on the ram air exhaust
(b) modulating vanes on the ram air inlet
(c) modulating vanes on the ACM outlet.

7. Satellite communication systems use a low earth orbit to:
(a) provide greater coverage
(b) maintain a geostationary position
(c) minimize voice delays.

8. Audio-video on demand (AVOD) entertainment enables passengers to:
(a) pause, rewind, fast-forward or stop a programme
(b) make phone calls via satellite communication
(c) ignore PA system voice announcements and chime signals.

9. Supplemental oxygen is generally required for flight operations above:
(a) 3000 ft
(b) 10,000 ft
(c) sea level.

10. The optical fibre receiver unit consists of a photodiode or phototransistor that passes a:
(a) relatively large current when illuminated and negligible current when not illuminated
(b) relatively small current when illuminated and high current when not illuminated
(c) relatively small current when illuminated and low current when not illuminated.

11. The gasper fan is used in cabin ventilation systems to:
(a) increase air pressure in the air outlets
(b) re-circulate filtered cabin air
(c) assist with ram air recovery.

12. Cabin altitude is typically
(a) cruising altitude
(b) sea level
(c) between 6–7000 feet.

13. The cargo compartment is normally pressurized to be:
(a) equal to cabin pressure
(b) above cabin pressure
(c) below cabin pressure.

14. Referring to Fig. 13.15, cladding is used in fibre optic cables to:
(a) protect the fibre
(b) prevent moisture ingress
(c) guide the light signal.

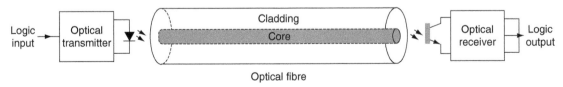

Optical fibre

Figure 13.15 See Question 14

15. Referring to Fig. 13.16, the iridium system transfers messages around the world:
(a) directly between satellites
(b) via gateways on the earth's surface
(c) directly between users.

16. The iridium system satellite orbits are:
(a) random
(b) equatorial
(c) polar.

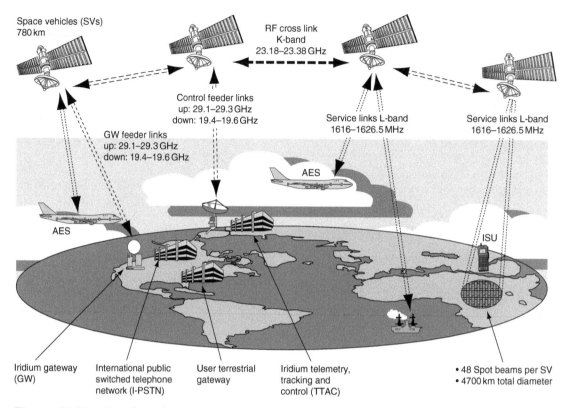

Space vehicles (SVs)
780 km

RF cross link
K-band
23.18–23.38 GHz

Control feeder links
up: 29.1–29.3 GHz
down: 19.4–19.6 GHz

GW feeder links
up: 29.1–29.3 GHz
down: 19.4–19.6 GHz

Service links L-band
1616–1626.5 MHz

Service links L-band
1616–1626.5 MHz

AES

AES

ISU

Iridium gateway
(GW)

International public
switched telephone
network (I-PSTN)

User terrestrial
gateway

Iridium telemetry,
tracking and
control (TTAC)

• 48 Spot beams per SV
• 4700 km total diameter

Figure 13.16   See Question 15

Airframe monitoring, control and indicating systems

Various sensors are needed for the monitoring and control of airframe systems. The indications and/or control circuit need to be informed of the position of a particular feature on the aircraft; this is achieved by a position sensor. Broadly speaking, the sensors can be considered as detecting one of two states (or 'conditions'), or a variable position. Two state conditions include: landing gear (up or down) or cabin doors (open or closed). Variable positions include: control surfaces and flap position. Two state position devices are detected by micro-switches and proximity sensors; variable position devices are detected by a variety of devices including synchros and variable resistors. This chapter reviews airframe systems such as landing gear control and indication, control surface position and indicating systems.

14.1 Landing gear

Three control and indicating system configurations are found on aircraft with retractable landing gear (or undercarriage):

- hydraulic control and operation, electrical indication of position
- electrical control and operation, electrical indication of position
- electro-hydraulic control and operation, electrical indication of position.

The landing gear comprises two or more wheels and shock absorbers, or **oleo legs**; these can be fixed in position or retractable. Large aircraft have hydraulic systems with electrical position sensors for gear up/down indications. General aviation (GA) aircraft normally have electrical control and operation, and electrical indication of position. Operation is via a series-wound split field electrical motor (Fig. 14.1) driving through to each undercarriage leg via a gearbox, torque tubes, cables and pulleys.

The 28 V DC power supply is connected to the **reversible** motor via a safety switch and the position selector switch. This switch operates the landing gear relay via up-lock and down-lock micro-switches. When the landing gear starts to retract, the down-lock switch contacts close. When the landing gear is fully retracted the up-lock switch contacts open, and this removes power from the motor. When the landing gear is selected down, the up-lock contacts close and the gear can be extended.

The landing gear safety switch senses when the aircraft is on the ground to prevent the landing gear from being retracted. Some aircraft have a **micro-switch** (or **proximity-sensor**) attached to the oleo leg; when the aircraft is on the ground, the oleo leg is compressed from the weight of the aircraft; this closes the switch contacts. The switch (or sensor) is now being used to detect the **weight on wheels (WoWs)**; this is sometimes referred to as a **squat** switch. When the aircraft takes off, the oleo leg extends when the aircraft weight is transferred to the wings and the WoW, or squat switch, opens its contacts. (Note that the switch can actually be configured as normally open or normally closed on the ground depending on the aircraft design.) WoW or squat switch contacts are also used by other systems that need to be set into air or ground modes, e.g. to prevent the pitot probes and ice detectors from being heated on the ground.

Key point

Two state conditions include landing gear (up or down) or cabin doors (open or closed).

Landing gear position indication is derived from simple micro-switches in two locations: gear up and gear down. Switches are operated by a cam or lever and this completes the circuit. The quantity of lights depends on the aircraft manufacturer and certification requirements of the aircraft type. The simple system shows if the nose and main gear are up or down with a single indication; it is more usual to

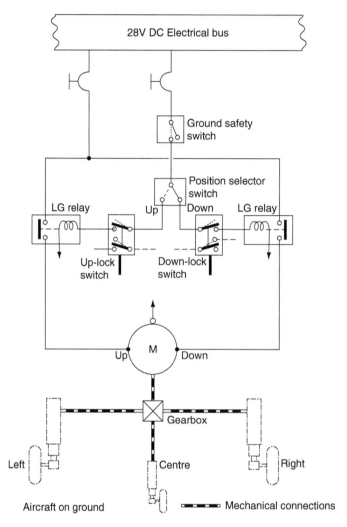

Figure 14.1 Landing gear control system

have an indication of each gear leg position. Landing gear indications sometimes include an audible warning (klaxon or horn) when the throttles are retarded and the gear is not locked down. The system configuration in widespread use has the following indications:

- gear down and locked (three green lights)
- gear up and locked (all three lights out)
- gear in transit (three red lights).

Figure 14.2(a) illustrates a typical landing gear position indication system; note that this system includes additional green lights that can be selected

in the event of lamp failure. The circuit is drawn with positions of switches for the aircraft on the ground.

With power applied, the three green lights are illuminated via the down-lock switch. When the gear is selected up, the down-lock switch opens, the green lights extinguish and the three red lights are turned on. When the gear is fully retracted and locked, the red lights are extinguished. When the gear is selected down, the up-lock switch closes, and three red lights come on. With the gear fully down and locked, the down-lock switch closes, the red lights are extinguished and the green lights are switched on. In the event that the throttles are closed (during an approach) and the gear is up, a warning horn will sound.

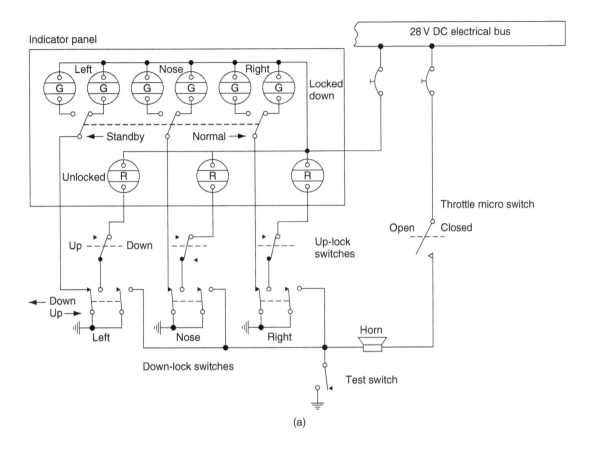

(a)

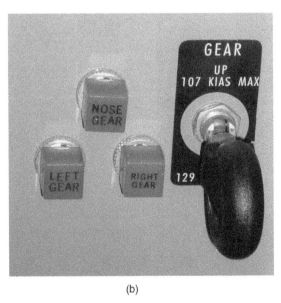

(b)

Figure 14.2 Landing gear position indication: (a) system schematic, (b) typical gear position selector and warning lights (general aviation)

Key point

Indications of landing gear fully down and locked are:

- red lights off
- green lights on.

Key point

When an electrically operated landing gear is fully retracted, the up-lock switch contacts open, thereby removing power from the motor

On smaller general aviation aircraft, the system is simplified by having three green lights for gear down and locked (Fig. 14.2(b)). In this example, the selector switch also has a placard stating the airspeeds for gear selection.

14.2 Trailing edge flaps

Three configurations are found on aircraft:

- hydraulic control and operation, electrical indication of position
- electrical control and operation, electrical indication of position
- electro-hydraulic control and operation, electrical indication of position.

The flap drive motor is a reversible DC motor that drives the flap control mechanism through a gearbox. GA aircraft have a simple three-position switch identified as OFF/UP/DOWN. The flaps on these aircraft are driven the full extent of travel until they operate up or down limit switches, see Fig. 14.3.

Selecting 'flaps up' applies power to the motor through the closed contacts of the up-limit switch; the motor operates and the flaps start to move. The down-limit switch changes over as shown by dotted lines, and the flaps travel to the fully retracted position; note that both the limit switches are closed during

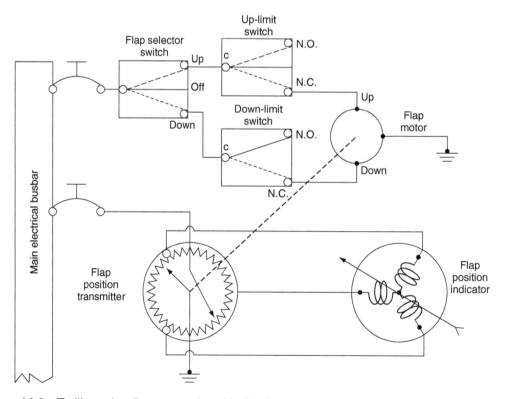

Figure 14.3 Trailing edge flaps control and indication

travel, and this allows the pilot to stop/start the flaps. Selecting 'flaps down' applies power through the down-limit switch and the motor runs in the opposite direction; the up-limit switch changes over to the position shown as solid lines in preparation for the next selection. The flaps continue to travel until they are fully extended and the down-limit switch is operated (solid line). This opens the supply to the motor. Selecting off at any time stops the motor at the given position.

The flap position indicator is an integral part of the system; a mechanical linkage is made between the flap control mechanism and position sensor. This varies the ratio of current in each of three coils in the indicator; flap position is displayed in degrees of flap movement. Flap position can be from a simple rheostat and ratiometer indication, or from a synchro system. Flaps on larger aircraft are normally driven by a hydraulic system; flap position is selected in specific settings referred to by simple up/down positions, or (on larger aircraft) by angular positions, see Fig. 14.4. The flap selection switch in this photograph is to the right of the pedestal and, in this example, are also referenced to the maximum airspeed for given flap settings.

Key point

An electrical flap drive system uses a reversible DC motor.

14.3 Control surfaces

Different aircraft types have a variety of control surface position indicators; the various flying control

Figure 14.4 Flap position selection (top right of photo)

surfaces on a large passenger aircraft are illustrated in Fig. 14.5. These control surfaces typically include:

- leading edge slats
- spoilers
- trailing edge flaps
- rudder(s)
- elevators
- ailerons.

The position of each control surface can be displayed by an analogue indicator (Fig. 14.6) or form part of an electronic display. Each of the control

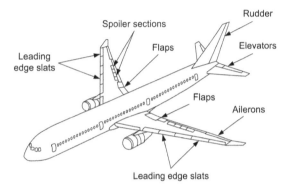

Figure 14.5 Flying control surfaces

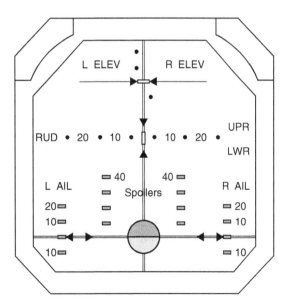

Figure 14.6 Control surface position indication

surfaces is fitted with a position sensor, typically a torque synchro.

Key point

Variable position features include control surface and flap position.

Key point

Two state position devices include micro-switches and proximity sensors.

14.4 Electronic indicating systems

The two systems used on the majority of passenger aircraft are EICAS and ECAM. The **Engine Indication and Crew Alerting System** (EICAS) is a Boeing developed system that provides all engine instrumentation and crew annunciations in an integrated format; this is covered in Chapter 10 (engine systems). The system used on Airbus aircraft is the **Electronic Centralized Aircraft Monitoring** (ECAM) system. The two systems operate on different philosophies; however their basic functions are to monitor aircraft systems and display relevant information to the pilots. Both systems produce warning, cautions and advisory messages that need to be evaluated by the crew; in certain cases, the system provides the procedures required to address the problem. Each colour display unit uses either an active matrix liquid crystal display (AMLCD) or a cathode ray tube (CRT), see Fig. 14.7.

ECAM provides the main features of EICAS but also displays corrective action to be taken by the crew as well as system limitations after the failures. Using a colour coded hierarchy, the crew can assimilate the information being presented and take the necessary corrective action. ECAM was first introduced in the Airbus A320 and provides the **paperless cockpit** since procedures for abnormal and emergency situations are presented on each of the displays.

Figure 14.7   ECAM displays

ECAM comprises a series of integrated systems that display information to the crew in an efficient way. Referring to Fig. 14.8, aircraft sensors are categorized into key monitoring functions; these sensors transmit data into two system data acquisition concentrators (SDACs). The data is processed and supplied into flight warning computers (FWCs). The FWCs are programmed to identify any inconsistencies in the data, and then output the data through three display management computers (DMCs).

If a system fault or event is detected, one of the FWCs generates the appropriate warning messages and aural alerts. Critical systems such as engine and fuel quantity are routed directly into the FWCs so that they can still be monitored in the event of both SDACs failing. (ECAM can tolerate the failure of one SDAC and one FWC and still continue to operate.)

The typical synoptic for an electrical system is shown in Fig. 14.9. In this illustration, the power sources supplying specific busbars is shown graphically, together with information such as battery and transformer rectifier unit (TRU) voltages and supply currents. Generator capacity is shown as a percentage of maximum, together with output voltage and frequency.

Aircraft system failures are prioritized as level 1, 2 or 3 failures for the upper and/or lower display. The warning and caution hierarchy is as follows.

Level 3 failures – red warnings: these are situations that require immediate crew action and indicate that the aircraft is in danger. Examples of level 3 failures include an engine fire or loss of cabin pressure. Level 3 system failures illuminate the red master warning light, a warning (red) ECAM

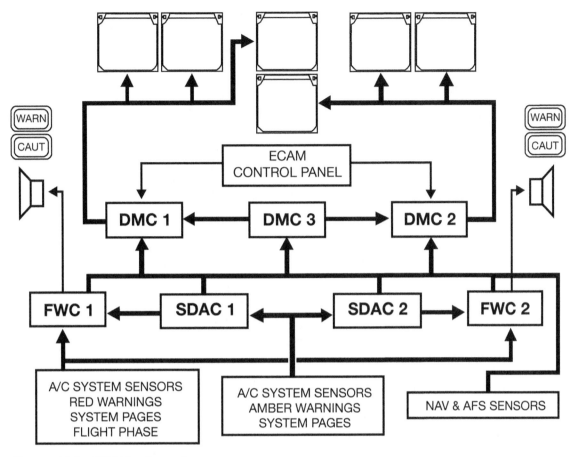

Figure 14.8 ECAM schematic

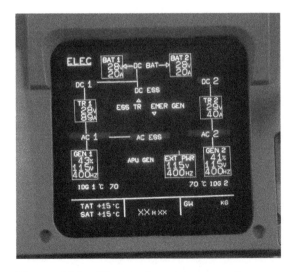

Figure 14.9 Electrical system synoptic

message and an aural warning; this can be a continuous repetitive chime, a specific sound or a synthetic voice. Pressing the master warning push button silences the aural warning.

Level 2 failures – amber cautions: these are situations that require crew attention but not immediate action. Examples of level 2 failures include bleed air failure or a fuel system fault. Level 2 failures have no immediate or direct impact on flight safety; cautions are displayed to the crew by an amber master caution light, an amber ECAM message and a single chime.

Level 1 failures – these are system failures and/or faults that could lead to a loss of system redundancy. Level 1 failures require monitoring but have no immediate impact on continued safe operation of the aircraft. Examples of level 1 failures include the loss of a fuel system temperature sensor. Level l failures are displayed to the crew by amber ECAM messages only (no aural warning).

14.5 Multiple choice questions

1. Indications of landing gear fully down and locked are:
   (a) red lights on, green lights off
   (b) red lights on, green lights on
   (c) red lights off, green lights on.

2. Two state conditions include:
   (a) flap position or doors (open or closed)
   (b) landing gear (up or down) or doors (open or closed)
   (c) landing gear (up or down) or control surface position

3. An electrical flap drive system uses a:
   (a) reversible DC motor
   (b) variable speed DC motor
   (c) unidirectional DC motor.

4. Level 3 ECAM failures are indicated by:
   (a) red warnings, requiring immediate crew action
   (b) amber cautions, requiring crew attention
   (c) red warnings, having no immediate impact on the aircraft.

5. Variable position features include:
   (a) doors (open or closed) and flap position
   (b) control surface and proximity sensor position
   (c) control surface and flap position.

6. When an electrically operated landing gear is fully retracted, the up-lock switch contacts:
   (a) open, thereby removing power from the motor
   (b) close, thereby removing power from the motor
   (c) open, thereby applying power to the motor.

7. Engine fire or loss of cabin pressure would be displayed on ECAM as:
   (a) level 3 failures
   (b) level 2 failures
   (c) level 1 failures.

8. Two state position devices include:
   (a) micro switches and variable resistors
   (b) synchros and proximity sensors
   (c) micro switches and proximity sensors.

9. Electrically driven flaps continue to travel until they are:
   (a) fully retracted and the down-limit switch is operated
   (b) fully extended and the down-limit switch is operated
   (c) fully extended and the up-limit switch is operated.

10. Variable position devices include:
    (a) synchros and variable resistors
    (b) micro switches and variable resistors
    (c) synchros and proximity sensors.

Chapter 15

Warning and protection systems

Various systems are installed on aircraft to protect them from a variety of hazards including: stalling, ice, rain, unsafe configuration during takeoff and skidding. Stall protection systems provide the crew with a clear and distinctive warning before the stall is reached. The primary sensor required for this protection system is the aircraft's angle of attack. Flying in ice and/or rain conditions poses a number of threats to the safe operation of the aircraft. Ice formation can affect the aerodynamics and/or trim of the aircraft. The anti-skid system (also called an anti-lock braking system: ABS) is designed to prevent the main landing gear wheels from locking up during landing, particularly on wet or icy runway surfaces. The configuration warning system (also known as a take-off warning system) provides a warning if the pilot attempts to take-off with specific controls not selected in the correct position, i.e. an unsafe configuration. This chapter describes a variety of protection systems found on a range of aircraft.

15.1 Stall warning and protection

Aircraft wings are in the form of an aerofoil designed to produce lift; the key features of an aerofoil together with specific terminology are shown in Fig. 15.1.

When the airflow passes over the wing without turbulence it is said to have a **streamline airflow**, see Fig. 15.2. The **angle of attack** (AoA) is the angle between the chord line of the wing and the vector representing the relative motion of the aircraft and the atmosphere. (The term angle of incidence is sometimes used instead of angle of attack.)

Basic aerodynamic principles state that increasing the angle of attack increases the lift produced by the wing; when the AoA reaches a certain angle, the airflow over the wing becomes turbulent and the lift is dramatically decreased. When this occurs, the aerofoil is in a **stall** condition as illustrated in Fig. 15.3. It can be seen that the streamlined airflow is breaking up and becoming turbulent.

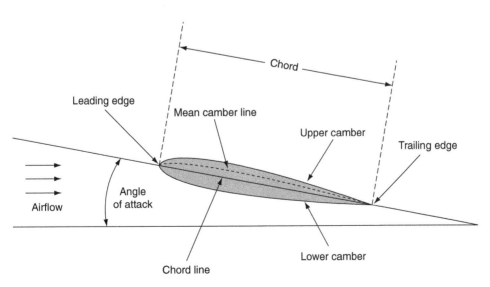

Figure 15.1   Aerofoil terminology

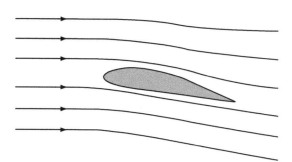

Figure 15.2   Streamline airflow

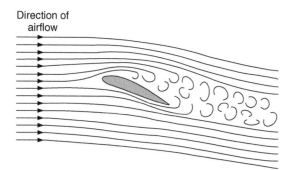

Figure 15.3   Aerofoil stall

It is important to note that an aerofoil stalls at a certain angle; the speed at which the aircraft is flying makes little difference to the stall angle. The relationship between speed and stalling can be related directly to the formula for lift:

$$\text{Lift} = \tfrac{1}{2}\rho V^2 S C_L$$

where ρ is the air density, V is the aircraft speed, S is the wing area, and C_L is the aerofoil's coefficient of lift.

Referring to Fig. 15.4, an aircraft can have a different angle of attack for a given pitch attitude. Aircraft pitch angle should not be confused with angle of attack. Relative airflow over the wing will change direction in flight in relation to the pitch angle of the aircraft.

The critical features of aerofoil performance are centred on the boundary layer as illustrated in Fig. 15.5. When the airflow divides between the upper and lower surfaces, the point of division is at the **stagnation point**. Airflow over the upper surface remains stable until, at higher AoA, it starts to detach; this occurs at the **transition point**, before giving rise to turbulence at the **separation point**. The turbulence behind the separation point increases with AoA.

The relationship between lift and stall is depicted in Fig. 15.6. The two axes on this chart are the lift coefficient (C_L) and angle of attack. It can be seen that lift

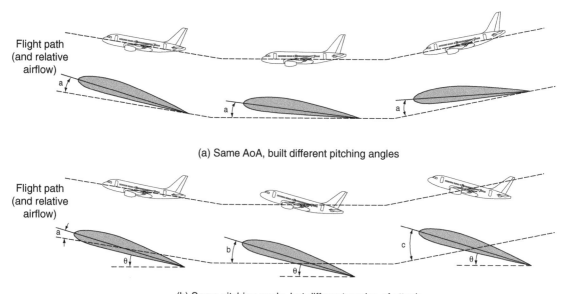

(a) Same AoA, built different pitching angles

(b) Same pitching angle, but different angles of attack

Figure 15.4   Angle of attack and attitude

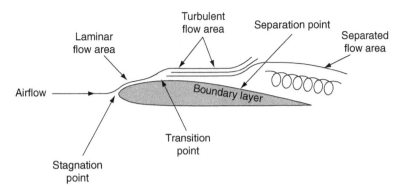

Figure 15.5 Boundary layer separation

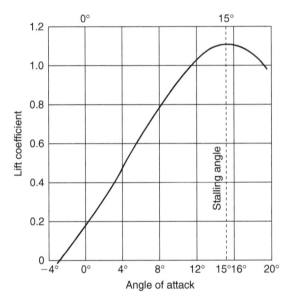

Figure 15.6 Angle of attack and lift

increases in proportion to the AoA over the ordinary angles of flight. At higher angles of attack, the stall condition is reached. Different aerofoil sections have different lift coefficients, and therefore different stall angles, but the principles remain the same.

Deploying lift augmentation devices, e.g. leading and/or trailing edge flaps, can change the wing area and also the wing camber, thereby changing the coefficient of lift. By reducing aircraft speed, the lift will decrease; for a given flap setting, the only other way of increasing lift is to increase the angle of attack. (Aerodynamic theory is covered in more detail in another book from the aircraft engineering series: *Aircraft Engineering Principles*.)

Key point

An aerofoil stalls at a certain angle; the speed at which the aircraft is flying makes little difference to the stall angle.

Warning systems are installed to provide the crew with clear and distinctive warning before the stall is reached. This can be a warning horn and/or warning light on smaller aircraft. On passenger aircraft, the stall warning is provided by a **stick shaker**. Warning systems require a basic sensor to detect this stall condition; typical sensors are:

- reed sensor
- vane sensor
- pressure sensor
- angle of attack sensor.

Key point

Stall warning systems provide the crew with a clear and distinctive warning before the stall is reached.

15.1.1 Reed sensor

Many GA aircraft are installed with a **reed sensor** (Fig. 15.7); this does not require any electrical power.

Airflow is directed from a scoop (Fig. 15.7, item 5) in the wing leading edge and into a reed (3) and horn assembly (2). In the pre-stall condition, the air pressure on the leading edge reduces (relative to cabin pressure)

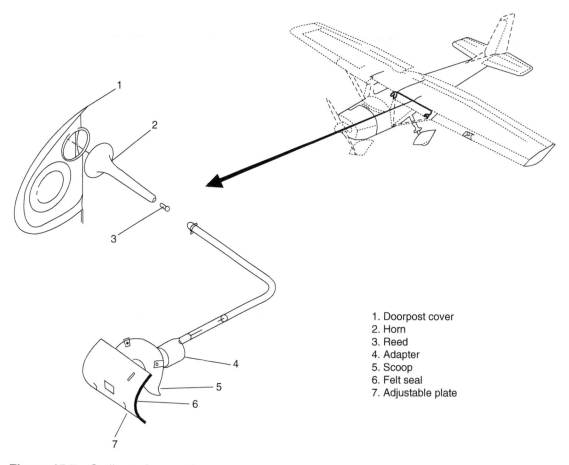

Figure 15.7 Stall warning: reed sensor

1. Doorpost cover
2. Horn
3. Reed
4. Adapter
5. Scoop
6. Felt seal
7. Adjustable plate

because the stagnation point has moved. This draws air through the reed, causing it to vibrate at an audible frequency (much like a musical instrument); this is amplified in an acoustic horn and the crew receive a tone. An adjustable plate (7) is used to alter the airflow intake such that the intake is aligned with stagnation point.

Key point

When the AoA reaches a certain angle, the airflow over the wing becomes turbulent and the lift is dramatically decreased.

15.1.2 Vane sensor

An alternative device used on GA aircraft is the **vane sensor** located in the wing leading edge as shown in

Fig. 15.8. As with the reed sensor, the vane is sensitive to the movement of the stagnation point. The vane is held in a forward position by an internal spring and connected to a micro-switch.

At normal angles of attack, the vane is held back by the airflow (against spring pressure) and the micro-switch is held in the open position, see Fig. 15.9. At higher angles of attack, airflow pressure on the switch is reduced and the vane eventually moves forward (by the spring force) thereby closing the micro-switch. This completes the warning circuit that activates a light and horn.

Key point

At normal angles of attack, the vane sensor is held back by the airflow against spring pressure.

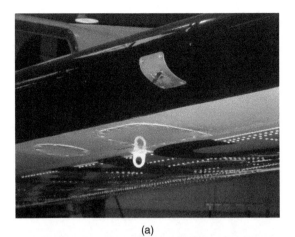

(a)

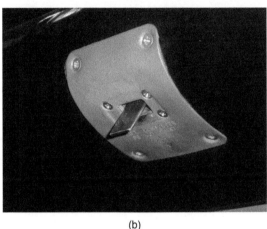

(b)

Figure 15.8 Stall warning: vane sensor installation

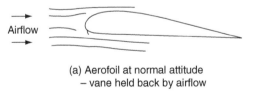

(a) Aerofoil at normal attitude
– vane held back by airflow

(b) Aerofoil in stall condition
– vane pushed forward by spirng

Figure 15.9 Stall warning: vane sensor principles

15.1.3 Pressure sensing sensor

Another angle of attack sensor technology is based on measuring the **pressure** at two points on the sensor housing, see Fig. 15.10. The conical housing rotates on its axis; slots A and B are connected to a pressure chamber that contains a pivoted vane. The sensor housing aligns with the angle of attack because slots A and B maintain equal pressures on the vane, and hence the conical housing.

If the angle of attack increases, slot A has increased pressure compared with slot B. The vane moves to re-align the housing with the airflow to equalize the pressures in both slots. Rotation of the housing is detected by a potentiometer; the centre contact picks off a signal voltage tapped from the resistance windings and this is used to measure the angle of attack.

Key point

When airflow passes over the wing without breaking up, it is said to have streamline airflow.

15.1.4 Sensor vane

The final type of stall warning sensor to be described is illustrated in Fig. 15.11. An angle of attack (AoA) **sensor vane** aligns itself with the prevailing airstream; this rotates a shaft inside the housing. The vane's shaft is connected to a synchro that provides an electrical output proportional to the angle of attack.

A **viscous damper** connected to the AoA vane stabilizes vane movements and reduces the effects of turbulence. The AoA sensor contains a heater that provides continuous de-icing/anti-icing, prevents condensation and reduces changes in damper fluid viscosity.

15.1.5 Stick shaker

On larger aircraft the stall warning system comprises an angle of attack vane and stick shaker. A motor is attached to one or both control columns; an out-of-balance weight is attached to the motor shaft so that it vibrates when the motor is running. The motor design is matched to the control column to provide a distinctive frequency of approximately 10–30 Hz together with physical movement of the controls.

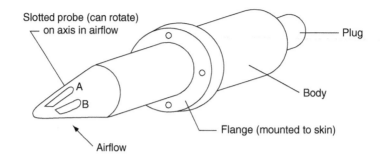

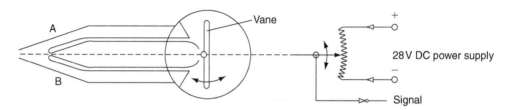

Figure 15.10   Stall warning: pressure sensor

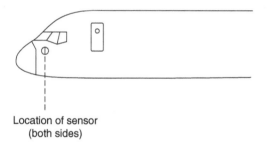

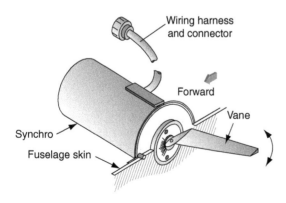

Figure 15.11   Stall warning: AoA sensor

Key point

An angle of attack (AoA) sensor vane aligns itself with the prevailing airstream.

15.1.6 Stall identification system

Certain aircraft types, typically **tee-tailed aircraft**, are installed with stall identification systems; this is normally integrated with the stall warning system. The tee-tailed aircraft can be vulnerable to a deep-stall as illustrated in Fig. 15.12. In addition to the wings (or main-planes) stalling, the deep stall condition also makes the elevators less effective; this compounds the problem, since the elevators are the means of pitching the aircraft nose down to recover from the stall. Furthermore, if the aircraft has rear-mounted engines, the airflow into the engine is turbulent, possibly causing a loss of performance.

The stall identification system contains an actuator that pushes the control column forward with a sharp

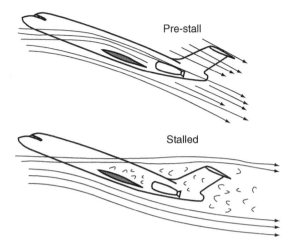

Figure 15.12   Deep stall

and positive action. The **stick push** is released once the aircraft is returned to an acceptable angle of attack. This system has to be highly reliable, both in terms of operating when needed and having low probability of false operation.

Key point

The stall identification system contains an actuator that pushes the control column forward.

15.2 Airframe ice and rain protection

Flying in ice and/or rain conditions poses a number of threats to the safe operation of the aircraft. Ice formation can affect the aerodynamics and/or trim of the aircraft. Icing occurs when **super-cooled water** makes contact with the airframe and/or ice particles. Super-cooled water exists at temperatures below freezing point; this is because water needs nuclei to form ice crystals. The water freezes on the surface of the nuclei, and then grows in size by forming layers. Water freezes when energy is given up to its surroundings. Ice accumulation on an aircraft can cause secondary damage by breaking off and being ingested

into the engine. Ice and rain also reduces visibility through the windscreen. External sensors and equipment can also be affected by the build-up of ice, e.g. pitot tubes that sense airspeed and angle of attack sensors, and these must not be allowed to freeze.

15.2.1 Ice detection

Various technologies are available to provide a warning to the crew of icing conditions. The simple method is to monitor the outside air temperature and atmospheric conditions. When flying at night, an ice inspection light (see Chapter 12) can be used to illuminate critical areas, e.g. wing leading edges. Automatic ice detectors are normally located at the front of the aircraft. Various technologies are employed; the function of the sensor is to detect **ice accretion** and provide a warning to the crew and/or turn on the ice protection systems.

One type of automatic ice detector consists of a motor-driven sensor that is located against a **knife-edge cutter**, see Fig. 15.13, at the front of the fuselage. As ice builds on the rotor, the gap closes with the knife-edge and torque is applied to the rotor. The body of the motor is held in position by springs, and as the torque increases, the motor starts to rotate in its mounting until a pre-determined point is reached and a micro-switch is closed, thereby operating a warning light.

An alternative technology uses an **ultrasonic** ice detector; see Fig. 15.14. The sensing probe is exposed to the airflow and is vibrated by an electromagnetic coil at a natural frequency of 40 kHz. The probe vibrates at a lower frequency when it accumulates ice due to its increased mass.

A logic unit within the detector housing (Fig. 15.15) determines when the probe vibration is less than 39,867 Hz; this occurs when a known mass of ice has been accumulated, and the cylinder is heated to melt the ice.

A timing schedule (Fig. 15.16) is programmed into the control unit and monitors the time taken for the frequency to change. The heater remains on until the probe vibration returns to its nominal 40 kHz value. Nominal heating time is six seconds; if the heater is on for more than 25 seconds, power is removed, and a fault condition is notified to the crew. The detector provides control functions for selection of both engine and wing ice protection.

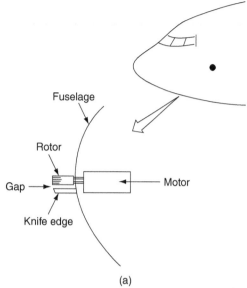

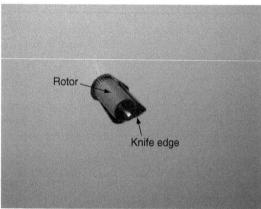

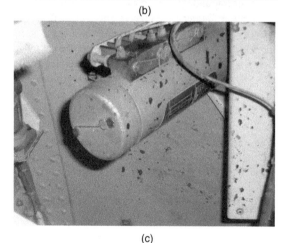

Figure 15.13 Ice detector: knife edge cutter.
(a) Schematic, (b) sensor, (c) motor

Key point

When the ultrasonic ice detector probe accumulates ice, it vibrates at a lower frequency.

Fault conditions are reported to the on-board maintenance system. Weight on wheels logic prevents the detector from being heated on the ground. Press to test (PTT) is used to check the detector's logic functions and heater continuity.

15.2.2 Ice protection

Two strategies are used for ice protection: de-icing and anti-icing. De-icing allows ice to form and then be removed on a periodic basis. The build-up of ice will have been investigated during type testing of the aircraft and the build-up removed before it poses a hazard. Anti-icing is when ice is not permitted to form at all. Three primary methods are used for both de-icing and anti-icing:

- fluid
- pneumatic
- thermal.

All three are controlled and operated by electrical systems; an aircraft can be fitted with any one method, or a combination of all three. Specific areas to be protected from ice formation are as follows:

Airframe
- leading edges
- control surfaces
- lift augmentation devices
- windscreen (or windshield)

Propulsion
- air intakes
- propellers

External components
- pitot tubes
- temperature sensors
- angle of attack sensors
- water drains.

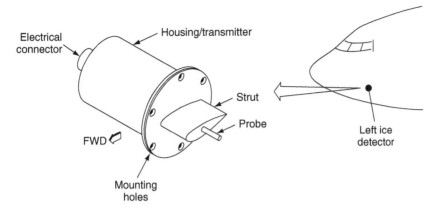

Figure 15.14   Ice detector: ultrasonic sensor

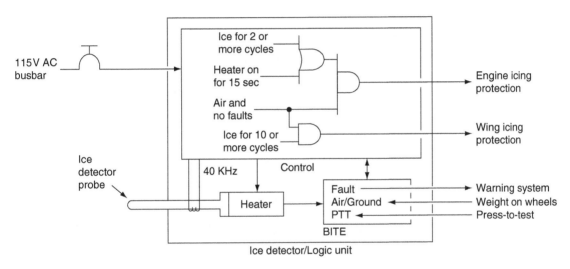

Figure 15.15   Ice detector: ultrasonic system

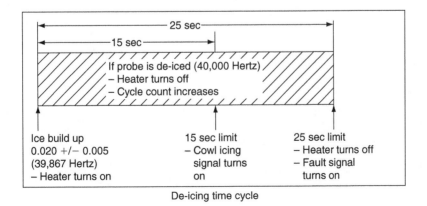

Figure 15.16   Ice detector: ultrasonic system timer

15.2.2.1 De-icing fluids

Fluids are typically used on wings, vertical stabilizer, horizontal stabilizer, propellers and windscreens. Onboard fluid protection is applied through pipes and/or small holes in the airframe. The fluid is transferred from a storage tank by electrical pumps and directed to the required zones by electrically operated valves.

The fluid can also be applied on the ground by specially equipped vehicles with booms to allow easy access to the entire aircraft. **De-icing fluids** are typically composed of ethylene glycol or propylene glycol, together with thickening agents, corrosion inhibitors, and coloured dye. The aircraft is sprayed with a fluid that melts any existing ice on the aircraft and also prevents ice formation prior to takeoff. The timing of when the aircraft is sprayed has to take into account the weather conditions (in particular ambient temperature, wind speed, precipitation, humidity) aircraft skin temperature, and the scheduled departure time.

Fluid performance is characterized by its **holdover time**: this is the period of time that an aircraft remains protected against ice prior to takeoff, and can be up to 80 minutes depending on conditions and fluid additives. The coloured dye is used so that treated areas can be readily identified.

Key point

De-icing allows ice to form and then be removed on a periodic basis.

15.2.2.2 Pneumatic ice protection

Pneumatic ice protection is used on the leading edge of wings and the vertical/horizontal stabilizers. Fabric reinforced rubber tubes (known as **boots**) are attached to the surfaces of leading edges of wings, tail planes and fins, see Fig. 15.17. These tubes are inflated and deflated with air on a cyclic basis to cause a slight deformation of the boot; this causes the ice to break off.

The air supply is from an electrically driven pump, or bled from the engine compressor via a regulator. Electrical solenoid valves direct the air into the boots either in sequence or simultaneously. The tubes are deflated and kept flush with the airframe when not in use; this is achieved by connecting the boots to a vacuum source. The system can be operated automatically or manually.

Key point

When not in use, pneumatic de-icing boots are deflated and kept flush with the airframe.

15.2.2.3 Thermal ice protection

Thermal ice protection can be achieved by one of two methods: hot air bled from the engine compressor, or electrical heaters. Hot air is directed to the required zone(s) by electrically operated valves controlled from the flight compartment or by a control unit. Temperature sensors are used as part of a control

(a)

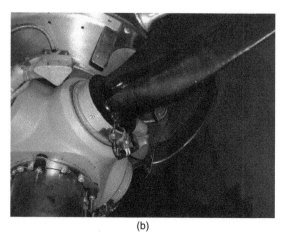

(b)

Figure 15.17   Ice protection: (a) wing leading edge, (b) electrical connections

and feedback system to provide overheat protection. Applications for **bleed air ice protection** include the leading edge of wings, vertical/horizontal stabilizers and engine air intakes.

Key point

Anti-icing is used where ice is not permitted to form at any time.

Electrical heating systems are used for both anti-icing and de-icing; heater mats are bonded to the:

- airframe
- engine intakes
- propellers
- rotor blades
- windscreens.

The heater mats are formed from fine-wire elements inside layers of insulation and protective materials. Typical heating elements are nickel, nickel-chrome or copper-nickel. Larger aircraft use a 115 V AC variable frequency power supply. Smaller aircraft use a 28 V DC power supply.

Propeller de-icing is achieved by electrical heating elements that are bonded to the leading edge of the blades, see Fig. 15.18. Some aircraft have heating mats on the inboard and outboard sections of the propeller blade. The elements are connected to the 28 V DC power supply via slip rings and brushes inside the hub. Typical current requirements for propeller heating are 15–20 A. Power is applied from a timer control circuit that alternates the heating of outer and inner mats on each of the propeller blades over a 30 second cycle.

Engine intake leading edges are continuously heated when the system is turned on (anti-icing). This applies to gas turbine engines and turbo-props. The inner and outer surfaces are heated via a cyclic timer to provide de-icing. Engine intakes are installed with mats that are shaped to provide breaker strips; this ensures that the ice breaks off in pre-determined sections in a controlled sequence.

Test your understanding 15.1

What is the difference between anti-icing and de-icing?

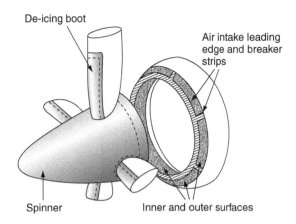

Figure 15.18   Propeller de-icing

Larger aircraft use engine bleed air for thermal anti-icing. Fig. 15.19 illustrates how ducts are located along the leading edge of the wing. Spray holes direct the air inside the leading edge before it is ejected overboard. Similar installations are used for tail plane and engine nacelles.

15.3 Windscreen ice and rain protection

Various methods of windscreen ice and rain protection are used on a range of aircraft types. One method uses a metallic film deposited onto the surface of the screen; this is connected to an electrical supply on either side of the panel. An alternative technology uses fine wire elements (Fig. 15.20) sandwiched within the laminated glass panel. Individual windscreen heaters use 4 kW of power, to keep window temperature at approximately 30°C.

Automatic cycling of power regulates the amount of heat being absorbed by the windscreen. Power is isolated from the heaters in the event of fault conditions; temperature sensors detect overheat and current transformers detect electrical overload conditions. Temperature sensors are monitored by a control unit; these can be simple thermostats or thermistors.

Key point

Windscreens are kept clear of rain by wipers, rain repellent (sprayed on the windows), or by treating the glass with a hydrophobic coating.

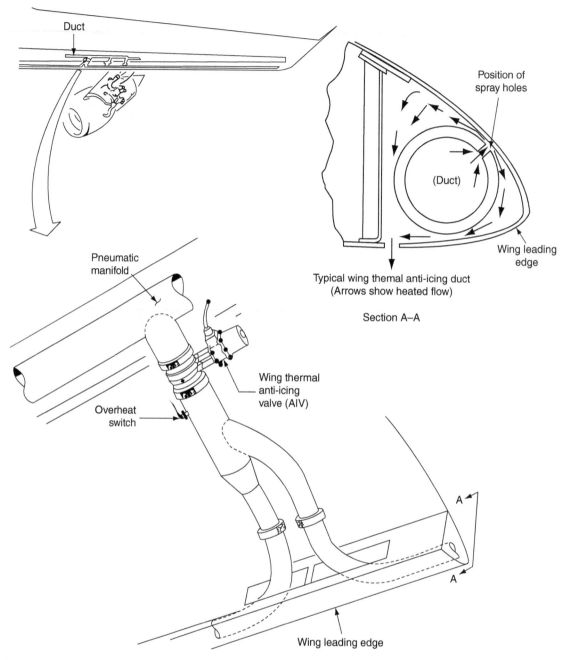

Figure 15.19   Thermal anti-icing

15.3.1 Windscreen wiper

The windscreen wiper system is based on 28 V DC variable speed motors; the rotary motion of the motor is changed by a gear mechanism in the converter to produce the sweeping motion of the wiper arm over the windscreen. A typical windscreen wiper circuit is shown in Fig. 15.21. The normal arrangement is to have one wiper assembly per screen to ensure that at least one pilot can keep a clear screen in the event of

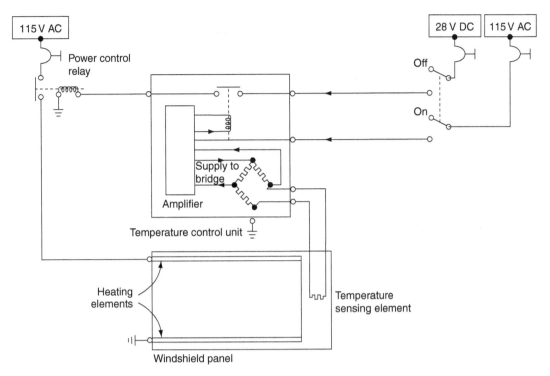

Figure 15.20   Windscreen de-icing circuit

failure. The motors are set by the control switch; typical wiper speeds are:

- **low**, 160 cycles per minute
- **high**, 250 cycles per minute.

A parking switch in the motor/converter sets the wiper blade to the park position when the wiper is selected off. In the off position, the park switch in each motor/converter closes; this causes the blades to position themselves at the bottom of each windscreen.

15.3.2 Rain repellent

This system is used to maintain a clear area on the windscreen during take-off and landing; a typical system is shown in Fig. 15.22. The rain repellent bottle is located inside the fuselage roof; this contains a pressure gauge, visual contents reservoir and a shut-off valve.

The control panel (normally combined with the wiper system) contains two switches to control the rain repellent system. Figure 15.23 illustrates a typical rain repellent system schematic. The repellent fluid is stored in a container; this is pressurized from an external air supply. Two electrically operated solenoid valves control the flow of repellent to the windscreens.

Nozzles on the fuselage (forward of the windscreen) direct the repellent spray onto the windscreens.

The rain repellent switches are momentary make type; pushing a switch once activates the respective solenoid valve. A time delay circuit in the valve controls the amount of repellent being released by closing the valve after a short time period, typically ½ second. A pressure gauge indicates when the container needs replacing.

15.4 Anti-skid

The anti-skid system (also called an anti-lock braking system: ABS) is designed to prevent the main landing gear wheels from locking up during landing, particularly on wet or icy runway surfaces. Traditional method used to operate brakes is from hydraulic pressure controlled from the brake pedals. This pressure can be varied by the pilot by increasing or decreasing the amount of force being applied to the pedals. If too much pressure is applied by the pilot, the wheels will lock-up and the aircraft will skid on the runway.

The anti-skid system ensures optimum braking under all conditions by modifying the pressure being exerted by the pilot on the brake pedals.

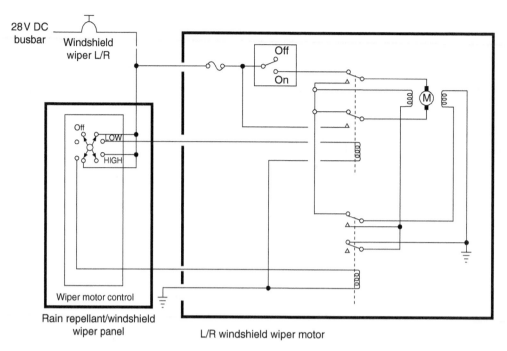

Figure 15.21   Windscreen wiper circuit

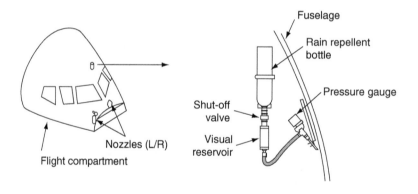

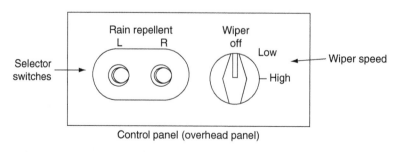

Figure 15.22   Rain repellent system overview

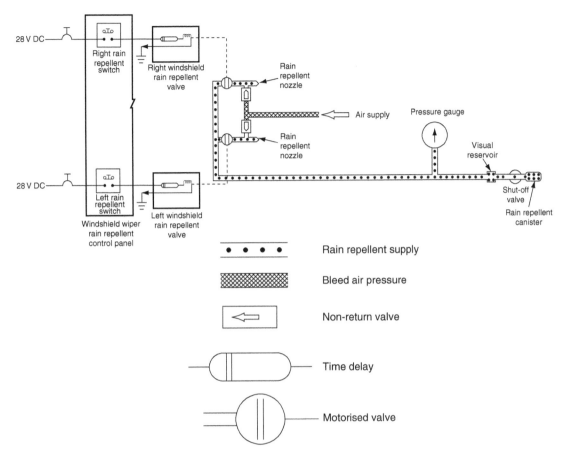

Figure 15.23   Rain repellent system schematic

The applied pressure is reduced before the wheels lock up, and then reapplied to continue the braking action; this occurs as a repeated on/off cycle. Modifying the applied pressure is achieved by modulation; this can vary the applied pressure in both time and pressure.

The system comprises speed transducers (or sensors) on each wheel, a control unit and electro-hydraulic control valves, see Fig. 15.24. During a skid condition, the wheel(s) experience a rapid deceleration. Typical speed sensors are based on the tachometer; this is a small AC generator with the stator formed with a permanent magnet. The rotor coil is built into the axle and turns inside the stator field. Referring to fundamental principles, when the coil is rotated inside the field, currents are induced and a voltage is generated; the tachometer's output is proportional to wheel speed. The tachometer AC output is rectified in the control unit; the DC output is monitored for rate of change to determine if the wheel is accelerating or decelerating. A comparator circuit generates an error signal which is amplified and used

to operate a control valve to modulate the applied brake pressure. Control laws are used to determine when the wheel is approaching a skid condition, this occurs when the wheel speed is decreasing at a given rate. Hydraulic pressure is decreased in the braking system and the wheel accelerates. Once the skid has been averted, hydraulic pressure can be reapplied.

Test your understanding 15.2

Explain how wheel speed is monitored to determine when an aircraft is approaching a skid condition.

15.5 Configuration warning

The system (also known as a **take-off warning** system) provides a warning if the pilot attempts to take-off with

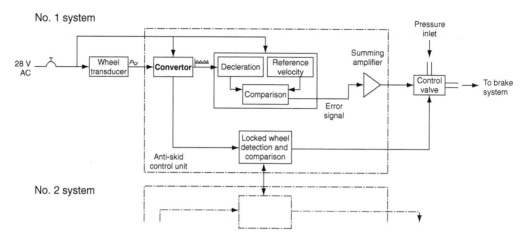

Figure 15.24   Anti-skid system

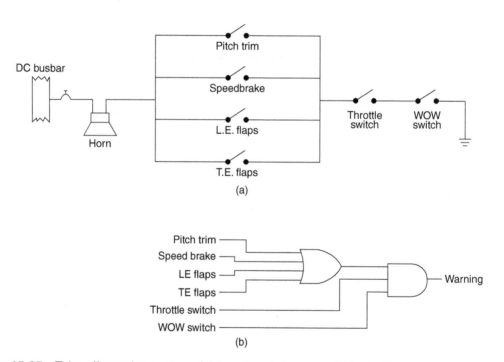

Figure 15.25   Take-off warning system: (a) functional diagram, (b) logic diagram

specific controls not selected in the correct position, i.e. an unsafe configuration. A simple take-off warning system is illustrated in Fig. 15.25(a). The type of system fitted depends on the aircraft type and size; typical parameters being monitored include (but are not limited to):

• pitch trim
• speed brake

• leading edge (LE) flaps
• leading edge (LE) slats
• trailing edge (TE) flaps

The position of each of these controls is monitored together with the squat switch (weight on wheels) and throttle position. If an unsafe take-off configuration is detected, a warning horn is sounded. This system can be viewed in combination logic terms as illustrated in

Fig. 15.25(b). On larger aircraft, more parameters are monitored, and more logic functions are required for configuration warning (Fig. 15.26).

15.6 Aural warnings

Flight compartment aural warnings typically include the:

- fire bell
- take-off configuration warning
- cabin altitude warning
- landing gear configuration warning
- mach/airspeed overspeed
- stall warning
- TAWS
- TCAS.

Typical external aural warnings are also provided for a fire bell in the wheel well and ground call horn in the nose wheel-well for an equipment bay overheat. Certain warnings can be silenced whilst the condition exists. (TCAS is an automatic surveillance system that helps aircrews and ATC maintain safe separation of aircraft. It is covered in a related book in this series: *Aircraft Communications and Navigation Systems*.

An aural warning system that combines many functions on has a **priority logic** based on alert levels to ensure that certain warnings, e.g. TAWS, are announced in a predetermined threat level. Examples of three priority levels are given in Table 15.1. Note that there could be 25 or more levels of aural warning system inputs.

Table 15.1 Aural warning system priority logic

Priority	Description
1	Windshear
2	Pull-up (sink rate)
3	Pull-up (terrain closure)

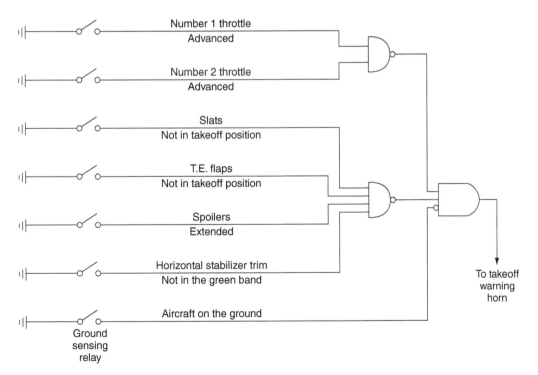

Figure 15.26 Configuration warning system

15.7 Multiple choice questions

1. Stall warning systems provide the crew with a clear and distinctive warning:
   (a) before the stall is reached
   (b) after the stall is reached
   (c) at all angles of attack.

2. When an ultrasonic ice detector probe accumulates ice, it vibrates at a:
   (a) higher frequency
   (b) lower frequency
   (c) constant frequency.

3. The stall identification system contains an actuator that:
   (a) maintains the angle of attack
   (b) pulls the control column rearward
   (c) pushes the control column forward.

4. When not in use, pneumatic de-icing boots are:
   (a) inflated and kept flush with the airframe
   (b) deflated and kept flush with the airframe
   (c) cycled on a periodic basis.

5. When the AoA reaches a certain angle, the airflow over the wing:
   (a) becomes turbulent and the lift is dramatically decreased
   (b) becomes streamlined and the lift is dramatically decreased
   (c) becomes turbulent and the lift is dramatically increased.

6. When a wheel is approaching a skid condition, this is detected when the speed is:
   (a) decreasing at a given rate
   (b) increasing at a given rate
   (c) constant.

7. At normal angles of attack, the vane sensor is held:
   (a) back by spring pressure against the airflow
   (b) forward by the airflow against spring pressure
   (c) back by the airflow against spring pressure.

8. De-icing:
   (a) allows ice to form, and then be removed on a periodic basis
   (b) prevents ice from forming at any time
   (c) directs the repellent spray onto the windscreens.

9. An angle of attack (AoA) sensor vane aligns itself with the:
   (a) boundary layer
   (b) prevailing airstream
   (c) stagnation point.

10. When airflow passes over the wing without breaking up, it is said to have a:
    (a) boundary layer
    (b) streamline airflow
    (c) stalled airflow.

Chapter 16 Fire and overheat protection

Fire on board an aircraft is a very serious hazard; all precautions must be taken to minimize the risk of a fire starting. In the event that a fire does occur, there must be adequate fire protection on the aircraft. Fire protection includes the means of both detecting and extinguishing the fire. The subject of fire protection theory is a branch of engineering in its own right. Basic fire protection theory is covered in this chapter to provide the reader with sufficient information to understand how this theory is applied through aircraft electrical systems. The chapter focuses on the equipment and systems used to detect fire and smoke together with the means of delivering the fire-extinguishing agent.

16.1 Overview

The type of fire protection systems and equipment fitted to an aircraft can be sub-divided into specific areas of the aircraft:

- engines/APU
- cargo bay
- passenger cabin.

The fire protection technologies used on aircraft depend on these areas and specified fire risk. In addition to the fire risk in these areas, high temperatures resulting from engine **bleed air leaks** can also be hazardous. Bleed air is tapped from the compressor stage of a gas turbine engine, distributed in the pneumatic system for a number of purposes including thermal anti-icing. The consequences of hot air leaks are less severe than fire, but overheats can weaken structure and damage components. The pneumatic system takes air from the engines and distributes it throughout the aircraft; typical areas that are fitted with dedicated overheat detection systems are the wing leading edges, in the wheel wells and under the cabin floor.

Up until the 1990s, the two agents used extensively in the aircraft industry were:

- bromochlorodifluoromethane, also known by the trade name Halon 1211, or BCF

- bromotrifluoromethane, also known by the trade name Halon 1301, or BTM.

These are both inert gases and are either applied locally (typically in hand-held extinguishers) or as total flooding applications (typically in cargo bays). Both these agents are very effective at extinguishing fires, however they are in a group of **halogenated hydrocarbons** that have been shown to contribute to depletion of the earth's ozone layer. Although Halon fire extinguishers are still being specified on aircraft, both gases come under the terms of the 1989 Montreal Protocol (and subsequent revisions), that prohibits the new production of these agents. The industry is supporting the supply of Halon 1211 and 1301 through **recycling** of existing Halon stocks for in-service and new production aircraft.

Alternative and replacement agents are being developed, and these are gradually being introduced throughout the fire protection industry. Examples of agents that do not deplete the ozone layer include:

- water
- dry powder
- vaporizing liquids (gaseous agents)
- Inert cold gas, e.g. Carbon Dioxide.

Combustion can be defined as a rapid and complex chemical reaction in which light and heat are evolved. All equipment in an aircraft is carefully designed and tested so that there is a low probability of starting and sustaining fire. Three factors are required to initiate and sustain combustion:

- fuel
- heat
- oxygen.

Fuels include solids, liquids and gases; each type of fuel requires a minimum temperature to be reached and maintained. Oxygen is normally available from the ambient air. The type of fire detector and extinguishing agent deployed is largely determined by which fuel is likely to combust. Detecting a fire

Table 16.1 Fire classifications and extinguishing strategies

Class of fire	Fuel/heat source	Appropriate extinguishing strategies
Class A	Solid/organic	Water, vaporizing gas
Class B	Flammable liquids	Carbon dioxide, dry powder, or vaporizing liquid
Class C	Flammable gases	Dry powder, vaporizing liquids
Class D	Metals	Dry powder
Class E	Electrical equipment	Carbon dioxide, dry powder, or vaporizing liquid
Class F	Cooking fat/oil	Carbon dioxide, dry powder, or vaporizing liquid

can be achieved by one or more of the following strategies:

- thermal sensors
- optical sensors
- smoke detectors.

Extinguishing a fire can be achieved by one or more of the following strategies:

- limiting or eliminating the fuel
- limiting or eliminating the oxygen
- reducing the temperature of the fire
- interfering with the chemical reaction.

Fires can be categorized by the types of fuel that are combusting; this in turn determines the detection and extinguishing strategy. There are several classifications of fire used in the USA, Europe and Australasia. Typical fire classifications and extinguishing strategies are illustrated in Table 16.1.

Primarily sources of risk/fuel in aircraft applications are:

- engines and APUs: class B
- cabin/flight compartment: class A
- cargo bay: class A.

Key point

Electrical fires are sometimes stated as Class E, or combined as Class A, B and D. The strategy adopted for this type of fire is to isolate electrical power and then deploy the agent.

Each of the fire extinguishing agents has different attributes. Water reduces the heat in a fire. It is often

mixed with an additive, glycol, to enable it to be used in sub-zero temperatures. Vaporizing liquids reduce heat, but their significant contribution is to interfere with the chemical reaction of the fire. Carbon dioxide (CO_2) and dry powder both reduce the fuel's temperature and reduce the oxygen levels.

Key maintenance point

Always consult with the appropriate authority before deploying fire extinguishers.

16.2 Engine/APU fire detection

Engine/APU fire detectors are designed to very high standards; they are needed when other systems, procedures and safety features have failed to prevent a hazardous situation developing.

In terms of fire protection, the engines and auxiliary power unit (APU) can be considered as having the same requirements. Fire detection within an aircraft engine can be achieved by thermal and/or optical methods. **Thermal** fire detectors sense heat energy that is radiated from the fire; they can be either unit detectors (sometimes referred to as point, or 'spot' detectors) or linear detectors. **Optical** fire detectors are based on sensing ultraviolet (UV) or infrared (IR) energy.

Whatever the sensor technology, the engine is fitted with sensors in positions that provide coverage of all areas where a fire could start; these positions are determined mainly from experience with an engine type. Ignition sources within the engine are electrical faults, and hot surfaces, e.g. exhaust ducts. Fuel and/or oil supply line failures in the engine will provide the means of sustaining the fire. There is a ready supply of oxygen in the ventilation air that is used to cool components.

When a fire is detected, the warning system activates an alarm bell and illuminates red warning lights in the flight compartment. The fire warning also illuminates a **master warning light** and another light (typically in the fire handle) to identify the affected engine(s). When the crew confirm that they have an engine fire, they operate the fire handle by first pulling and then twisting the handle. The action of pulling the fire handle activates micro-switches that shuts down the engine by:

- closing the fuel shutoff valves
- opening the field circuit of that engine's generator(s)
- closing the bleed air supply from the engine into the pneumatic system
- closing the hydraulic systems engine driven pump shutoff valve.

If the fire warning continues for a prescribed time after these actions, the fire handle is twisted; this closes micro-switches that activate the engine fire extinguisher.

16.2.1 Thermal fire detection

There are various techniques used to detect fire and overheat by including unit (or spot detectors), linear detectors and optical sensors.

16.2.1.1 Unit fire detectors

These are effectively thermostatic overheat switches based on the **bimetallic principle**. Figure 16.1 illustrates the external features and internal view and electrical schematic of a typical overheat switch. The two contacts are attached to spring bows; these are compressed during assembly and held apart by the outer barrel of the switch. As the detector is heated, it expands and the two contacts close, thus completing the warning circuit. An adjusting screw is found on the end-cap of some switches to allow for minor adjustments of operating temperature.

This type of detector can be used in engine fire detection systems; alternatively it can be used in the

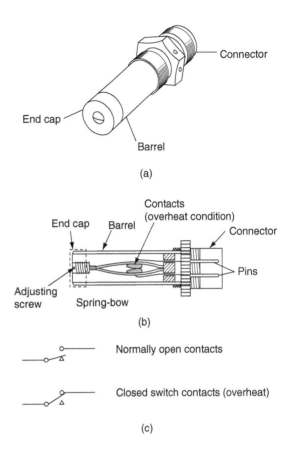

Figure 16.1 Overheat switch: (a) external features, (b) internal features, (c) electrical schematic

bleed air overheat detection system. The detector can only sense fire/overheat in a localized volume of the installation; it is more likely that multiple detectors are used to provide increased detection probability. A small gas turbine engine would typically be fitted with six of these detectors.

The wing leading of a large aircraft could be fitted with over 20 overheat switches. The switches are connected in parallel as shown in Fig. 16.2; this forms a **detection loop**. If any one of the switches closes due to a fire or overheat being detected, the alarm circuit is activated thereby illuminating a system warning light. A simple test circuit allows some of the circuit to be tested from the flight compartment. Some aircraft are fitted with a dual loop of detectors to provide a back-up system in the event of loop failure, e.g. open circuit wiring.

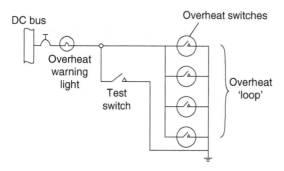

Figure 16.2   Overheat switches connected in parallel

Figure 16.3   Linear sensor installation

Test your understanding 16.1

How would the crew know if more than one switch in a multiple detection system went into alarm?

The actual alarm temperature of the unit fire detector is set by the amount of compression on the bow springs. During the aircraft design phase, the normal ambient temperature of the zone being protected is determined. The alarm temperature setting of the fire detector has to be carefully selected by the system designers. If it is too close to the nominal ambient temperature, there is a risk of false warnings. An operating temperature far higher than the zone's nominal ambient temperature will cause delays in the switch operating (if at all). Aircraft specifications and designs vary; however, a margin of approximately 50°C between ambient and operating temperature is typical. Detectors with different alarm temperatures are sometimes fitted in the same loop; this takes into account varying ambient temperatures in different zones. Typical settings are 230°C in the cooler forward sections of the engine nacelle and 370°C in the aft sections of the nacelle.

16.2.1.2 Linear fire detectors

Greater coverage of the engine nacelle can be achieved with linear fire detectors. Three types of **linear fire detectors** are used on aircraft; these are based on averaging or discrete detectors. The alarm temperature of **averaging** linear detectors depends on the length of element heated. **Discrete** linear detector systems are essentially independent of the length of element heated to provide an alarm. The response time for any fire detector is critical to the safety of the aircraft; the standard test for a linear fire detector is for it to respond to a 1100°C flame within five seconds. Actual response time will depend on many factors including the convection of heat energy; this is why it is so important to route the fire detector in the critical areas of the engine nacelle. Linear fire detectors are manufactured in lengths of up to three metres; some types can be joined in series to so that all these areas can be covered in convenient lengths. Installing the detector in manageable lengths facilitates installation and removal during maintenance.

Averaging linear detectors are based on either thermistor or pneumatic principles. Both detectors consist of a thin (typically 2 mm diameter) sensor that is routed around the engine. The sensor is attached to the engine using clips and bushes, see Fig. 16.3. The **thermistor** type fire detector is constructed in a coaxial form with a corrosion-resistant (typically stainless steel) outer sheath, nickel-chrome centre wire and temperature-sensitive separating, or insulating material, typically very fine granules of a silicon compound. Figure 16.4 illustrates the detector's physical features, a detector loop and the resistance temperature characteristics. This forms a complete electrical circuit, or **fire detector loop**.

Referring to Fig. 16.4(c), when the sensor is at low ambient temperature (t_1), the resistance (R_1) between the outer sheath and centre wire is very high. The silicon insulating material has a negative temperature

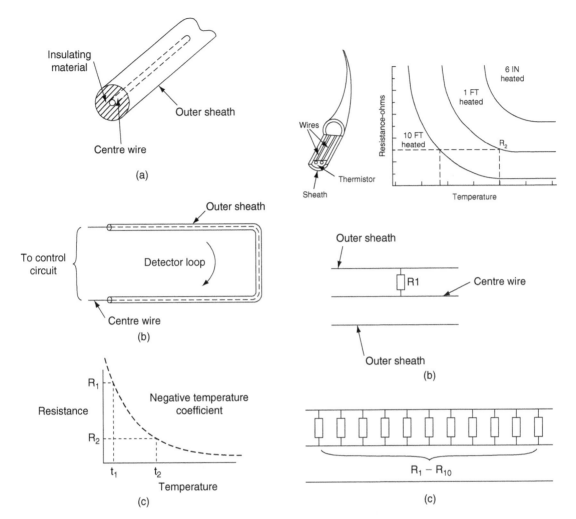

Figure 16.4 Linear fire detector principles: (a) external features (b) detector loop, (c) resistance/temperature characteristics

Figure 16.5 Thermistor characteristics: (a) effect of length being heated, (b) short length heated, (c) longer length heated

coefficient, i.e. when the sensor is heated, the resistance of the insulating material decreases.

When this temperature reaches the alarm point (t_2), the control unit senses a predetermined value, typically $1\,k\Omega$. A relay contact in the control unit closes and this activates the alarm circuit. When the fire has been extinguished, the sensor cools down, causing the resistance to increase; the relay opens and the alarm circuit is deactivated.

The thermistor fire detector is an **averaging** type, responsive to the length of element heated, see Fig. 16.5. It can be seen that if the alarm is tripped at an equivalent resistance indicated by R_2, a shorter length

operates at a higher temperature. The thermistor fire detector can be thought of as an infinite number of resistors connecting the centre wire to the outer sheath.

If only one of these imaginary resistors is heated, i.e. a short length of element, R_1 has to reach a high temperature before its resistance drops to the alarm point. If a large number of these imaginary resistors is heated, i.e. a longer length of element, since the resistors R_1 through R_{10} are in parallel, the alarm point will be reached at a lower temperature.

There are certain fault conditions that can lead to false fire warnings; these can be caused by moisture ingress into the sensor and/or short circuit of the sensor wiring. Control units have been developed with the ability to discriminate between fire and fault conditions. Moisture ingress into the connectors is prevented by hermetically sealed terminal lugs.

Testing of the system is via a test switch located in the flight compartment. When the test circuit is activated, the following aspects are checked:

- continuity of centre element
- resistance of the insulating material
- alarm circuit.

Key maintenance point

It is very important that the sensing element and wiring does not come into contact with the engine structure.

Test your understanding 16.2

Why are averaging fire detectors dependent on the length of element heated to trigger an alarm?

Key maintenance point

Abrasion of the outer sheath will expose the insulating material leading to moisture ingress, decreased resistance and potential false warnings.

Key maintenance point

A short circuit on the sensor wires could be interpreted by the control unit as a fire condition.

Eutectic salt overheat detectors are constructed from a thin (typically 2 mm diameter) tube filled with a salt compound. They are particularly effective at detecting overheats when only a short length of element is heated. Referring to Fig. 16.6, the insulating material in this type of detector is a salt compound that melts at a specific temperature. All of the constituents crystallize simultaneously at this temperature and the insulation changes state into a molten liquid solution.

As with the thermistor system, a control unit monitors the resistance between the inner conductor and the outer sheath. When the salt melts, the resistance drops very rapidly and the control unit signals a warning. The resistance between the outer sheath and inner conductor decreases and the control unit senses this as a fire and/or overheat condition.

When the temperature is reduced, the salt compound solidifies and the resistance increases, thereby resetting the alarm output. This type of detector is used in both engine fire detection and bleed air overheat detection applications. The discrete detector is ideally suited for the detection of hot air leaks in:

- wing leading edges
- wheel wells
- under floor locations.

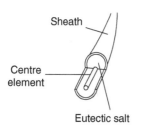

Figure 16.6   Eutectic salt overheat

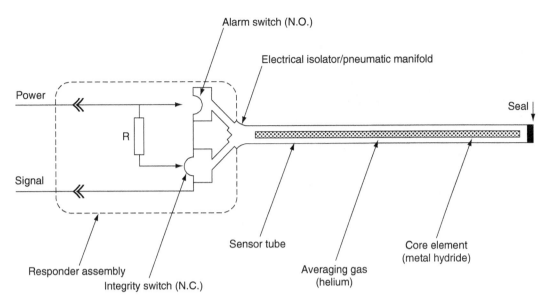

Figure 16.7 Pneumatic detector schematic

Pneumatic fire detectors are constructed from a thin (typically 2 mm diameter) sensor tube filled with helium, and a gas-charged metal-hydride centre core. The tube is sealed at one end and attached to a pressure-sensing device (the **responder)** at the other end. The responder assembly is electrically isolated from the sensing tube and contains two pressure switches, a resistor and electrical connector, see Fig. 16.7.

The normally open (NO) **alarm switch** is closed when a fire or overheat causes the helium to expand. A normally closed (NC) **integrity switch** in the responder is held closed by the helium pressure; the internal resistor and switch contacts form the warning circuit. Measurement of the resistor value by an external circuit provides an indication of fire, overheat or loss of integrity. In the non-alarm condition, a monitoring circuit detects the known resistor value. When large sections of the sensor tube are heated, the increased pressure of the helium causes the alarm switch to close; this bypasses the resistor and the monitoring circuit signals an alarm condition. When the overheat condition is removed, the gas pressure decreases and the switch opens.

If only a small section of sensor tube is heated, there will be insufficient helium gas pressure to operate the alarm switch. To accommodate this localized heating, the centre core is formed with a **metallic hydride compound** which liberates hydrogen when

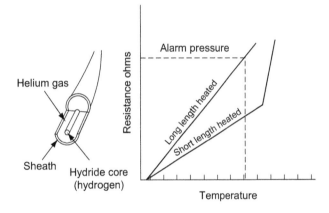

Figure 16.8 Pneumatic detector characteristics

rapidly heated; this now provides sufficient gas pressure to operate the alarm switch. When the sensor cools, the hydrogen is re-absorbed back into the centre core, pressure reduces and the alarm switch opens. Should a leak develop in the sensor or responder, the integrity switch opens. The control unit detects an open circuit condition from the change of resistance and this is signalled as a fault condition. The characteristics of a pneumatic detector are shown in Fig. 16.8. The entire assembly is hermetically sealed; the responder can be located in the engine fire zone as illustrated in Fig. 16.9.

Figure 16.9 Responder located in the engine fire zone

Key point

Thermal fire and overheat detectors are often installed as **dual loops** to ensure that fire detection is available even if one system is inoperative.

Key point

Dual loop fire detector systems can also be configured to only go into alarm if both loops sense fire, thereby reducing the possibility of a false warning.

16.2.2 Optical fire detection

Optical sensors sense the light emitted from a flame in much the same way that a person recognizes fires. Our brains can distinguish between the light energy from a fire and the light energy from another source, e.g. a light bulb. The optical fire detector (or sensor) also has to be able to make this distinction. Some aircraft use optical fire detectors in place of thermal detector elements to simplify the installation. Depending on the size of engine, several optical sensors may be required. The sensors are easier to install and maintain than linear detector elements. Optical sensors offer the advantage of being able to monitor a specific volume of engine nacelle.

The spectral analysis of a hydrocarbon fuel fire reveals peaks of energy in the **infrared** (IR) and ultraviolet (UV) frequency bands. Optical fire detectors are designed to detect radiation in one or both of these frequency bands. The type of detection technology used depends on a number of factors including the:

- speed of response to a fire
- ambient temperature for where the sensor is located
- likely source of potential false alarms.

One of the characteristics of burning fuel in an aircraft engine is the emission of IR radiation. A particular feature of a **hydrocarbon** fire is that it emits a high energy level at a nominal 4.4 micron band of the infrared radiation, referred to as the CO_2 spike, see Fig. 16.10(a). This is caused by the emission of energy from excited CO_2 molecules burning in the fuel.

Detecting a fire is one thing; being able to discriminate against other light sources is a vital part of the optical detector's performance. Extraneous light sources that are capable of causing a nuisance alarm emit very low levels of radiation in this range. These sources include:

- lightning
- arc welding
- x-rays
- sunlight
- hot surfaces.

An IR sensor designed for this specific frequency band provides a high level of reliable fire detection, while being relatively immune to nuisance alarm signals.

The detection cell incorporates a **pyro-electric cell** and an optical filter; the latter only transmits radiation within the wavelength band of 4.2 to 4.7 microns. This is packaged in a one-inch-diameter, three-inches-long cylindrical housing with an optical window at one end and electrical connector at the other. The **pyro-electric** detection cell responds to a fire by generating a signal when 4.4 micron radiation energy is detected. The optical fire detector has a **cone of vision** as illustrated in Fig. 16.10(b); 100% represents the maximum detection distance for a given fire. The sensitivity of the detector increases as the angle of incidence decreases.

As with any fire detector (whether thermal or optical) the **response time** of the detector depends on the:

- size of the fire
- rate of propagation
- type of fuel burning
- distance from the detector.

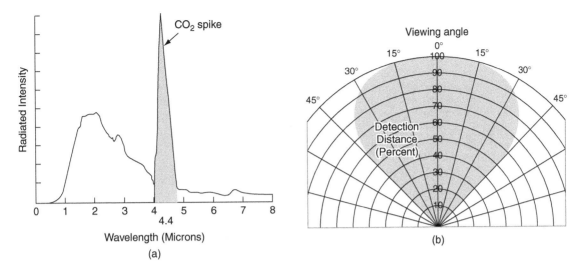

Figure 16.10 Optical fire detection: (a) hydrocarbon fire and 4.4 micron CO_2 'spike', (b) optical fire detector viewing angle

The average **response** time of aircraft engine fire detectors is less than five seconds; this is in response to a standard 1100°C, six-inch-diameter flame.

A typical turbine engine fitted to a medium-sized general aviation (GA) aircraft is installed with three IR optical detectors in each engine nacelle, a control amplifier, warning lights and test switch. The detectors are positioned to receive direct and indirect radiation from flames, thus providing optimum coverage of the nacelle.

Key point

Increased temperature as a result of a fire is not a factor in the operation of an optical sensor.

Key point

The IR cell's output is an electrical signal proportional to the amount of IR energy falling onto it.

16.3 Cargo bay/baggage area

The most likely source of fire in the aircraft cargo bay or baggage compartment is from organic materials; the detection method in widespread use is the **smoke detector**. The primary design consideration for this application is ensuring that the smoke reaches the detector in a timely way. **Open area** smoke detectors rely on the transfer of particulate matter from the source of fire to the detector by **convection**. Smoke from burning materials in cargo/baggage areas often stratifies; in the early stages of a cargo/baggage fire there is insufficient heat energy to cause the smoke to rise through convection. Smoke will eventually accumulate in the roof area of the cargo/baggage area and so this is where the detectors are located. In larger cargo/baggage areas several detectors will be installed to speed up the detection time, which can be up to 60 seconds. In some aircraft, air samples are taken from numerous locations and supplied to a centralized **ducted** smoke detector.

16.3.1 Smoke detector principles

There are two basic principles used in smoke detectors: ionization and photoelectric.

16.3.1.1 Ionization smoke detectors

The **ionization** type monitors ionized products of combustion as they pass through a charged electrical field. This type of smoke detector can detect particles of smoke that are often too small to be visible. The

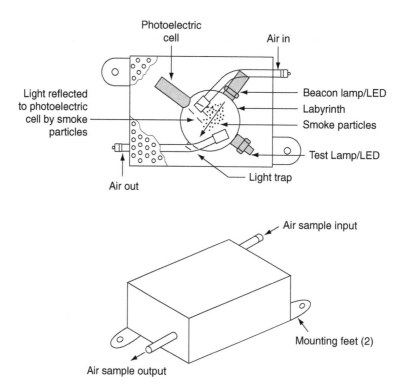

Figure 16.11 Optical smoke detector principles

detector uses a very small amount of **radioactive isotope** that is a source of **alpha radiation**. The radiation passes through an air-filled chamber between two electrodes; this radiation allows a low constant current to flow between the electrodes. Smoke entering the chamber absorbs the alpha particles, thereby reducing the ionization and interrupting the current; this interruption is used to trigger a warning circuit. Ionization smoke detectors can be manufactured at a lower cost compared with optical detectors; however they have several disadvantages:

- they will also go into an alarm if dust and/or fine water droplets enter the chamber
- the detector's sensitivity changes with pressure, i.e. altitude changes and also with age.

16.3.1.2 Photoelectric smoke detectors

Photoelectric smoke detectors measure light attenuation and/or reflection within a chamber. An air sample is passed through the detector via ducts, or pipes, see Fig. 16.11. This air sample passes trough a labyrinth chamber. A photoelectric cell is located at right angles to the light beam; when smoke particles enter the chamber, some of the light is reflected and scattered off the particles, causing energy to fall on the second cell.

Early types of photoelectric smoke detectors used a low-voltage incandescent light bulb projecting light across the chamber with a photosensitive cell on the other side of the chamber. Incandescent lamps do not have a very long life; this reduces the overall reliability of the detector. Modern aircraft smoke detectors now use a light-emitting diode (LED) as the light source. The smoke detector can be tested by illuminating an internal LED that simulates the deflected light.

Some aircraft use **open area** detectors where the smoke samples pass directly into the chamber. Open area cargo bay smoke detectors are arranged through the aircraft as shown in Fig. 16.12.

Main deck freighter aircraft typically use ducted smoke detectors installed as shown in Fig. 16.13. This configuration has a pair of smoke detectors per zone connected via orifices connected via manifolds. A centralized vacuum manifold draws air through all of the smoke detectors. Alarms are given on a zoned

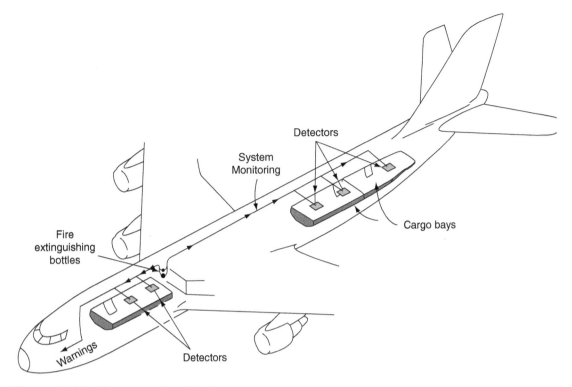

Detectors

System
Monitoring

Cargo bays

Fire
extinguishing
bottles

Warnings

Detectors

Figure 16.12   Cargo smoke detectors

basis when smoke enters through any of the sampling orifices.

A classification process approach is used to develop the fire protection strategy for cargo bay/baggage areas. The safety requirements are specified according to each class of compartments (A through E).

Class A cargo or baggage compartments are compartments where:

- the presence of a fire would be easily discovered by a crew member while at his/her station; and
- each part of the compartment is easily accessible in flight.

Class B cargo or baggage compartments are typically used on aircraft carrying a combination of cargo and passengers on the same deck. In these compartments:

- there is sufficient access during flight to enable crew members to reach any part of the compartment with the contents of a hand fire extinguisher
- when the access provisions are being used, no hazardous quantity of smoke, flames, or extinguishing agent will enter any compartment occupied by the crew or passengers

- there is a separate approved smoke detector or fire detector system to give warning at the pilot or flight engineer station.

Class C cargo or baggage compartments are typically compartments below the main deck of an aircraft, or on the same deck as passengers, but are not readily accessible. In these compartments:

- there is a separate approved smoke detector or fire detector system to give warning at the pilot or flight engineer station
- there is an approved built-in fire extinguishing or suppression system controllable from the cockpit
- there are means to exclude hazardous quantities of smoke, flames, or extinguishing agent from any compartment occupied by the crew or passengers
- there are means to control ventilation and drafts within the compartment so that the extinguishing agent used can control any fire that may start within the compartment.

Class D cargo or baggage compartments were typically compartments below the main deck of an aircraft, or on the same deck as passengers, not readily

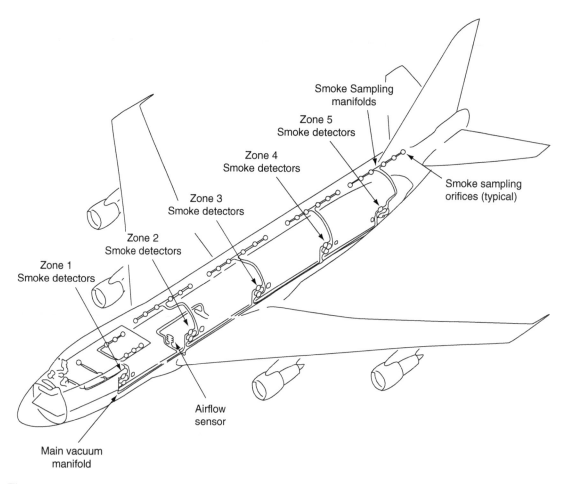

Figure 16.13   Main deck/freighter smoke detectors

accessible and were designed to use oxygen starvation to contain fire. This class of cargo or baggage compartments is being phased out due to incidents involving uncontained fires. They must now meet the same standards as Class C cargo or baggage compartments.

Class E cargo or baggage compartments are compartments on aircraft used only for the carriage of cargo and do not carry passengers. In these compartments:

- there is a separate approved smoke or fire detector system to give warning at the pilot or flight engineers station
- there are means to shut off the ventilating airflow to or within the compartment, and the controls for these means are accessible to the flight crew in the crew compartment

- there are means to exclude hazardous quantities of smoke, flames, or noxious gases from the flight crew compartment
- the required crew emergency exits are accessible under any cargo loading condition.

16.4 Fire extinguishing

16.4.1 Overview

Extinguishers are used in the engines, APUs, cargo bay, cabin and flight compartment. The type of fire extinguishing agent used on an aircraft, and the method of activating the extinguisher, depends on its application. The type of agent used depends primarily on the type of fire being addressed; other factors to be considered include toxicity and storage temperature. Extinguishers are constructed from a metal casing that contains the agent and a means of expelling the agent into the fire. Some agents, e.g. water and dry powder, are expelled from the container by high-pressure gas; some agents, e.g. carbon dioxide, are gaseous at room temperature, and can be stored as liquids under pressure. The container is fitted with a discharge head; this contains a valve that is opened to release the agent. The discharge valve can be opened manually, electrically or automatically by exposure to heat.

16.4.2 Cabin/flight compartment

Hand-held fire extinguishers are portable and operated by first removing a safety pin and then squeezing the handle. This instinctive squeezing action punctures a diaphragm in the discharge head and the agent is expelled under pressure through the aperture left by the ruptured diaphragm.

Waste bins located in the aircraft cabin washrooms are vulnerable from fire started by passengers disposing of cigarettes; this threat is largely eliminated by no-smoking flights; however, self-contained extinguishers are still used. The extinguisher is fitted with a pipe that is plugged by a fusible material; when the tip of the pipe is exposed to a predetermined temperature (typically 75°C), the plug melts and the extinguishing agent is released automatically into the fire.

16.4.3 Engine/APU and cargo bay

Engine/APU and cargo bay fire extinguishers are secured to the aircraft or engine structure outside of their respective fire zones; the agent is directed into the fire zone via a pipe. In some aircraft this pipe feeds into a manifold to distribute the agent. The containers (often referred to as **bottles**) are either cylindrical or spherical in shape; a typical engine fire 'bottle' installation is shown in Fig. 16.14.

This type of extinguisher is filled with a vaporizing liquid agent pressurized with dry nitrogen to 450 psi at 20°C. Figure 16.15 illustrates a two-outlet fire extinguisher. The discharge head contains a **cartridge** unit formed by a small **pyrotechnic** device (sometimes called a **squib)** fitted with a heating coil.

When the fire switch is operated, current is passed through the heater coil. This causes the pyrotechnic charge to ignite, creating a rapid increase in pressure within the discharge head, thereby rupturing a **frangible diaphragm**. The agent is expelled under pressure

Figure 16.14 Engine fire extinguisher installation

through the aperture left by the ruptured diaphragm. When the agent has been expelled, a discharge plug moves to the end of the discharge head; this causes an indicator pin to become visible. On the aircraft fuselage, a polythene plug is forced into the position shown in Fig. 16.16; this provides additional confirmation that the extinguisher has been fired. If the fire extinguisher is exposed to high temperatures outside of its specified range, a pressure relief valve is opened and the contents discharged overboard.

The engine/APU fire extinguisher cartridge (or squib) is energized when the pilot operates the engine fire handle as previously described. Many aircraft are installed with extinguishers that have two or three outlets; a dedicated cartridge operates each outlet. These types of extinguishers are referred to as two-shot or three-shot bottles depending on the construction. The fire extinguisher outlets are connected to two separate engines; this provides the opportunity to discharge a second fire extinguisher into one engine in the event that the first discharge did not extinguish the fire, see Fig. 16.17. The so-called **two-shot fire extinguisher** is controlled from the fire handle as before; the first shot is discharged when the handle is turned in one direction, the second shot is discharged by twisting the handle in the opposite direction.

Key maintenance point

A discharge plug on the aircraft fuselage provides confirmation that the fire extinguisher has been fired.

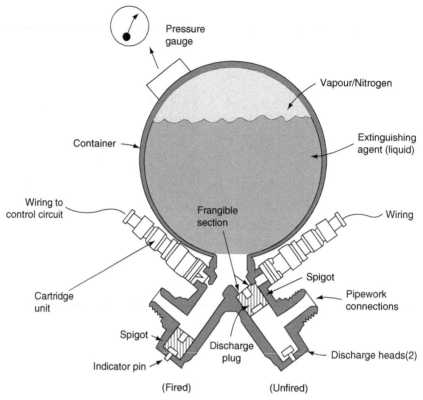

Figure 16.15   Engine fire extinguisher features

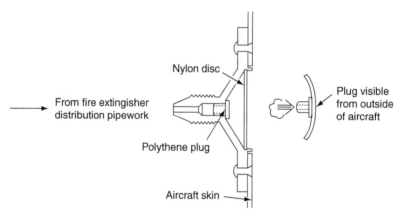

Figure 16.16   Fire extinguisher discharge plug

Cargo bay fire extinguishers are operated from a control panel selector. A typical cargo bay extinguishing system is illustrated in Fig. 16.18; this system features two fire extinguishers. The first bottle provides a high rate discharge to 'knock down' the fire; the second provides a low rate discharge to prevent re-ignition.

16.4.4 Fire extinguisher maintenance

Depending on the construction of the fire extinguisher, some of its contents could escape through the discharge head. Some extinguishers are **hermetically** sealed, some are constructed with gaskets. It is

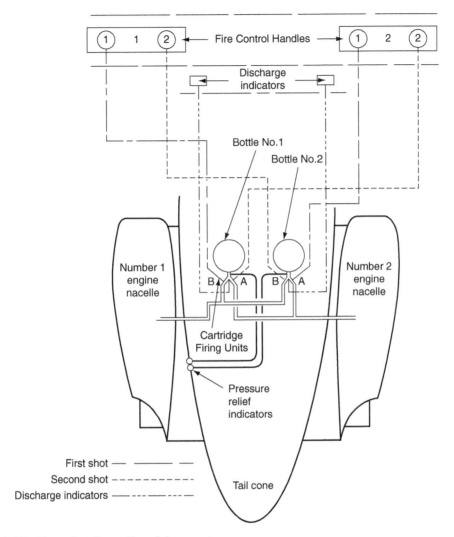

Figure 16.17   Two-shot fire extinguisher system

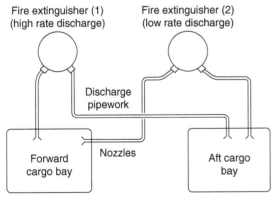

Figure 16.18   Cargo fire extinguisher system

essential that any loss of gas pressure and/or extinguishing agent be established. One way of ascertaining this is to weigh the extinguisher on a periodic basis; the disadvantage of this method is the cost of removal/installation together with the logistic costs of transporting the item back to a workshop, together with the additional spares requirements. Most of these fire extinguishers are fitted with a means of indicating the **pressure** of the contents. This can either be a direct reading gauge, typically using a Bourdon tube, or a pressure switch that is linked via an indicating system to provide a warning to the crew that they have an extinguisher with low pressure; either from a leak or when the extinguisher has been discharged.

16.5 Multiple choice questions

1. The action of twisting a fire handle closes micro-switches that:
   - (a) activate the engine fire extinguisher
   - (b) cancels the alarm
   - (c) shuts off the fuel.

2. When eutectic salt melts, the resistance between the centre wire and outer sheath:
   - (a) drops very rapidly and the control unit resets the warning
   - (b) increases very rapidly and the control unit signals a warning
   - (c) drops very rapidly and the control unit signals a warning.

3. Open area smoke detectors rely on the transfer of particulate matter from the source of fire to the detector by:
   - (a) convection
   - (b) radiation
   - (c) conduction.

4. Dual-loop fire detector systems can be configured to only go into alarm if:
   - (a) both loops sense fire, thereby reducing the probability of a false warning
   - (b) either loop senses fire, thereby reducing the probability of a false warning
   - (c) both loops sense fire, thereby increasing the probability of a false warning.

5. Multiple bimetallic overheat detectors are connected in parallel to provide:
   - (a) variable alarm temperatures
   - (b) increased detection probability
   - (c) linear fire detection.

6. A discharge plug on the aircraft fuselage provides confirmation that the fire extinguisher:
   - (a) is fully pressurized
   - (b) squib is still intact
   - (c) has been fired.

7. When a pneumatic fire detector is rapidly heated; hydrogen is liberated causing:
   - (a) sufficient gas pressure to close the alarm switch
   - (b) the integrity switch to open
   - (c) the alarm switch to reset.

8. Photoelectric smoke detectors measure:
   - (a) light attenuation and/or reflection within a chamber
   - (b) radiated heat within a chamber
   - (c) absorption of alpha particles within a chamber.

9. When a thermistor sensor is heated, the resistance of the insulating material:
   - (a) decreases because of its negative temperature coefficient
   - (b) decreases because of its positive temperature coefficient
   - (c) increases because of its negative temperature coefficient.

10. To operate an engine fire extinguisher, the fire handle is:
    - (a) twisted
    - (b) twisted and then pulled
    - (c) pulled and then twisted.

Chapter 17 Terrain awareness warning system (TAWS)

During the 1970s, studies were carried out by accident investigators and regulatory authorities to examine one of the most significant causes of aircraft accidents of the time: **controlled flight into terrain** (CFIT). This can be defined as an accident where a serviceable aircraft, under the control of a qualified pilot, inadvertently flies into terrain, an obstacle or water. CFIT accidents usually occur during poor visual conditions, often influenced by other factors, e.g. flight crew distraction, malfunctioning equipment or air traffic control (ATC) miscommunication. With CFIT, the pilots are generally unaware of this situation until it is too late. The outcome of these investigations was that many CFIT accidents could be avoided with a ground proximity warning system (GPWS). A system was developed in 1967 to alert pilots that their aircraft was in immediate danger of CFIT. This system was further developed into the enhanced ground proximity warning system (EGPWS) by adding a forward-looking terrain avoidance (FLTA) feature, made possible via global positioning system technology. This chapter describes the generic name given to this type of protection: terrain awareness warning system (TAWS).

17.1 System overview

The terrain awareness warning system comprises aircraft sensors, a computing function and warning outputs. The computer receives numerous inputs from on-board sensors and processes these to determine if a hazardous situation is developing. The TAWS computing function can either be from a self-contained unit; alternatively, it can be integrated within another aircraft system.

Early ground proximity warning systems were based on aircraft sensors that detected:

- barometric altitude
- vertical speed
- radio altitude.

These early systems warned the crew of excessive descent rates or if they were flying towards rising terrain. TAWS can be considered as a **ground proximity warning system** (GPWS) combined with **forward-looking terrain avoidance** (FLTA), see Fig. 17.1. The combined systems provide a significant

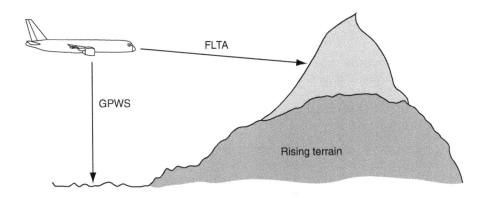

TAWS – Terrain Awareness Warning System
GPWS – Ground Proximity Warning System
FLTA – Forward Looking Terrain Awareness

Figure 17.1 TAWS overview

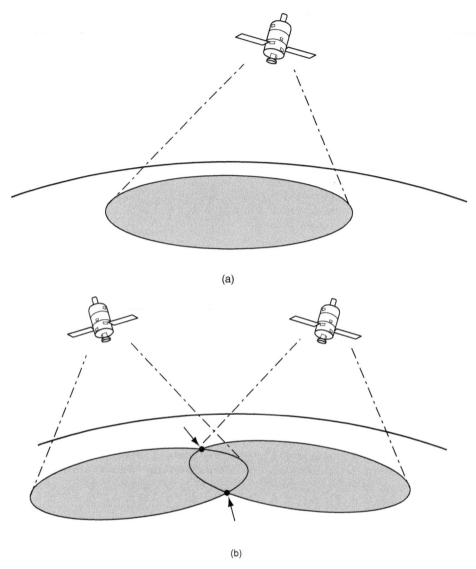

Figure 17.2 GPS principles: (a) Single satellite describes a circle on the earth's surface, (b) Two satellites define two unique positions

increase in safety margins by combining accurate position information (via the GPS) together with a terrain database.

GPS is an artificial constellation of navigation aids that was initiated in 1973 and referred to as Navstar (navigation system with timing and ranging). The global positioning system was developed for use by the US military; the first satellite was launched in 1978 and the full constellation was in place and operating by 1994. GPS is now widely available for use in many applications including aircraft navigation; the system calculates the aircraft position by triangulating the distances from a number of satellites, see Fig. 17.2.

Three satellites are required for the receiver to calculate a unique **lateral** position; a fourth satellite provides altitude, i.e. a **three-dimensional** position.

The availability of GPS technology thereby provides FLTA capability; it can look ahead of and

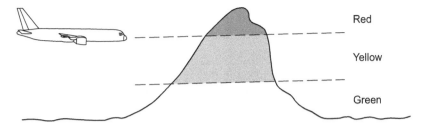

Figure 17.3   Obstacle clearance/colour coding

below the aircraft flight path. This can be depicted on a display that is colour-coded (Fig. 17.3) to show the required obstacle clearance:

- **red** is used to indicate terrain above the aircraft's current altitude
- **yellow** is dependent on the flight phase, i.e. en route, terminal areas or approach
- **green** areas on the display are safe in terms of required terrain clearance.

The TAWS computer function creates a four-dimensional position comprising: latitude, longitude, altitude and time. It compares this position with the on-board database that contains details of terrain, obstacles and runways to determine any conflicts. (More information on GPS is provided in a related book in the series, *Aircraft Communications and Navigation Systems*.)

Key point

TAWS can be considered as a ground proximity warning system (GPWS) combined with forward-looking terrain avoidance (FLTA).

17.2 System warnings and protection

Cautions and warnings are provided to the crew depending on the level of CFIT threat; these include visual indications and audible announcements. TAWS equipment is designed to provide the flight crew with adequate time to take appropriate action. Equipment is classified either as Class A or B depending on the level of protection that the system provides.

In general terms, passenger carrying aircraft require **TAWS-A**, smaller aircraft require **TAWS-B**; see Tables 17.1 and 17.2 for a summary of features.

Cautionary alerts are given by TAWS that require immediate crew awareness; subsequent corrective action will normally be necessary. Warning alerts are given for a terrain threat that requires immediate crew action. Aural alerts are synthesized voice messages that are produced through the audio panel into crew headsets and through the flight compartment speakers. Messages are presented on a suitable electronic display in text format, typically on a multi-function or other navigation display. Electronic navigation displays provide situational awareness, i.e. direction of flight and the location of nearby airports (given as four-letter codes). Local terrain features can be **overlaid** onto this presentation. Messages are either red or amber, depending on the threat level. Master caution and GPWS lights are illuminated depending on the threat level.

We have seen from the above that TAWS can be considered as a ground proximity warning system (GPWS) and forward-looking terrain avoidance (FLTA). Forward-looking terrain avoidance produces alerts from obstacles ahead of the aircraft using a terrain data base, GPS position and other reference data, e.g. aircraft speed, etc. The ground proximity warning system produces alerts from the terrain below the aircraft and configuration inputs, e.g. gear and flap settings during approach and landing. FLTA features in Class A and B TAWS are identical and provide the same level of safety for the aircraft. The primary differences between Class A and B TAWS are in the GPWS features and terrain awareness display. Class A TAWS can be considered as being **autonomous**, i.e. all the aircraft reference inputs are from independent sources. Class B TAWS derives much of its references from within the TAWS computer (or TAWS section of

Table 17.1   TAWS: features, purpose and application (Class A or Class B)

Feature	Purpose	A	B
Forward-looking terrain avoidance (FLTA)	The system looks ahead of and below the aircraft's lateral and vertical flight path to provide a suitable alert if a potential threat exists. FLTA sub-functions are detailed below	✓	✓
FLTA reduced required terrain clearance alerts	Generated when the aircraft is currently above the terrain of the projected flight path of the aircraft and terrain clearance ahead of the aircraft is considered unsafe	✓	✓
FLTA imminent terrain impact alert	Generated when the aircraft is currently below the elevation of local terrain along the lateral flight path and when the projected vertical flight path is considered unsafe	✓	✓
FLTA high terrain impact alert	Generated when the terrain ahead of the aircraft is higher than the projected vertical path	✓	✓
FLTA flight path intent advisory alerts	Generated when the terrain ahead of the aircraft conflicts with the flight plan	✓	✓
Premature descent alert	The system compares the aircraft's current lateral and vertical position with the proximity of the nearest airport. Predicted flight path information and details from the database are used to determine if the aircraft is below the normal approach path for the nearest runway	✓	✓
Attention alerts	The system provides appropriate visual and audio alerts for both cautions and warnings	✓	✓
Terrain awareness display	The system provides terrain information to a suitable display system, e.g. a multi-function display (MFD)	✓	

Table 2   TAWS ground proximity warning Modes (Class A and Class B)

Mode	Description	A	B
1	Excessive descent rate	✓	✓
2	Excessive terrain closure rate	✓	
3	Negative climb rate or altitude loss after take-off	✓	✓
4	Flight into terrain when not in landing configuration	✓	
5	Excessive downward deviation from an ILS glide slope	✓	
6	Altitude callout at 500 feet	✓	✓
7	Wind shear	✓	

another computer). In summary, the following sensor inputs and displays are required by Class A TAWS, but not Class B:

- radar altitude
- air data
- gear position
- map display.

A terrain awareness display is only required for Class-A TAWS, although the display is found in many Class-B installations as an option. The graphical

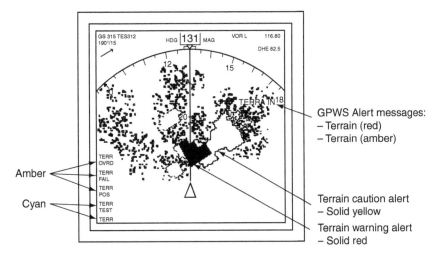

Figure 17.4 MFD/navigation display

display of terrain relative to the aircraft position and projected flight path is a major enhancement to **situational awareness**. Terrain can be displayed on general purpose multi-function displays (MFDs) or on a navigation display (ND), see Fig. 17.4. Colour-coding and intensity are used to depict terrain relative to the aircraft.

Displays can be either a profile view or plan view. In addition to the terrain map, text messages are produced depending on system status and the terrain awareness threat level. When integrated with an electronic display, the surrounding terrain can be viewed relative to the aircraft position; this provides strategic terrain information up to 30 minutes in advance of a potential terrain conflict.

Test your understanding 17.1

Define the meaning of CFIT.

Key point

A terrain awareness display is only required for Class-A TAWS.

The two classes of TAWS are aimed at different types of aircraft; in broad terms Class A is intended for larger passenger aircraft, Class B is for general aviation aircraft. By deriving critical parameters such

as synthetic radar altitude, the cost of Class B TAWS is substantially reduced in terms of both equipment and certification costs.

17.3 External references

The ground proximity modes are derived from a number of signal inputs to the computer; each input is received from a variety of aircraft system outputs including:

- barometric air data
- radio altitude
- instrument landing system
- attitude and heading reference system.
- global navigation satellite system.

The source of these inputs to the TAWS computer will vary depending on aircraft type.

Barometric air data includes altitude and vertical rate, or vertical speed. These two parameters are usually derived from an air data computer **(ADC)**. This computer combines the functions of individual instruments, and can be used to determine data from the aircraft's pitot-static system:

- altitude
- vertical rate, or speed
- calibrated airspeed
- mach number.

Air data computers usually have an input of total air temperature to enable computation of static air temperature and true airspeed.

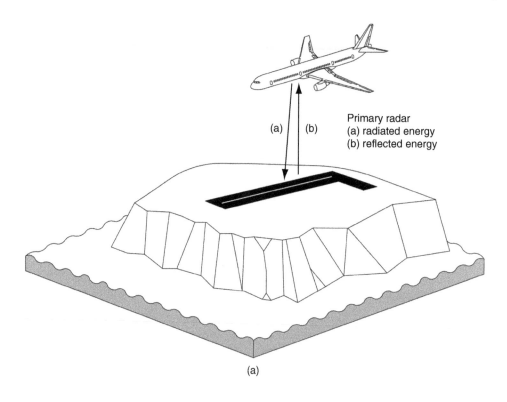

(a)

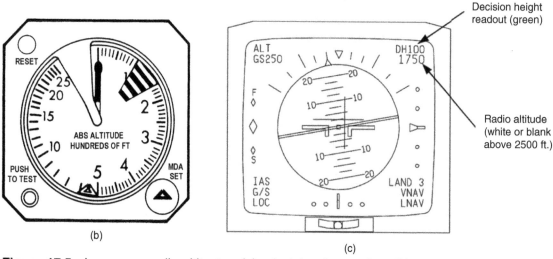

(b)

(c)

Figure 17.5 Low-range radio altimeter: (a) principle of operation, (b) analogue indication, (c) electronic indication

The **low-range radio altimeter** (LRRA) is a self-contained vertically directed primary radar system operating in the 4.2 to 4.4 GHz band. Airborne equipment comprises a combined transmitting and receiving antenna, LRRA transmitter/receiver and a flight deck indicator. Most aircraft are fitted with two independent systems. Radar energy is directed via a transmitting antenna to the ground (Fig. 17.5(a)); some of this energy is reflected back from the ground and is collected in the receiving antenna. This is

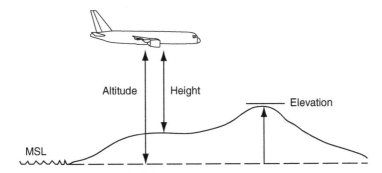

Figure 17.6 Elevation, altitude and height

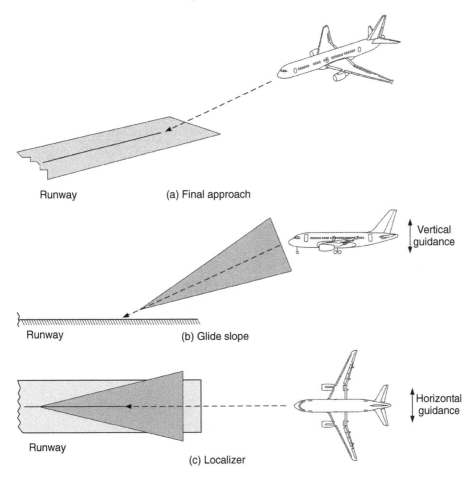

Figure 17.7 Instrument landing system

displayed to the crew on an dedicated display (Fig. 17.5(b)) or an electronic instrument (Fig. 17.5(c)).

For aircraft operations, and in particular for TAWS, it is important to differentiate between elevation, altitude and height. Referring to Fig. 17.6, the basic reference point is mean sea level (MSL). Terrain that

rises above MSL is measured as **elevation**. The aircraft's **altitude** is measured (normally in feet) above MSL, whereas the aircraft's **height** is measured above the terrain.

The **instrument landing system** is used during the final approach and is based on directional beams

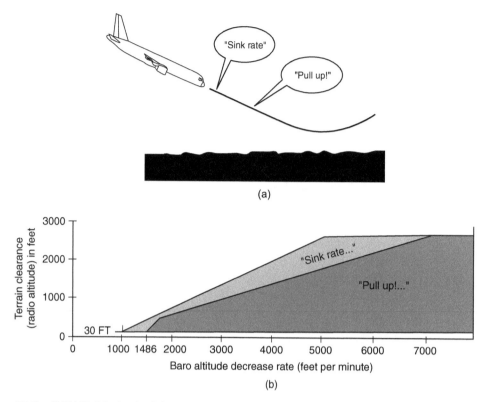

Figure 17.8 GPWS Mode 1: (a) excessive descent rate profile, (b) excessive descent rate envelope

propagated from two transmitters at the airfield; see Fig. 17.7. One transmitter (the localizer) guides the aircraft in the horizontal plane. The second transmitter (the glide slope) provides guidance in the vertical plane and has a range of approximately 10 nm.

The **attitude** of an aircraft (pitch and roll) is sensed by gyroscopes. These sensors provide reference outputs that are processed to develop navigation and attitude data. Larger passenger aircraft derive this data from an inertial reference system (IRS). Developments in micro-electromechanical systems (MEMS) technology have led to silicon accelerometers that are more reliable and can be manufactured onto an integrated circuit. MEMS is the integration of mechanical elements, sensors and electronics on a common silicon substrate through micro-fabrication technology. This technology is being introduced onto general aviation aircraft for attitude and heading reference systems (AHRSs).

Key point

Warning alerts are given for a terrain threat that requires immediate crew action.

Key Point

First-generation GPWS technology used the radio altimeter to provide warning of terrain clearance. GPWS would not detect an aircraft flying across level ground, toward a vertical cliff.

Vertical position for TAWS may come from a barometric source such as an altimeter, air data computer, or from a geometric source, such as GNSS provided that accuracy requirements are met through equipment approvals, and certification specifications.

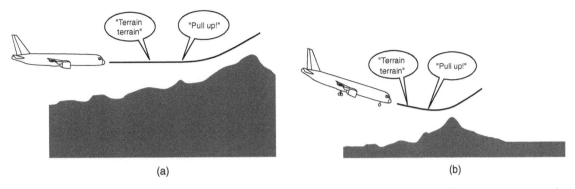

Figure 17.9 GPWS Mode 2 profiles: (a) excessive terrain closure rate, (b) excessive terrain closure rate, flaps and gear down

17.4 Ground proximity modes

17.4.1 Mode 1: excessive descent rate

Referring to Fig. 17.8, Mode 1 GPWS detects excessive sink (or descent) rate. The computer will provide a repeated aural alert of 'SINK RATE'. This is followed by a repeated aural warning of 'PULL UP' if the excessive descent rate continues, together with illumination of the **ground proximity warning light**. Amber and red text messages consistent with the aural messages are given on the navigation display. The computer derives excessive descent rate alert and warning from the following primary inputs:

- radio altitude
- vertical speed
- barometric altitude.

There are two envelopes within the TAWS software algorithms. The first envelope provides the 'SINK RATE' caution, if the aircraft flight path continues into the second envelope, the 'PULL UP' warning is produced. Mode 1 is inhibited below a minimum height, typically 30 feet; it is also independent of aircraft configuration, e.g. flaps and/or undercarriage positions.

17.4.2 Mode 2: excessive terrain closure rate

This mode provides protection in one of two ways (see Fig. 17.9): excessive terrain closure rate with flaps not in landing position (Mode 2A) and excessive terrain closure rate with flaps and gear in the landing position (Mode 2B).

Entering the Mode 2A envelope (see Fig. 17.10) will cause the repeated 'TERRAIN-TERRAIN' aural alert to sound. This is followed by the 'PULL UP' warning if the closure rate is too large. Airspeed schedules further refine the Mode 2A profile.

The height of the Mode 2B envelope floor varies dependent upon barometric descent rate and airspeed (Fig. 17.11). The cautionary voice message is accompanied with an amber text message consistent with the aural message. If the terrain closure rate penetrates the warning zone, an aural 'PULL UP' voice message is produced together with a red text message consistent with the aural message. Mode 2 is inhibited below a minimum terrain clearance, typically 30 feet.

17.4.3 Mode 3: negative climb rate or altitude loss after take-off or go around

During an approach, in the landing configuration, Mode 3 arms when the aircraft descends below a specified altitude as illustrated in Fig. 17.12. The mode becomes active if either gear or flaps are retracted.

The cautionary alerts are 'DON'T SINK' or 'TOO LOW TERRAIN' (depending on what has been programmed into the computer during installation) together with an amber text message consistent with the aural message. Mode 3 does not have a warning zone; it is inhibited below a minimum terrain clearance and barometric altitude.

17.4.4 Mode 4: flight into terrain when not in landing configuration

This mode is programmed in one of three ways; see Fig. 17.13. Mode 4A is for unsafe terrain clearance

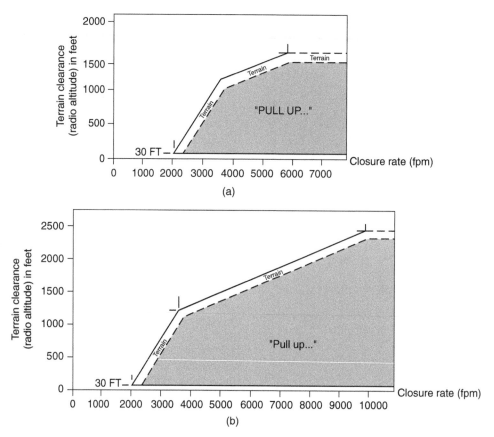

Figure 17.10 GPWS Mode 2A envelopes: (a) excessive terrain closure rate (airspeed < 220 knots), (b) excessive terrain closure rate (airspeed > 220 knots)

with flaps in the landing position, but the gear is up; this will initially generate a 'TOO LOW TERRAIN'. This is followed by a 'TOO LOW GEAR' voice message with an amber text message consistent with the aural message.

Alternatively, if the flaps are up and the gear is down (Mode 4B), the voice message 'TOO LOW FLAPS' is produced. The third condition (Mode 4C) is caused by unsafe terrain clearance after take-off or go around; the will produce a 'TOO LOW TERRAIN' caution. The various envelopes associated with Mode 4 are illustrated in Fig. 17.14.

17.4.5 Mode 5: excessive downward deviation from an ILS glide slope

Mode 5 provides protection on the glide slope in both hard and soft alert areas (Fig. 17.15(a)). This mode becomes active when the number one

ILS receiver has a valid signal, gear is down and radio altitude is between the limits shown in Fig. 17.15(b).

Descent below the glide slope (G/S) causes 'GLIDE SLOPE' voice message and an amber text message. Mode 5 can be inhibited by selecting the GPWS G/S push button switch when below 1000 ft RA. The mode automatically reactivates for a new envelope penetration.

17.4.6 Mode 6 Altitude callouts

When the aircraft descends to 500 feet above runway elevation, a voice message of 'Five hundred' is given; there are no text messages generated with this Mode. Within Mode 6, some manufacturers offer predetermined altitude callouts, see Fig. 17.16.

This mode can also incorporate other optional features such as bank angle protection, see Fig. 17.17.

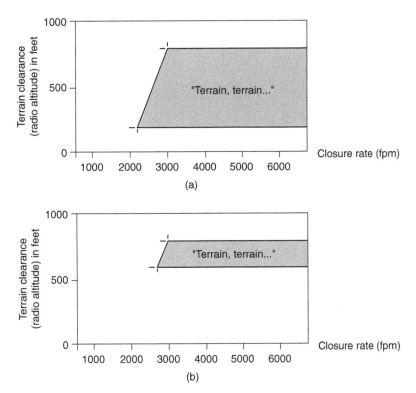

Figure 17.11 GPWS Mode 2B envelopes: (a) excessive terrain closure rate – flaps down (barometric rate < 400 f.p.m.), (b) excessive terrain closure rate envelope – flaps down (barometric rate > 1000 f.p.m.)

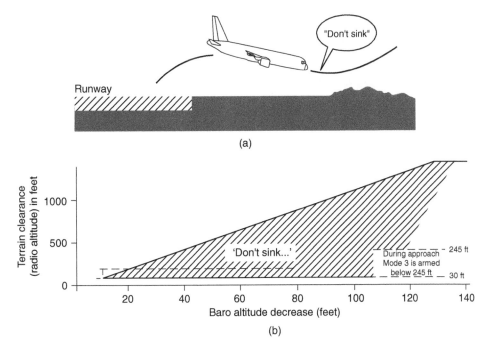

Figure 17.12 GPWS Mode 3: (a) altitude loss after take-off profile, (b) altitude loss after take-off envelope

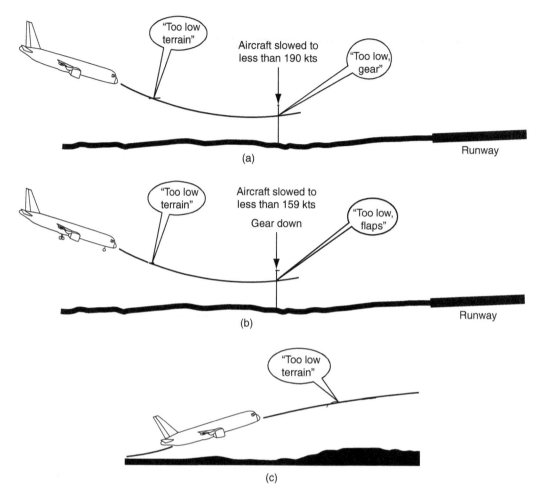

Figure 17.13 GPWS Mode 4 profiles: (a) flight into terrain when not in landing configuration – gear up, (b) flight into terrain when not in landing configuration – gear down/flaps up, (c) flight into terrain when not in landing configuration (take-off or go-around)

17.4.7 Mode 7 wind shear

Additional features on larger aircraft include a **wind shear** functionality within the TAWS computer. This is incorporated as **ground proximity Mode 7**. Wind shear can occur in both the vertical and horizontal directions; this is particularly hazardous to aircraft during take-off and landing. Specific weather conditions known a **microbursts** cause short-lived, rapid air movements from clouds towards the ground. When the air from the microburst reaches the ground it spreads in all directions, this has an effect on the aircraft depending on its relative position to the microburst.

17.4.7.1 Microburst

Referring to Fig. 17.18, when approaching the microburst, it creates an **increase** in headwind causing a temporary increase of airspeed and lift; if the pilot were unaware of the condition creating the increased airspeed; the normal reaction would be to reduce power.

When flying through the microburst, the headwind reduces and the aircraft is subjected to a **downdraught**. As the aircraft exits the microburst, the downdraught now becomes a tail wind, thereby **reducing** airspeed and lift. This complete sequence of events happens very quickly, and could lead to a

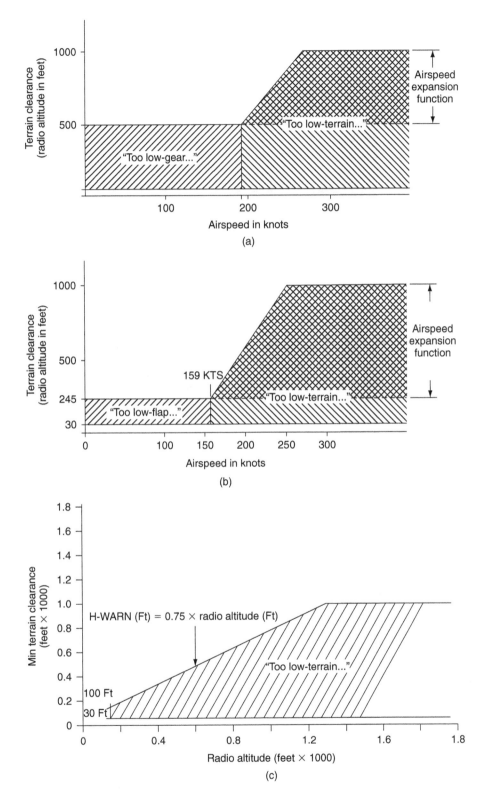

Figure 17.14 GPWS Mode 4 envelopes: (a) flight into terrain when not in landing configuration – gear up, (b) flight into terrain when not in landing configuration – gear down/flaps up, (c) flight into terrain when not in landing configuration (take-off or go-around)

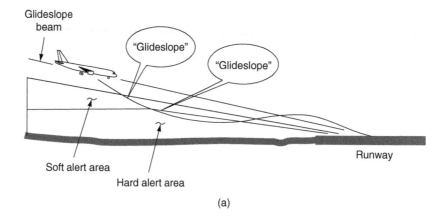

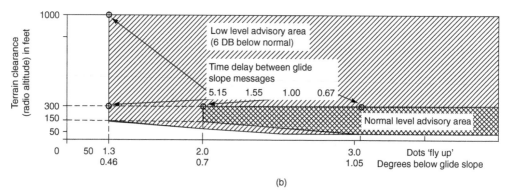

Figure 17.15 GPWS Mode 5: (a) excessive downward deviation from an ILS glide slope profile, (b) excessive downward deviation from an ILS glide slope envelope

sudden loss of airspeed and altitude. In the take-off and climb-out phase of flight, an aircraft is flying just above stall speed; wind shear is a severe threat. During approach and landing, engine thrust will be low; if a microburst is encountered, the crew will have to react very quickly to recognize and compensate for these conditions.

17.4.7.2 Doppler radar

Modern weather radar systems are able to detect the horizontal movement of droplets using **Doppler shift** techniques. Doppler is usually associated with self-contained navigation systems. (This subject is described further in *Aircraft Communications and Navigation Systems.*) The Doppler effect can be summarized here as: 'the frequency of a wave apparently changes as its source moves closer to, or farther away from an observer'. This feature allows wind shear

created by microbursts to be detected. Referring to Fig. 17.19, the microwave energy pulses from the antenna are reflected by the water droplets as in the conventional weather radar system.

Using the Doppler shift principle, the frequency of energy pulse returned by droplets (B) moving towards the aircraft will be at a higher frequency than the transmitted frequency. The frequency of energy pulse returned by droplets moving away from the aircraft (A) will be at a lower frequency than the transmitted frequency. These Doppler shifts are used to determine the direction and velocity of the air movement resulting from a microburst.

17.4.7.3 Predictive wind shear

Visual and audible warnings of **predictive** wind shear conditions are provided to the crew. The visual warnings are given on the multi-function or naviga-

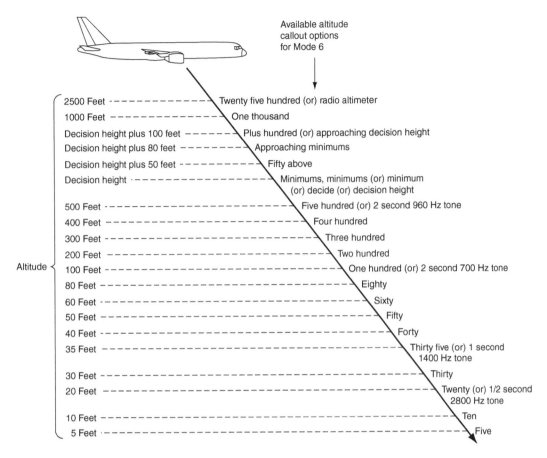

Figure 17.16 GPWS Mode 6 altitude callouts

Figure 17.17 GPWS Mode 6 bank angle

tion displays using a wind shear icon and message, together with warning lights on the glare shield. Audible warnings are provided as computer generated voice alerts over the cockpit speakers, typically 'WIND SHEAR WIND SHEAR'. The system automatically configures itself for the phase of flight; it is normally inhibited below 50 feet radio altitude during take-off and landing. During an approach, Mode 7 is activated below 2500 feet radio altitude and has typical envelope incorporated as illustrated in Fig. 17.20.

Key point

When approaching a microburst, it creates an increase in headwind, causing a temporary increase of airspeed and lift.

17.5 Forward-looking terrain avoidance (FLTA)

This TAWS feature uses aircraft position, altitude and **flight path** information together with details of

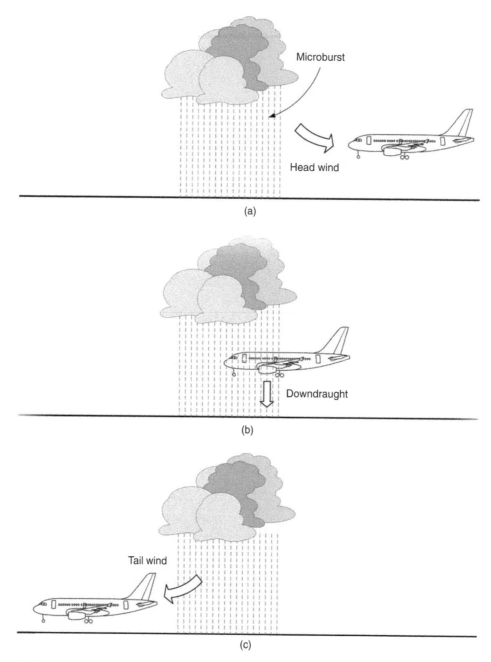

Figure 17.18   Microbursts

terrain, obstacles and known runways in the database. Referring to Fig. 17.22, the system provides an envelope of protection by comparing the flight path with the details contained in the database. There are two **look-ahead** distance features; these are functions of terrain clearance floor height and ground speed.

Referring to Fig. 17.21, alert areas are projected along the flight path in three levels:

- warning
- alert
- caution.

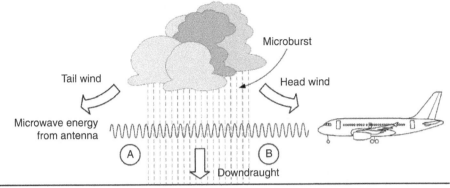

Doppler frequency shift between A and B indicates a wind shear condition

Figure 17.19 Doppler radar

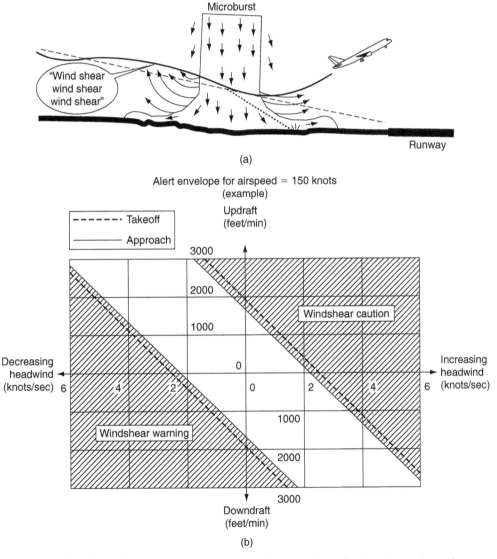

(a)

Alert envelope for airspeed = 150 knots
(example)

(b)

Figure 17.20 GPWS Mode 7: (a) wind sheer detection profile, (b) wind sheer detection envelope

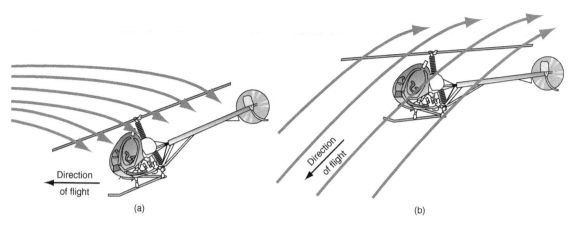

Figure 17.21   Rotorcraft autorotation: (a) normal powered flight, (b) autorotation

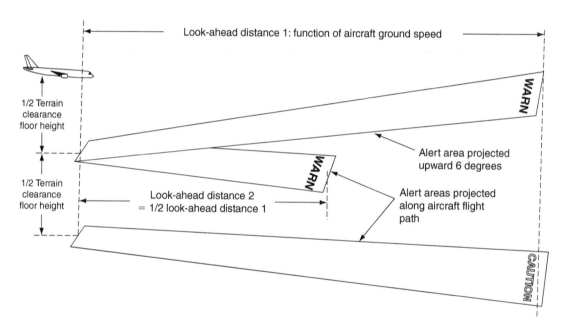

Figure 17.22   FLTA flight path

When a runway has been selected in the navigation system, a terrain clearance profile comes into effect as shown in Fig. 17.22. This profile take into account distance to the runway and height above ground level.

17.6 Rotorcraft TAWS

Rotorcraft, or helicopters, require additional TAWS considerations as result of their unique flying char-

acteristics and operational roles. In addition to the functions already described, systems installed into rotorcraft need to include:

- excessive bank angle protection
- tail strike protection
- autorotation.

During **autorotation**, the relative upward airflow causes the main rotor blades to rotate at their normal speed; the blades are 'gliding' in their rotational plane, see Fig. 17.21. Autorotation facilitates a con-

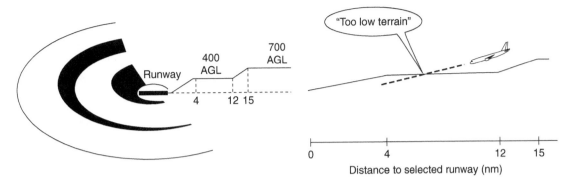

Figure 17.23 FLTA runway

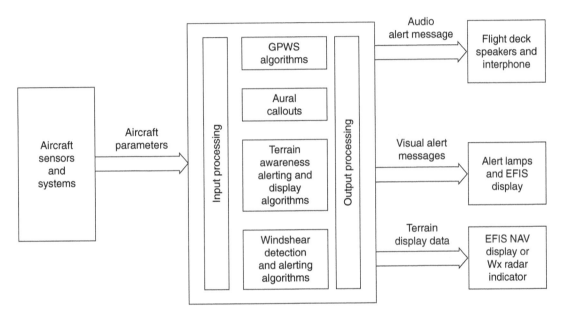

Figure 17.24 TAWS architecture

trolled descent through to an emergency landing in case of power plant failure; this flight regime has to be recognized by TAWS.

Many national aviation authorities including the FAA and CAA recommend the use of TAWS in helicopters performing specialized roles, e.g. emergency medical services and oil-pipe inspections these often involve low-level operations at night or in low visibility.

17.7 Architecture and configurations

Typical Class A system architecture is illustrated in Fig. 17.24. Larger aircraft require TAWS Class A, smaller aircraft require TAWS Class B; this is reflected in the complexity (hence cost) of the hardware and certification requirements. The various TAWS configurations also take into account the type of interfaces that TAWS requires, e.g. whether the existing aircraft sensors are digital or analogue.

Smaller general aviation aircraft are adopting a simple terrain only feature. This is normally a sub-feature of a GPS navigation system, referred to as **Class C TAWS**. Typical features include:

- terrain data base
- terrain elevations relative to the aircraft

- visual alerting
- reduced terrain clearance avoidance
- imminent terrain impact avoidance
- premature descent alert.

17.8 Future developments

TAWS is specified in transport category aircraft as a result of mandatory rulemaking. Two main categories of general aviation (GA) aircraft are not included in these rules, however: non-turbine, fixed-wing aircraft (particularly privately owned aircraft), and rotorcraft. Both of these categories fall under the voluntary Class C TAWS arrangement. Although there are no firm indications of general aviation's adoption of Class C TAWS equipment, the functionality is often included within existing equipment, and many aircraft owners would prefer some level of protection.

Some manufacturers are developing an **assisted recovery** concept, where the aircraft's automatic flight control system (AFCS) automatically performs the pull-up manoeuvre following a TAWS alert. The system could also steer the aircraft away from prohibited airspace for security purposes.

Performance-based systems are being developed that utilize data from an active aircraft database that stores the aircraft's performance characteristics under varying conditions of engine thrust, aircraft weight, altitude, temperature and other inputs. When a TAWS 'PULL UP' warning is given, the system determines if the aircraft can actually climb at the required rate to avoid impacting the terrain ahead. If the system predicts that this is not possible, the crew receive an 'AVOID TERRAIN' command, and the system indicates an alternate manoeuvre to fly the aircraft into safer areas.

To minimize nuisance warnings, the system limits its look-ahead terrain search profile to a narrow sector either side of the aircraft's projected track. Terrain advisory lines are indicated on a wider scan either side of the projected track to visually alert the crew of potential obstructions. When a terrain avoidance manoeuvre is initiated, the system switches to a wider scanning sector and anticipates the rate of turn while searching the terrain ahead from the original track. Terrain that the aircraft can safely climb over is then depicted with appropriate graphics on the navigation display.

The ground collision avoidance module (GCAM) is a predictive alerting technique derived from a terrain-following/terrain-avoidance system used by military aircraft. GCAM uses a collision prediction and alerting (CPA) function that combines terrain and airport databases with a Model of the aircraft's climb capability. The CPA function predicts potential terrain conflicts from the terrain environment and the aircraft's predicted flight path.

The former correlates precise aircraft position information from the GPS and/or other navigation sources with the digital database. The flight path prediction uses current aircraft flight path data to project the flight path more than 2 minutes ahead of the aircraft, and provides visual cautions up to 4 minutes ahead. Alerts are generated when the CPA system predicts the flight path could intersect the terrain, providing **terrain avoidance** rather than just a warning.

17.9 Multiple choice questions

1. A terrain awareness display is only required for:
   (a) Class A TAWS
   (b) Class B TAWS
   (c) GPWS.

2. Mode 1 terrain awareness cautions are given for:
   (a) negative climb rate or altitude loss after take-off
   (b) excessive descent rate
   (c) altitude callout at 500 feet.

3. When approaching a microburst, it creates:
   (a) a decrease in headwind causing a temporary increase of airspeed and lift
   (b) an increase in headwind causing a temporary decrease of airspeed and lift
   (c) an increase in headwind causing a temporary increase of airspeed and lift.

4. Warning alerts are given for a terrain threat that requires:
   (a) immediate crew awareness
   (b) confirmation with air traffic control
   (c) immediate crew action.

5. Forward-looking terrain avoidance (FLTA) looks:
   (a) ahead of and below the aircraft's lateral and vertical flight path

(b) ahead of and above the aircraft's lateral and vertical flight path

(c) either side of and below the aircraft's lateral and vertical flight path.

6. Premature descent alert compares the aircraft's:
   (a) ground speed with the proximity of the nearest airport
   (b) lateral and vertical position with the proximity of the nearest airport
   (c) lateral and vertical position with the proximity of high terrain.

7. Red areas are used on TAWS displays to indicate terrain that is:
   (a) above the aircraft's current altitude
   (b) level with the aircraft's current altitude
   (c) safe in terms of required terrain clearance.

8. The low-range radio altimeter (LRRA) is a:
   (a) self-contained vertically directed primary radar system
   (b) self-contained horizontally directed primary radar system
   (c) self-contained vertically directed secondary radar system.

9. Mode 7 (wind shear) is normally inhibited:
   (a) above 50 feet radio altitude during take-off and landing
   (b) below 50 feet radio altitude during take-off and landing
   (c) during take-off and landing.

10. The TAWS computer function creates a four-dimensional situation comprising:
    (a) latitude, longitude, heading and time
    (b) latitude, speed, altitude and time
    (c) latitude, longitude, altitude and time.

11. Referring to Fig.17.25, this illustrates a:
    (a) FLTA profile
    (b) GPWS Mode 2 profile
    (c) GPWS Mode 1 profile.

12. Referring to Fig. 17.26, this illustrates a:
    (a) GPWS Mode 6 caution
    (b) GPWS Mode 3 caution
    (c) GPWS Mode 1 warning.

13. Referring to Fig.17.27, this is the warning envelope for:
    (a) GPWS Mode 5
    (b) GPWS Mode 2
    (c) GPWS Mode 1.

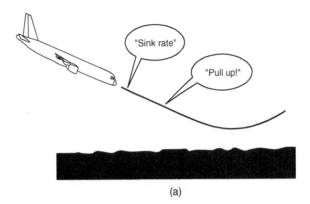

(a)

Figure 17.25 See Question 11

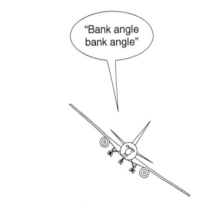

Figure 17.26 See Question 12

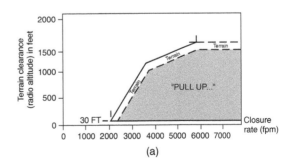

(a)

Figure 17.27a See Question 13

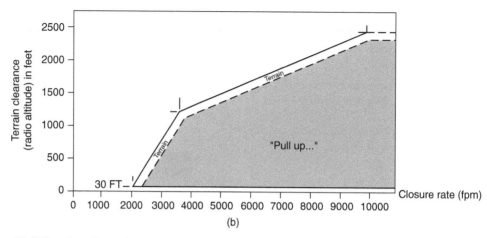

Figure 17.27b   See Question 13

14. GPWS Mode 3 is defined as:
    (a) Excessive terrain closure rate
    (b) Negative climb rate, or altitude loss after take-off or go around
    (c) excessive descent below the glide slope

15. Flight into terrain whilst not in the landing configuration is GPWS Mode:
    (a) one
    (b) three
    (c) four

Chapter 18

Flight data and cockpit voice recorders

Flight data recorders used for accident investigation are mandatory items of equipment in commercial transport aircraft. Efforts to introduce crash-survivable flight data recorders can be traced back to the 1940s; the FDR has now been supplemented with the **cockpit voice recorder** (CVR). Recorders are used after an incident or accident as an integral part of the investigators' efforts to establish the cause(s).

Data recorders can also be used to indicate trends in aircraft and engine performance. Algorithms are established for healthy and normal conditions during the aircraft's flight-testing programme and the early period of service. These algorithms include engine parameters such as engine exhaust temperature, oil pressure and shaft vibration for given speeds and altitudes. These parameters are then monitored during the aircraft's life; any deviations from the norm are analysed to determine if the engine requires inspection, maintenance or removal. This chapter reviews the range of FDR/CVR technologies that are employed for both accident investigation and trend monitoring.

18.1 Flight data recorder history

A series of fatal accidents involving the De Havilland DH106 Comet in 1953 and 1954 led to the grounding of the entire Comet fleet pending an investigation. There were no survivors or witnesses to these accidents and an exhaustive investigation at the Royal Aircraft Establishment at Farnborough, UK, eventually concluded that metal fatigue was the primary cause of the accidents. One of the ideas that came out of these investigations was for the development of a **crash-survivable** device to record the flight crew's conversations and other aircraft data. Doctor David Warren of the Aeronautical Research Laboratories of Melbourne, Australia, was on one of the investigation committees; he concluded that such recorders would be of tremendous benefit in determining the likely cause(s) of an accident. The idea was that design

improvements and/or operating procedures would be introduced to eliminate avoidable accidents from the same cause. The proposed recorder would have to survive both the high g forces and the potential fire resulting from an aircraft crash; the technology required for such a device did not become available until 1958.

Key point

Flight data recorders (FDRs) are often referred to in the press as the 'black box', even though the item is painted bright orange.

Test your understanding 18.1

Describe how data recorders can also be used to indicate trends in aircraft and engine performance.

18.1.1 Scratch foil

The technology used in first-generation flight data recorders was based on a roll of **steel foil tape** embossed with five separate items of recorded data (parameters), see Fig. 18.1; these parameters were:

- heading
- altitude
- air speed
- vertical acceleration
- time.

The foil is made from a high nickel content steel; several hundred feet of foil, approximately 125 mm wide, is rolled onto a spool. When the recorder is switched on, the foil is transferred from a feed spool onto the take-up spool after passing under five recording needles (hardened or diamond-tipped styli) that move across the foil in response to their respective inputs. With the foil tape removed from the recorder, the various traces (Fig. 18.2) are compared with a transparent overlay with markings to measure the recorded parameters.

This early 'scratched-foil' technology gave the investigators more data than had been previously available; however, it was soon realized that five parameters did not provide sufficient data for meaningful accident investigation.

18.1.2 Magnetic recording

Second-generation recorders were introduced onto aircraft during the 1960s based on one of two recording media; **steel wire** or **magnetic tape**. When the recorder is switched on, the wire is transferred onto another drum after passing over record/replay heads. Steel wire recording technology was used by the telegraphy industry from the late 1890s; it was subsequently developed for other industrial applications and home entertainment until the introduction of magnetic tape recorders in the late 1940s. Even with

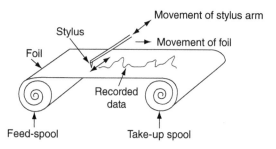

Figure 18.1 Scratched foil recorder principles

very fine diameter of approximately 100 micrometers, steel wire is intrinsically more crash-survivable than magnetic tape, and this medium was used for recording **analogue data** from the 1950s through to the 1970s. Wire recorders have a high media density; this is made possible by the solid metal medium and relatively fast speed (typically 24 inches per second) over the record/replay heads. One hour of recorded data could be achieved with approximately 7000 feet of wire wound onto a three-inch-diameter drum.

The technology required to protect magnetic wire or tape was eventually developed; the first application of magnetic tape recorders was to record the last thirty minutes of crew voice communications and noise within the flight compartment environment. The cockpit voice recorder (CVR) required substantial packaging to enable it to be crash-survivable. This magnetic tape technology (recording data in a **digital format**) was then applied to FDR applications. This third-generation FDR facilitated the recording of many more flight parameters from the airframe and engines while meeting higher crash and fire protection requirements. Inside the write head is a small coil; when voltages are applied to the coil, an electromagnetic field is created; particles in the wire or tape are aligned on a micro-level with the field as it passes over the head.

To analyse electromagnetic recorded tape, an electronic process is required such that digitally coded data can be converted back (decoded) into its analogue form required for presentation and/or display purposes. To read the data stored on the wire/tape, it is passed over a reading head; this is similar in physical construction to the writing head. As each section of wire or tape passes over the head, the particles are aligned in

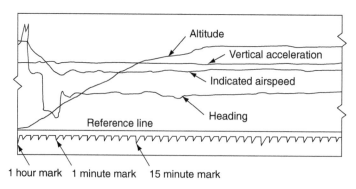

Figure 18.2 Scratched foil recorded data

specific directions, inducing small voltages into the coil. The polarity of the voltages indicates whether a binary one or zero is stored. The voltage signals are processed electronically to produce analogue signals for use in displays or charts.

18.1.3 Solid state data recorders

Solid-state digital flight data recorder (DFDR) technology became commercially viable in the 1990s. Data is stored in semiconductor memory via integrated circuits. Solid-state memory does not require any servicing, maintenance or overhaul. Furthermore, retrieval and interpretation of the data via personal computer-based software is a more efficient process.

Recorders are now also used for reasons other than accident investigation, e.g. to indicate trends in aircraft and engine performance. Algorithms are established for healthy conditions during the aircraft's flight-testing programme and the early period of service; these include engine parameters such as oil temperature, oil pressure and shaft vibration for given speeds and altitudes. These parameters are then monitored during the aircraft's life and any deviations from the norm are analysed to determine if the engine requires inspection, maintenance or removal. Low-cost, high-performance, rugged solutions are required for the processing and data recording. Equipment such as the S3DR product range developed by Specialist Electronics Services (SES) Ltd has many applications in the aerospace and defence industry. The S3DR product range developed and fielded by SES is designed with flexibility and expandability in mind.

The S3DR-F is a panel-mounted data recorder providing a recording solution where space and mass are at a premium. It is specified where access to the removable media is required from within the flight compartment. The unit features a high-speed recording function that is able to write data to a removable solid state memory module from multiple monitored interfaces at speeds up to 35 Mbits per second (MBPS). Memory modules are available in many sizes up to 16 GBytes. A Pentium® compatible processor, allowing application specific data gathering regimes to be accommodated, provides processing functions.

The Solid State Data Recorder-Expandable (S3DR-E) is a high-speed, high-capacity expandable data recorder with extensive data acquisition and communications flexibility. The recorder is built from the concept of multiple recording, data acquisition and communication modules that are assembled together to provide a system that meets the specific bespoke requirements of a customer application.

The recorder contains a master processor module which provides control and configuration of any other modules contained in the system. This master processor provides a high-performance general purpose processor that controls all other data acquisition, recording and communication modules. The processor module also provides multiple high-speed USB 2.0, Giga Ethernet and VGA interfaces.

The baseline S3DR-E recorder is built from a master processor combined with a recording module and an integrated power supply. Alternative recorder configurations can be built from any combination of up to seven data acquisition, communications and recording modules. Each recording module can be built from either a dual PCMCIA recorder with a dedicated I/O processor, or a high-capacity (100 byte), high-speed (up to 480 MBPS) dedicated recording module.

18.2 Mandatory equipment requirements

In the late 1960s, the UK Civil Aviation Authority (CAA) stipulated a need for additional parameters to be recorded; FDRs using scratched foil were developed to mark both sides of the foil. This allowed for the recording of additional parameters, e.g. pitch, roll and flap position. Although this provided additional information for the accident investigator, it increased the complexity of the recorder resulting in lower reliability. Furthermore, the additional information contained on the scratch-foil was becoming harder to read and required high skill levels to interpret the data.

Larger, more complex aircraft were introduced in the 1960s/70s. These aircraft were flying faster, higher and farther than had been previously possible. They had reduced flight crew; the navigator and wireless operators were being replaced by computerized (albeit analogue) avionics bringing reduced operating costs. High-density passenger operations, reduced flight crew and longer flights were becoming the norm. This created new problems for the accident investigators; more data was required to identify the probable cause of an accident. Flight data acquisition units (FDAUs) were introduced to provide an interface between aircraft systems and the FDR. The FDAU collects, or acquires, a variety of analogue signals from around the aircraft, converts them into a single digital data stream to the recorder. In the late 1970s and early 1980s the

digital flight data recorder (DFDR) could record 32 parameters for a 25 hour period. (This represented the round-trip time between cities in the USA, Europe and Asia.)

Key point

The FDAU collects, or acquires, a variety of analogue signals and converts them into a digital data stream to the recorder.

The specific parameters to be recorded are stipulated by the regulatory authorities. In Europe, commercial air transport (fixed wing and rotary wing) aircraft are regulated under JAR OPS 1 and 3 respectively. For flight data recorders, these regulations take into account when the aircraft was first issued with an individual Certificate of Airworthiness. Aircraft operating under JAR OPS 1 are required to install a **digital flight data recorder** (DFDR). The DFDR has to be able to retain the recorded data for a minimum of the last 25 hours of its operation (10 hours for aircraft below 5700 kg maximum take of mass). The flight data recorder must start to record data automatically prior to the aircraft being capable of moving under its own power. The recorder must stop automatically after the aircraft is incapable of moving under its own power. The flight data recorder must have a device to assist with locating that recorder if it is submerged in water.

The mandatory parameters required for an aircraft depend on the size of the aircraft and the prevailing regulatory rules applied to that aircraft. The mandatory parameters required for multi-engine turbine powered with maximum approved seating configuration for more than nine passengers and maximum take-off mass under 27,000 kg are shown in Table 18.1. (Note that all parameters must be referenced to a timescale or relative time count.)

For aircraft over 27,000 kg maximum take-off mass, additional mandatory requirements are required; these are listed in Table 18.2. Aircraft with an electronic flight instrument system (EFIS) are required to have the additional parameters listed in Table 18.3.

Electronic flight instrument system (EFIS) technology was introduced during the 1980s based on the cathode ray tube (CRT); the established technology is now liquid crystal displays (LCDs). The attitude director indicator (ADI) and horizontal situation indicator (HSI) are included within a basic EFIS installation;

Table 18.1 JAR OPS 1 mandatory FDR parameters

- Time or relative time count
- Pressure altitude
- Indicated airspeed
- Heading
- Normal acceleration
- Pitch attitude
- Roll attitude
- Manual radio transmission keying
- Propulsive thrust/power on each engine and thrust/power lever position (if applicable)
- Trailing edge flap or cockpit control selection
- Leading edge flap or cockpit control selection
- Thrust reverser status
- Ground spoiler position and/or speed brake selection
- Total or outside air temperature
- Autopilot and autothrottle mode and engagement status
- Longitudinal acceleration (body axis)
- Lateral acceleration[1]
- Angle of attack[2]

[1] Lateral acceleration on aircraft with 5700 kg maximum take-off mass.
[2] Angle of attack on aircraft with less than 5700 kg maximum take-off mass.

other indications and displays are incorporated through various configurations.

There are some exceptions to the above allowed under JAR OPS 1, that have to be agreed and authorized, e.g. if the required transducer is not available; or if the aircraft system or equipment producing the data needs to be modified; or if the signals are incompatible with the recording system. It is a fundamental requirement that the parameters being recorded are obtained from sources that enable accurate correlation with the information displayed to the flight crew.

Helicopters used for commercial transportation (with maximum certificated take-off mass over 3175 kg) also require a flight data recorder. This must be capable of recording data during at least the last eight hours of its operation; mandatory parameters are shown in Table 18.4. Helicopters over 7000 kg require the additional parameters shown in Table 18.5.

Table 18.2 JAR OPS 1 Mandatory FDR parameters (aircraft over 27,000 kg)

- Primary flight controls – control surface position and/or pilot input (pitch, roll, yaw)
- Pitch trim position
- Radio altitude
- Vertical beam deviation (ILS glide path or MLS elevation)[1]
- Horizontal beam deviation (coupled approaches: ILS localizer or MLS azimuth)
- Marker beacon passage
- Warnings (specified on the aircraft installation)
- Optional navigation receiver frequency selection
- Optional DME[2] distance
- Landing gear squat switch status or air/ground status
- Terrain awareness warning system (TAWS)
- Angle of attack
- Low-pressure warning (hydraulic and pneumatic power)
- Groundspeed
- Landing gear or gear selector position

[1]Instrument landing and microwave landing systems (ILS and MLS) are used for precision approach and landing.
[2]Distance measurement equipment (DME) is a short/medium-range navigation system based on secondary radar principles. (These systems are described in a related book in the series: *Aircraft Communications and Navigation Systems*.)

Table 18.3 Additional parameters for aircraft with electronic instruments

- Selected barometric setting
- Selected altitude
- Selected speed
- Selected mach
- Selected vertical speed
- Selected heading
- Selected flight path
- Selected decision height
- EFIS display format
- Multi function/engine/alerts display format

Table 18.4 Mandatory FDR parameters for commercial transport helicopters (less than 7000 kg)

- Time or relative time count
- Pressure altitude
- Indicated airspeed
- Heading
- Normal acceleration
- Pitch attitude
- Roll attitude
- Manual radio transmission keying
- Power on each engine
- Engine power control position
- Main rotor speed
- Rotor brake (if installed)
- Primary flight controls
- Collective pitch
- Longitudinal cyclic pitch
- Lateral cyclic pitch
- Tail rotor pedal
- Controllable stabilator
- Hydraulic selection
- Warnings
- Outside air temperature
- Autopilot engagement status
- Stability augmentation system engagement

Table 18.5 Additional parameters for helicopters over 7000 kg

- Main gearbox oil pressure
- Main gearbox oil temperature
- Yaw rate or yaw acceleration
- Indicated sling load force (if installed)
- Longitudinal acceleration (body axis)
- Lateral acceleration
- Radio altitude
- Vertical beam deviation (ILS glide path or MLS elevation)
- Horizontal beam deviation (ILS localizer or MLS azimuth)
- Marker beacon passage
- Warnings
- Optional navigation receiver frequency selection
- Optional DME distance
- Optional navigation data
- Landing gear or gear selector position

Key point

The flight data recorder must stop automatically after the aircraft is incapable of moving under its own power.

Key point

Parameters being recorded are obtained from sources that are accurately correlated with the information displayed to the flight crew.

Figure 18.3 Underwater location transmitter

18.3 Flight data recorder (FDR) specifications

Initial FDR technology required that the unit be able to withstand 100 g of impact; the equipment was often installed alongside other equipment in the avionics bay. As more experience was gathered in the operational use of FDRs, the impact requirements were increased to 1000 g. It was also learned through experience that the optimum location for the recorder was at the rear of the aircraft. This rationale is based on the statistical probability that, following initial impact, the rear structure of the aircraft will remain intact. When the front of the aircraft is crushed, it effectively reduces the shock impact that reaches the rear of the aircraft. Evidence from aircraft accidents shows that the rear fuselage and empennage are most likely to survive intact and therefore provide the optimum location for the recorder. It is imperative that, in the event of an accident, the data recorder be located as swiftly as possible. For this reason they are painted bright orange and are fitted with an **underwater location transmitter**, see Fig. 18.3.

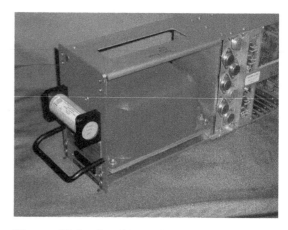

Figure 18.4 Crash-proof recorder

portion of the recorder (see Fig. 18.4) is designed and tested for a variety of severe conditions derived from aircraft crash data; the most significant requirements being:

Crash impact:	3400 g deceleration for 6.5 ms.
Crushing:	5000 pounds pressure applied against all six sides of the recorder
Piercing:	500 pounds mass with a hardened steel pin dropped from ten feet
Fire:	1100°C for 60 minutes
Water:	20,000 ft (submerged)

Key point

The underwater location transmitter is sometimes called an underwater locator beacon (ULB).

To ensure survivability of a flight data recorder, modern products are typically double-wrapped in strong corrosion-resistant stainless steel or titanium, with high-temperature insulation. The crash-survivable

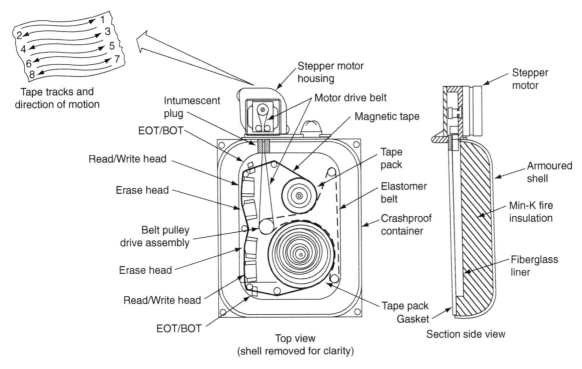

Figure 18.5 Crash-proof recorder tape drive

These are specific requirements for the **crash-survivable** section of the recorder. This part of the recorder fully encloses the tape and transport mechanism (Fig. 18.5); the drive motor is contained externally. Generic requirements for the entire product include resistance to a number of environmental conditions including:

- hydraulic fluids
- fuel de-icing fluids
- fire extinguishing agents, etc.

Key point

The mandatory parameters required for an aircraft DFDR depend on the size of the aircraft and the prevailing regulatory rules applied to the operation of that aircraft.

18.3.1 Recording tape

Typical DFDR recording tape is 0.25 in. wide made of Kapton or Mylar; these are the trade names for self-lubricating polyamide films which can remain stable in a wide range of temperatures from −269°C to +400°C. Up to 500 feet of tape is arranged by eight bi-directional tracks with sequential switching.

The tape is transferred from one reel to another by a motor-driven belt, and transported over the write/read heads. The motor is a reversible stepper motor, located outside of the crash-survivable enclosure rotating at a nominal rate of 700 steps per second. Odd-numbered tracks are used when the tape is travelling in the nominal forward direction; even-numbered tracks are used in the reverse direction, see Fig. 18.5. When an entire track has been used, the direction of travel is switched and the next track utilized. At the end of recording on track eight, the recording will be switched back to track one, and old data is thereby overwritten. The tape travels at speeds up to five inches per second on an incremental basis when data is being written, or five inches per second continuously when data is being read. Beginning of tape (BOT) and end of tape (EOT) sensors are used to reverse the direction of travel and switch to the next track.

Key point

The DFDR on large aircraft has to be able to retain the recorded data for a minimum of the last 25 hours of its operation.

Test your understanding 18.2

What are the five significant requirements that a crash-survivable recorder has to withstand?

Two four-channel write/read heads are used for recording and playback from the tape. One head is enabled when the tape moves in one direction, and writes the odd numbered tracks. When the tape direction is reversed, the other write/read is enabled and writes to the even numbered tracks. Data is written at a rate of 11.4 kilo bits per second (kbps). Before writing new data, the old data are removed on the engaged track by dedicated **erase heads**.

18.3.2 Data acquisition

Modern flight data recorders are based on digital technology to enable more parameters to be recorded. Digital flight data recorders (DFDRs) installed on aircraft equipped with a distributed digital data bus (Fig. 18.6) are integrated as one of the many avionic systems.

There are many aircraft, however, that have limited or no digital equipment. Depending on the size and category of aircraft, e.g. if it is used for public trans-

port, it still has to have a DFDR. The DFDR will be used for recording both mandatory and domestic parameters. The latter is used for the benefit of the operator; it might only have a 30 minute duration, sufficient for monitoring specific systems, e.g. to evaluate the performance of engines or automatic landings. This type of data is often recorded on a cassette termed a **quick access recorder** (QAR); the QAR is removed at the convenience of the operator and information is analysed via customized software packages.

Assuming that the aircraft has limited digital capability, parameters required for both mandatory and domestic recording are gathered in analogue form and converted into digital form via an **aircraft integrated data system** (AIDS). This system also allows the digital information to be retrieved and formatted for analogue purposes, e.g. for display on a monitor or printed on a chart; Fig. 18.7 illustrates a typical system.

The acquisition units sample the data to be recorded and converts them into digital form suitable for recording. The logic unit also inhibits the recording (REC) function until the aircraft is capable of moving under its own power; this switch is controlled by one of many possible options, e.g. when ground power is removed, when an alternator comes on line or when ground speed increases over a specified value. The TEST switch allows the system to be tested on the ground by overriding the REC logic.

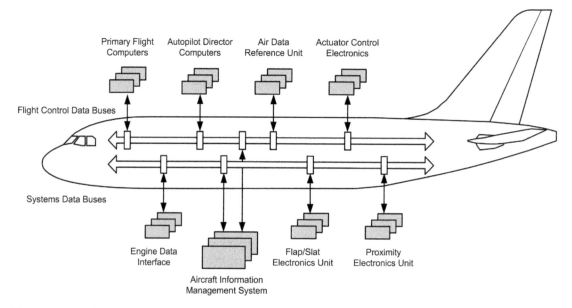

Figure 18.6 Digital data bus

Test your understanding 18.3

What is the purpose of a QAR?

The DFDR control display panel is used to enter the aircraft flight number (or some other identification, weight, date, etc). Once the system is in operation, specific parameters being recorded can be displayed on the panel. The control panel often has a push-button momentary switch that is used to mark a particular event. (Pushing the event button puts a discrete signal on the recording to highlight the timing of the event.)

18.3.3 Digital data recording formats

Typical data output to the DFDR comprises a continuous stream of digital data formed into frames, see Fig. 18.8. Each frame is divided into four sub-frames, typically of one-second duration each; the sub-frame is formed by 64 12-bit words.

The first word in each sub-frame is a synchronizing word; the other three contain data. Each word contains 12 bits of digital information; the total number of bits in a sub-frame is therefore 768. Various formats are used to form the digital bits of logic one and logic zero, see Fig. 18.9. These formats include:

- non-return-to-zero (NRZ)
- bipolar return-to-zero (BPRZ)
- Harvard bi-phase format.

Non-return-to-zero (NRZ) logic one is formed by a 5 V DC level; logic zero is indicated by 0 V DC. The two logic levels are therefore represented by one of two significant conditions, with no other neutral or rest condition. A clock waveform is required to distinguish between bits. Two wires are used to carry the signal, together with a third wire for a clock reference.

Bipolar-return-to-zero (BPRZ) describes a code in which the signal drops (returns) to zero between each pulse. This takes place even if a number of consecutive logic zeros or ones occur in the signal. Logic one is indicated by a 0.5 V DC level, and logic zero by a ±0.5 V DC level. The signal returns to 0 V DC in the second half of each bit. The signal is self-clocking; therefore separate clock pulses are not required alongside the signal.

In the **Harvard bi-phase** format, each bit changes state at its trailing edge; either from high to zero or

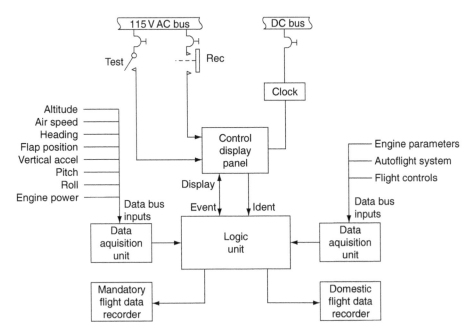

Figure 18.7 AIDS overview

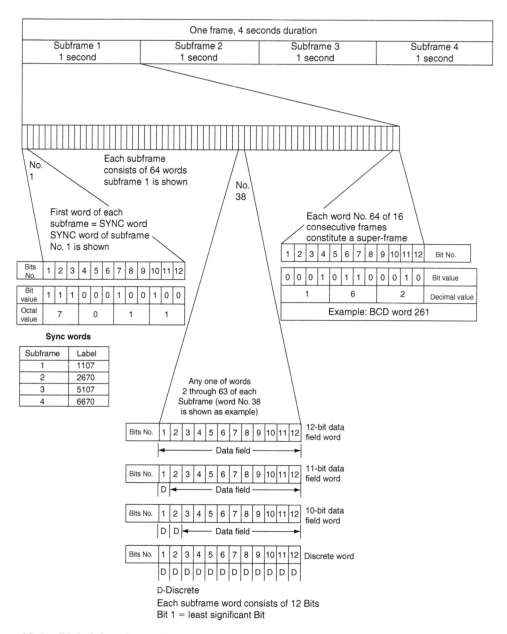

Figure 18.8 Digital data formed into frames

zero to high independently of its value. A logic one is indicated by a mid-bit change of state; a logic zero is indicated by no mid-bit change of state.

The 64th word of 16 consecutive frames are combined into a super-frame. These words are formed in **binary coded decimal** (BCD) format; each word comprises four bits, used to represent the denary

numbers zero to nine, see Table 18.6. This illustrates how binary numbers and then BCD represent the denary (or decimal) numbers 0–20.

The advantage of BCD is that it allows conversion to decimal digits for printing or display and faster decimal calculations. This is particularly useful where a numeric value is to be displayed, e.g. recording

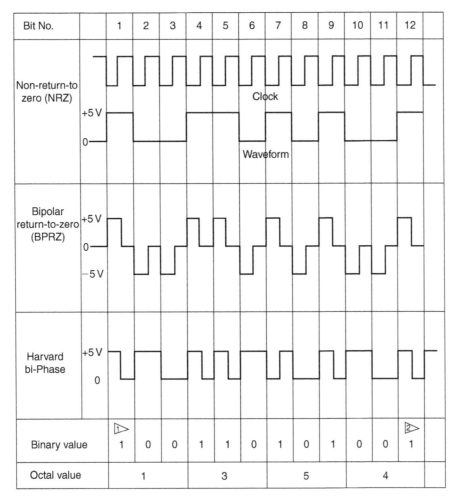

Flight data recorder system codes

▷ Least significant bit
▷ Most significant bit

Figure 18.9   Digital bits of logic one and logic zero

systems consisting solely of digital logic. Manipulation of numerical data for the recording readout can be simplified by treating each digit as a separate single sub-circuit. This matches the physical reality of data displayed to the pilot and recorded data. The disadvantage of BCD is the complexity of circuits needed to implement mathematical operations; BCD is a relatively inefficient technique for encoding, i.e. it occupies more memory than a binary representation.

Test your understanding 18.4

How would the following denary numbers be represented in BCD: 27, 3, 1954?

The data transmitted by these super-frames includes frame count, error codes and voltage levels used in the calibration of analogue signals. The DFDR continuously verifies the data integrity from the tape; this

Table 18.6 Binary coded decimal (BCD)

Decimal	Binary	Binary coded decimal	
		A	*B*
0	00000	0000	0000
1	00001	0000	0001
2	00010	0000	0010
3	00011	0000	0011
4	00100	0000	0100
5	00101	0000	0101
6	00110	0000	0110
7	00111	0000	0111
8	01000	0000	1000
9	01001	0000	1001
10	01010	0001	0000
11	01011	0001	0001
12	01100	0001	0010
13	01101	0001	0011
14	01110	0001	0100
15	01111	0001	0101
16	10000	0001	0110
17	10001	0001	0111
18	10010	0001	1000
19	10011	0001	1001
20	10100	0010	0000

is accomplished by backing up the tape after a write operation and then reading the data.

A typical DFDR receives data from an acquisition unit in Harvard bi-phase code. This data is then processed and stored in solid-state **non-volatile memory** (NVM). The data is received at the interface card and is transferred to the power supply card. Arinc 573 data transmitters and receivers interface with a compression/storage acquisition processor; this processor stores the data in a buffer memory before sending it to a flash memory unit in the crash-survivable portion of the recorder (Fig. 18.10).

The compression/storage acquisition processor also monitors the system for faults via built-in test routines. If any of these tests should fail, appropriate warning lights or LEDs are illuminated on the front of the unit.

18.4 Cockpit voice recorders

Commercial aircraft are required to carry a cockpit voice recorder (CVR). This unit captures and stores information derived from a number of the aircraft's audio channels that may later become invaluable in the event of a crash or serious incident. The CVR can provide valuable information that can later be analysed. The voice recorder preserves a continuing record of typically between 30 to 120 minutes of the most recent flight crew communications and conversations. The storage medium used with the CVR fitted to modern aircraft is usually based on one or more solid state memory devices whereas, on older aircraft, the CVR is usually based on a continuous loop of magnetic tape. The CVR storage unit must be recoverable in the event of an accident. This means that the storage media must be mounted in an enclosure that can withstand severe mechanical and thermal shock as well as the high pressure that exists when immersed in deep water.

The CVR control panel is usually fitted with a test switch, headphone jack and status light; the recorder itself is fitted with an underwater locator beacon (ULB) to facilitate undersea recovery. The ULB is a self-contained device (invariably attached to the front of the CVR) that emits an ultrasonic vibration (typically at 37.5 kHz) when the water-activated switch is immersed in water. A label on the ULB indicates the date by which the internal battery should be replaced. The specification for a typical ULB is shown in Table 18.7. An external view of a CVR showing its externally mounted ULB is shown in Fig. 18.11.

The audio input to the CVR is derived from the captain, first officer and observer position(s) and also from an open area microphone in the flight compartment which is usually mounted in the overhead panel and thus collects audio input from the entire flight-deck area.

In order to improve visibility and aid recovery, the external housing of the CVR is painted bright orange. The unit is thermally insulated and hermetically sealed to prevent the ingress of water. Because of the crucial nature of the data preserved by the flight, the unit should only be opened by authorized personnel following recovery from the aircraft.

Magnetic tape CVRs use a multi-track tape transport mechanism. This comprises a tape drive, four recording heads, a single (full-width) erase head, a monitor head and a bulk erase coil. The bias generator

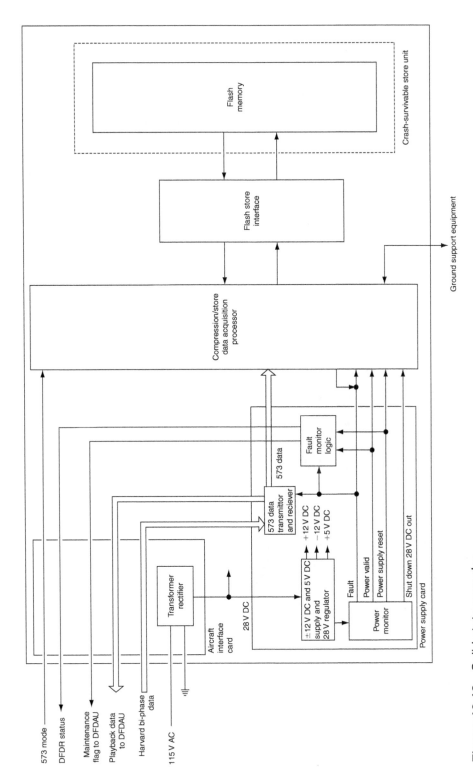

Figure 18.10   Solid state recorder

Table 18.7 Typical ULB specification

Parameter	Specification
Operating frequency	37.5 kHz ($\pm$1 kHz)
Acoustic output	160 dB relative to 1 μPa at 1 m
Pulse repetition rate	0.9 pulses per second
Pulse duration	10 ms
Activation	Immersion in either salt water or fresh water
Power source	Internal lithium battery
Battery life	6 years standby (shelf-life)
Beacon operating life	30 days
Operating depth	20,000 ft (6096 m)
Housing material	Aluminium
Length	3.92 in. (9.95 cm)
Diameter	1.3 in. (3.3 cm)
Weight	6.7 oz (190 g)

Figure 18.11 CVR/ULB installation

usually operates at around 65 kHz and an internal signal (at around 600 Hz) is provided for test purposes. **Bulk erase** can be performed by means of an erase switch; this is interlocked so that erasure can only be performed when the aircraft is on the ground and the parking brake is set. The erase current source is usually derived directly from the aircraft's 115 V AC

400 Hz supply. The magnetic tape (a continuous loop) is typically 308 feet in length and ¼ in. wide.

18.5 Health and usage monitoring system (HUMS)

18.5.1 Introduction

HUMS was originally pioneered to monitor helicopters servicing the oil platforms in the North Sea. The health and usage monitoring system (HUMS) monitors a variety of aircraft parameters, including aircraft vibration, engine and structural health. Ground support equipment and software is designed to decrease maintenance costs and increase operational readiness. The system includes on-board sensors and data processing to calculate specific maintenance recommendations. HUMS provides engine health and usage monitoring, exceedance monitoring, performance trends, operational usage monitoring and flight data monitoring and recording. Systems developed for helicopters now include rotor track and balance. All the required airborne data acquisition and processing, including crash survivable cockpit voice recorders (CVRs) and flight data recorders (FDRs), are often combined into a single line replaceable unit.

18.5.2 System overview

Typical system component locations are illustrated in Figure 18.12.

Vibration sensors, or accelerometers, are formed from piezoelectric crystals; a small electrical charge is created when the crystal is vibrated. The sensor's output is fed into the HUMS processor. Vibration sensing provides a very good diagnostic tool for health monitoring. One of the most demanding locations to place a sensor is the main gearbox (MGB) epicyclic module. Typical MGB failure modes include, but are not limited to:

- Defective bonding between hard metal and coating
- Improper coating of hard metal (carbide grains size, porosity, coating thickness, etc)
- Machining defects
- Permanent distortion (creep) of casings
- Seizure of roller bearing(s)
- Sub-surface cracks
- Wear due to load variations/movements.

The MGB epicyclic stage is a challenging location for mounting vibration sensors for a variety of reasons:

- Large rotating metallic components/high stored energy
- Limited space
- Risk of damage to gears
- Temperature
- Vibration.

Order analysis is a technique for analysing vibration signals from rotating or reciprocating components within engines, compressors, turbines, and pumps. Each component contributes a unique vibration pattern versus vibration pattern of the whole assembly. With order analysis, these vibration patterns can be identified and isolated as part of the subsequent analysis of each individual component. One of the most common methods for analysing vibration signals is fast Fourier transform (FFT) analysis.

The principles of Fourier analysis are illustrated in Figure 18.13; vibration waveforms from the various sensors are complicated. In FFT, these waveforms are broken down into a series of discrete sine waves, referred to as separate order harmonics (one through five in the illustration); each waveform can be evaluated individually in terms of frequency and amplitude. The fundamentals of vibration are covered in part of the book series.

Key Point

Fourier analysis breaks down, or separates, a complex waveform into sine waves of different frequencies and amplitudes that represent the original waveform.

The engines (power turbines/gas generators) vibration spectrum is monitored during run-up for first and second order harmonics (SO1 and SO2). The engine to main gearbox (MGB) transmission input drive shafts are monitored for imbalance and/or misalignment. MGB components (gears, shafts and bearings) are monitored for meshing frequencies, gear teeth, bearing wear, and epicyclic gear indicators (SO1 and SO2). The accessory gearbox components (gears, shafts and bearings) are monitored for gear meshing frequencies, gear tooth and bearing wear indicators (SO1 and SO2). The tail rotor drive shaft hanger bearings are monitored for imbalance, misalignment and bearing wear. Intermediate and tail gearboxes (gears, shafts and bearings) are monitored for gear meshing frequencies, and gear tooth indicators (SO1 and SO2). Oil cooler drive shafts and (if applicable) NOTAR circulation control blower drive shafts are monitored

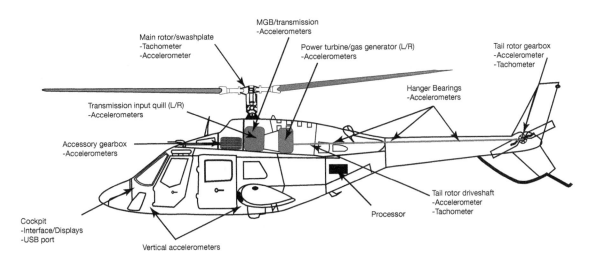

Figure 18.12 HUMS component locations

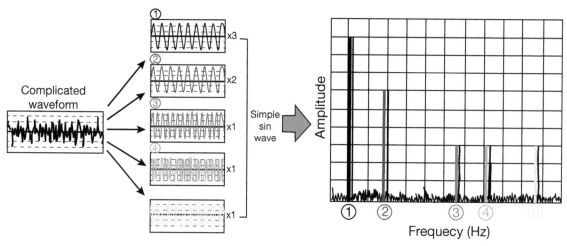

Figure 18.13   Fourier analysis principles

for bearing wear (SO1 and SO2). The main rotor blade track and balance swashplate is monitored for bearing wear indicators. Tail rotor (and if applicable) Fenestron® blade track and balance swashplates are monitored for bearing wear indicators.

Main rotor blades should all travel in the same plane and maintain equidistant angular spacings during flight. Smooth rotor operation is achieved when each blade tracks the same path as it rotates. Tracking can be achieved using vibration sensors, because the track of the rotor directly relates to vibrations in the airframe. Monitoring of each blade is performed with the use of accelerometers, typically located in the cockpit. The accelerometers are synchronized with a tachometer and a tracking device, e.g. a camera or strobe. The monitoring system collects the accelerometer, track, and phase information to provide solutions for reducing vibration and/or blade track differences. Accelerometers for HUMS applications are typically based on micro-electromechanical sensors (MEMS) technology; described in a previous chapter of this book.

Key Point

Although not shown as part of the HUMS, torque (Chapter 10) is an important part of engine performance; this could form part of any flight data recorder analysis, or HUMS algorithms.

18.6 Multiple choice questions

1. The flight data recorder must stop automatically:
   (a) after the aircraft is incapable of moving under its own power
   (b) before the aircraft is incapable of moving under its own power
   (c) after landing.

2. Parameters being recorded on the FDR are obtained from sources that are:
   (a) not displayed to the flight crew
   (b) independent of information displayed to the flight crew
   (c) accurately correlated with the information displayed to the flight crew.

3. Inputs form the audio system are recorded on the:
   (a) FDR
   (b) CVR
   (c) ULB.

4. The mandatory parameters required for an aircraft DFDR depend on the:
   (a) speed and weight of the aircraft
   (b) maximum weight of the aircraft
   (c) size of the aircraft and the prevailing regulatory rules applied to that aircraft.

5. The DFDR requires an electronic process such that:
   (a) digitally coded data can be converted back (decoded) into its analogue form

(b) digitally coded data can be converted back (encoded) into its analogue form

(c) analogue data can be converted back (decoded) into its digital form.

6. In binary coded decimal (BCD) format, each word comprises:
   (a) four bits, used to represent the denary numbers zero to nine
   (b) four bits, used to represent the binary numbers zero to nine
   (c) four bits, used to represent the denary numbers zero to ten.

7. The DFDR on large aircraft has to be able to retain the recorded data for a minimum of the last:
   (a) 30 minutes of its operation
   (b) 25 hours of its operation
   (c) 25 flights.

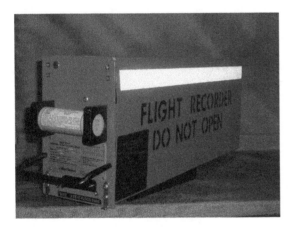

Figure 18.14   See Question 11

8. Lateral acceleration and radio altitude are typical parameters recorded on the:
   (a) FDR
   (b) CVR
   (c) ULB.

9. The FDAU collects, or acquires, a variety of:
   (a) analogue signals and converts them into a digital data stream for the recorder
   (b) analogue signals and outputs for the recorder
   (c) digital signals and converts them into an analogue data stream for the recorder.

10. The flight data recorder must start to record data automatically:
    (a) after the aircraft is capable of moving under its own power
    (b) after take-off
    (c) prior to the aircraft being capable of moving under its own power.

11. Referring to Fig. 18.14, the underwater locator beacon is activated when immersed in water:
    (a) after a time delay
    (b) immediately
    (c) at a specified depth.

12. Referring to Fig. 8.15, data bits that change at the trailing edge are:
    (a) Harvard bi-phase
    (b) Non-return-to-zero
    (c) Return-to-zero.

13. Referring to Fig. 18.16, what is the purpose of EOT/BOT sensors?
    (a) convert digital data into a numerical display
    (b) convert analogue data in digital format
    (c) reverse the direction of tape travel and switch tracks.

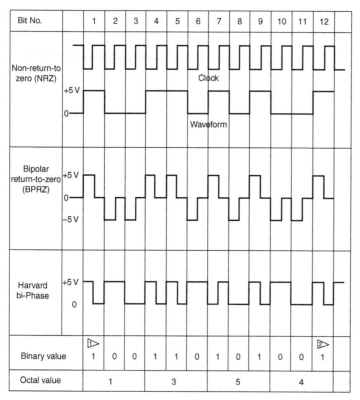

Bit No.		1	2	3	4	5	6	7	8	9	10	11	12	
Non-return-to zero (NRZ)	+5 V 0						Clock Waveform							
Bipolar return-to-zero (BPRZ)	+5 V 0 –5 V													
Harvard bi-Phase	+5 V 0													
Binary value		1	0	0	1	1	0	1	0	1	0	0	1	
Octal value		1			3			5			4			

Flight data recorder system codes

▷ Least significant bit
▷ Most significant bit

Figure 18.15 See Question 12

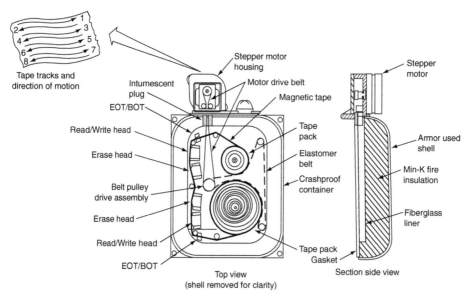

Figure 18.16 See Question 13

Chapter 19

Electrical and magnetic fields

One of the consequences of operating electrical and electronic equipment is the possibility of disturbing, or interfering with, nearby items of electronic equipment. The term given to this type of disturbance is electromagnetic interference (EMI). Placing a portable radio receiver close to a computer and tuning through the radio's wavebands can illustrate this effect. The computer will radiate electromagnetic energy; this is received by the radio and heard as 'noise'. Radio equipment is designed to receive electromagnetic energy; however, the noise in this simple experiment is unwanted. Electrical or electronic products will both radiate and be susceptible to the effects of EMI. This is a paradox since many principles of electrical engineering are based on electromagnetic waves coupling with conductors to produce electrical energy and vice versa (generators and motors). Furthermore, systems are specifically designed to transmit and receive electromagnetic energy, i.e. radio equipment. The problem facing aircraft electrical and electronic systems is the unwanted noise; in the case of the computer/radio experiment, this unwanted noise is no more than a nuisance. In complex avionic systems, the consequences of EMI can be more serious. The ability of an item of equipment to operate alongside other items of equipment without causing EMI is electromagnetic compatibility (EMC).

Modern digital equipment operates at very high speed and relatively low power levels. In addition to EMI, high-intensity radiated fields (HIRFs) are received from the external environment, e.g. from radio and radar transmitters, power lines and lightning. The high energy created by these radiated fields disrupts electronic components and systems in the aircraft. (This effect is also referred to as high-energy radiated fields – HERFs.) The electromagnetic energy induces large currents to flow, causing direct damage to electronic components together with the secondary effects of EMI.

19.1 Electromagnetic interference

Electromagnetic interference (EMI) can be defined as the presence of unwanted voltages or currents that can adversely affect the performance of an electrical/ electronic system. The effects of EMI include:

- errors in indications
- unwanted noise on audio signals
- random patterns on electronic displays
- repetitive 'buzzing' on intercom and cabin phone systems
- desensitizing of radio and radar receivers
- false indications in radar and navigation equipment
- nuisance triggering of alarms.

Consider the current-carrying conductor illustrated in Fig. 19.1. It can be seen that the field strength is a measure of **flux density** at any given point. From first principles, the field strength B is proportional to the applied current and inversely proportional to the distance from the conductor. If an alternating current is applied to the conductor, this field will build up and collapse around the conductor. If a second conductor is now placed alongside the first conductor, the alternating field from the first conductor will induce currents in the second conductor. These induced currents will be superimposed onto any current flowing in the second conductor. The first conductor is **radiating** EMI; the second conductor is **susceptible** to EMI.

The total EMI effect depends primarily on the amount of current in the first conductor, rate of change of current (i.e. alternating, digital or switched currents) and the distance between the two conductors. Alternating currents, or digital signals being carried by a conductor, will set up alternating magnetic fields thus causing EMI as described. In summary, the amount of electromagnetic field radiated from a conductor depends on amount of current in the conductor and the rate of change of a magnetic field from the conductor.

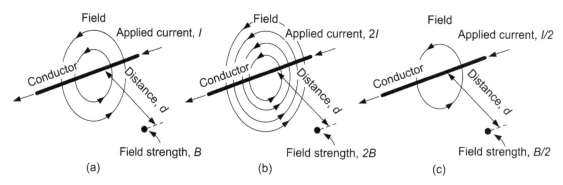

Figure 19.1   Current-carrying conductor

Key maintenance point

EMI can also be created if a conductor carrying direct current is vibrated close to another conductor. The net result is relative movement of a magnetic field over a conductor with the consequences of induced currents. Wiring looms must be secured to prevent them vibrating excessively.

19.1.1 Shielding

For a given current, the distance between conductors and the provision of shielding (or screening) are the main considerations when trying to design and install wiring and equipment to minimize EMI. From Fig. 19.1 we can see that the distance between conductors has a direct effect on the amount of unwanted current induced into the second conductor. The addition of shielding on conductors limits the coupling of electromagnetic fields, see Fig. 19.2. The electromagnetic field created by current I_1 in the first conductor induces a current I_2 in the second conductor.

With one conductor shielded, some of the electromagnetic field created by current I_1 induces current in the shielding; this current is taken away to a ground connection at one or both ends of the shielding. The shielding effectively absorbs some of the electromagnetic field created by I_1; the remaining (weaker) field induces a reduced current I_2 in the second conductor. If both conductors are shielded, most of the weaker field created by I_1 is absorbed into the shielding of the second conductor; this further reduces the current I_2 in the second conductor.

Shielding can reduce the coupling of electromagnetic fields between conductors. The amount of reduction depends upon the screening material used,

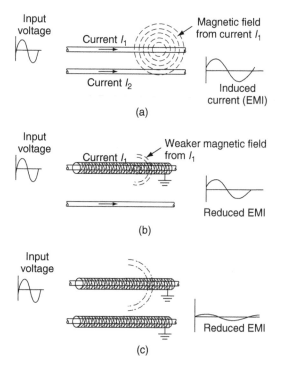

Figure   19.2 Shielding   principles: (a) conductors unshielded, (b) one conductor shielded, (c) both conductors shielded

its thickness, and the frequency of the applied current. Typical materials used for shielding are metallic mesh, or braiding formed on the outside of the conductor's protective sheath; this ensures that the wire or cable retains some degree of mechanical flexibility. The shielding impedes the radiation of signals from the first conductor, and also minimizes signals from being induced into the second conductor.

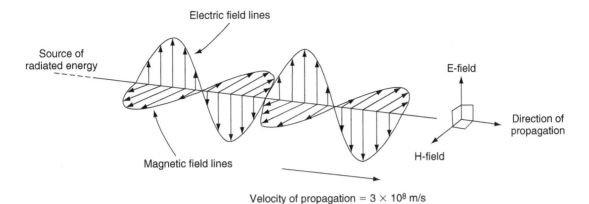

Figure 19.3   E and H lines mutually at right angles

19.1.2 Electromagnet waves

We have been using the term electromagnet wave
to develop the concept of EMI. The electromagnet
wave actually comprises two components. As with
light, electromagnet waves propagate outwards from
a source of energy (transmitter) and comprise **electric** (E) and **magnetic** (H) fields at right angles to
each other. These two components, the E-field and the
H-field, are inseparable. The resulting wave travels
away from the source with the E and H lines mutually at right angles to the direction of propagation, as
shown in Fig. 19.3.

   Electromagnet waves are said to be **polarized** in
the plane of the electric (E) field. Thus, if the E-field
is vertical, the signal is said to be vertically polarized
whereas, if the E-field is horizontal, the signal is said
to be horizontally polarized.

   In the case of intentional propagation of electromagnet waves, i.e. radio transmitters and receivers
(Fig. 19.4), the electric E-field lines are shown in
the space between a transmitter and a receiver. The
transmitter aerial (or antenna) is supplied with a
high-frequency alternating current. This gives rise
to an alternating electric field between the ends of
the aerial/antenna and an alternating magnetic field
around (and at right angles to) it. The direction of the
E-field lines is reversed on each cycle of the signal as
the wavefront moves outwards from the source.

   The receiving aerial/antenna intercepts the moving
field, and voltage and current is induced as a consequence. These voltages and currents are similar (but
of smaller amplitude) to those produced by the transmitter. Note that in Fig. 19.4 (where the transmitter
and receiver are close together) the radiated E-field

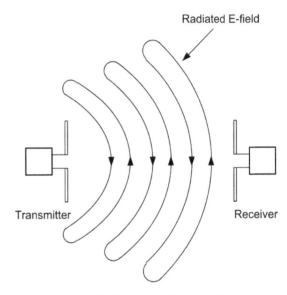

Figure 19.4   Radio transmitters and receivers

is shown spreading out in a spherical pattern (this is
known more correctly as the near field). The magnetic field (not shown) will be perpendicular to the
E-field.

   In practice, there will be some considerable distance between the transmitter and the receiver and
so the wave that reaches the receiving aerial/antenna
will have a plane wavefront. In this far field region,
the angular field distribution is essentially independent of the distance from the transmitting antenna.
(Radio wave theory is addressed in more detail in a
related book in the series, *Aircraft Communications
and Navigation Systems*.)

A simple wiring installation connects the shielding to ground, thereby 'soaking' away the unwanted currents, rather than inducing them into the second conductor. Although shielding provides some protection against EMI, it also adds cost and weight to the installation. Furthermore, the currents being carried away in the shielding radiate their own fields that can cause secondary EMI. This unwanted current is referred to as a **ground loop**; current flows in a conductor connecting two points that are intended to be at the same potential, e.g. ground potential. (The fact that current is flowing can only occur because they are actually at different potentials.) As with many engineering situations, solving one problem often comes at a cost together with the introduction of new problems!

Test your understanding 19.1

Explain the difference between EMI and EMC.

19.1.3 Twisted pair

Another technique used to minimize EMI in wiring is the **twisted pair**. This a form of wiring in which the two conductors are wound together to cancel out electromagnetic interference (EMI) from external sources, and minimize **cross-talk** between adjacent pairs of wires. Cross-talk is the consequence of EMI between one electrical circuit to another, i.e. when a signal transmitted in one circuit creates an undesired effect in another circuit. A twisted pair cable consists of two independently insulated wires twisted around one another.

Twisting the cables forms repetitive loops; each twist reverses the polarity of the loop, therefore magnetic fields can only couple into each loop and not the entire cable length. When current is supplied through a loop, a magnetic field is set up; it can be seen that the flux is concentrated in the centre of the loop as illustrated in Fig. 19.5. The twisted pair also provides a physical means of minimizing EMI in wires carrying digital signals; each wire is positioned alternatively next to the source of interference; the net effect is to cancel out any differential between the wires.

Twisting a pair of wires to form a cable is an extremely effective way of transmitting high-speed signals because:

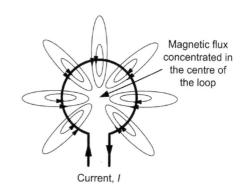

Current, I

Figure 19.5 Magnetic field

- most of the electrical noise entering into and/or radiating from the cable can be eliminated
- cross-talk (signals 'leaking' between wires in the cable) is minimized.

In addition to the electrical energy flowing down each wire, energy can also be coupled between wires due to electrostatic and magnetic effects. For the electrostatic effects, the insulation between the two conductors is the dielectric of a capacitor. More surface area is created by longer cables; this leads to increased inter-wire capacitance. Higher frequency signals can lead to cross-talk between wires as a result of reduced capacitance.

Typical applications of signals transmitted down a twisted pair of wires include Arinc 429 signals as illustrated in Fig. 19.6. Any other system wires adjacent to this pair will be affected by cross-talk.

Two wires (A and B) each carry a ± 5 volt digital signal arranged as bipolar return-to-zero (BPRZ). The extent of the cross-talk is equal to the sum of the digital signals; if this sum is zero (or nearly zero) then the affects of cross-talk are eliminated.

Consider three wires in a cable as illustrated in Fig. 19.7. Wires A and B are formed as a twisted pair carrying a digital signal. If the signal sent through wire A is ± 10 volts with respect to wire B (a reference voltage of zero) then wire C wire picks up cross-talk noise. If the digital signals are arranged as per the second illustration, the opposite polarity signal of ± 5 volts on wires A and B cancel each other out, the cross-talk effect wire C is eliminated.

Key maintenance point

Twisting wires together also provides a neater installation; the number of twists per given length

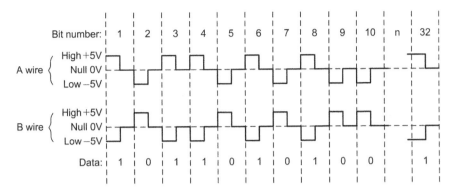

Figure 19.6   Digital signal

should be specified in order to achieve the desired EMI protection.

Key maintenance point

In some installations, further EMI protection is given by the addition of shielding over the twisted pair of wires.

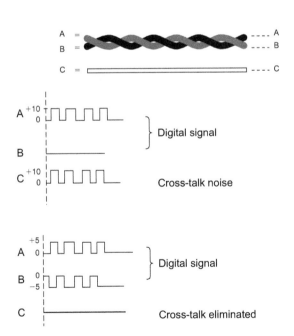

Figure 19.7   Twisted pairs and cross-talk

19.1.4 Bandwidth

Bandwidth is the difference between the upper and lower cut-off frequencies of analogue amplifying circuits; the unit of **bandwidth** is hertz (Hz). The characteristics of bandwidth is illustrated in Fig. 19.8. This is a central concept in electronics, radio-frequency (RF) systems and signal processing. It also has to be considered in the context of EMI; high frequencies should be filtered out of a circuit wherever possible, without compromising its functionality.

In computers, digital bandwidth refers to the rate at which data is transmitted/received; this is measured in **bits per second (BPS)**. A digital communication network has a given bandwidth in terms of its overall channel capacity and throughput (consumption). Channel capacity (in BPS) is proportional to the analogue bandwidth in hertz (Hz); this is the maximum amount of error-free digital data that can be transmitted via a communication link with a specified bandwidth in the presence of EMI. The deeper theory associated with this subject is beyond the scope of this book; students wishing to research this subject in more detail should refer to the **Shannon–Hartley theorem**.

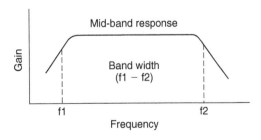

Figure 19.8   Bandwidth

Test your understanding 19.2

Why do higher-frequency signals lead to increased cross-talk?

Test your understanding 19.3

How does twisting pairs of wiring reduce cross-talk?

19.1.5 Radiated EMI

There are many sources of EMI throughout the aircraft. Those sources known to **radiate** EMI include:

- fluorescent lights
- radio and radar transmitters
- power lines
- AC powered window heat controllers
- motors/generators
- switching and light dimming circuits
- microprocessors
- pulsed high-frequency circuits
- data bus cables (but not fibre optic cables)
- static discharge and lightning.

The energy generated by these sources is radiated as an electromagnetic field. From first principles, we know that the coupling only takes place when there is a relative movement of electromagnet field and conductor; digital circuits are (by definition) switching currents on/off with fast and short pulse rise/fall times. Unless adequate precautions are taken to eliminate the interference at source and/or to reduce the equipment's radiation of EMI, the energy can then become coupled into other circuits. In electromagnetic field radiation, energy is transmitted through electrically nonconductive paths, such as air, plastic materials, or fibreglass.

19.1.6 EMI susceptibility

There are many systems on the aircraft that may be **susceptible** to electromagnetic interference. These include:

- radio and radar receivers
- microprocessors and other microelectronic systems
- electronic instruments
- control systems
- audio and in-flight entertainment systems (IFEs).

Whether a system will have an adverse response to **electromagnetic interference** depends on the type and amount of emitted energy in conjunction with the susceptibility threshold of the receiving system. The threshold of susceptibility is the minimum interference signal level (conducted or radiated) that results in equipment performance that is indistinguishable from the normal operation. If the threshold is exceeded then the performance of the equipment will become degraded. Note that, when the susceptibility threshold level is greater than the levels of radiated emissions, electromagnetic interference problems do not exist. Systems to which this applies have **electromagnetic compatibility** (EMC). In other words, the systems will operate as intended and any EMI generated is at such a level that it does not affect normal operation.

19.2 EMI reduction

Planning for electromagnetic compatibility must be initiated in the design phase of a device or system. If this is not satisfactorily addressed, interference problems may arise. The three factors necessary to produce an EMI problem are:

- source(s) of interference (sometimes called noise)
- a means of coupling (by conduction or radiation)
- susceptible components or circuits.

To reduce the effects of EMI, or **electrical noise**, at least one of these factors must be addressed. The following lists some techniques used for EMI reduction to tackle these three factors (note that some techniques address more than one factor).

1. Suppress the interference at source
   - Enclose the interference source(s) in a screened metal enclosure and then ensure that the enclosure is adequately grounded
   - Use transient suppression on relays, switches and contactors
   - Twist and/or shield bus wires and data bus connections
   - Use screened (i.e. coaxial) cables for audio and radio-frequency signals
   - Keep pulse rise times as slow and long as possible
   - Check that enclosures, racks and other supporting structures are grounded effectively.

2. Reduce noise coupling
   - Separate power leads from interconnecting signal wires
   - Twist and/or shield noisy wires and data bus connections
   - Use screened (i.e. coaxial) cables for audio and radio-frequency signals
   - Keep ground leads as short as possible
   - Pay close attention to potential ground loops
   - Filter noisy output leads
   - Physically relocate receivers and sensitive equipment away from interference sources.
3. Increase the susceptibility thresholds
   - Limit the bandwidth of circuits wherever possible
   - Limit the gain and sensitivity of circuits wherever possible
   - Ensure that enclosures are grounded and that internal screens are fitted
   - Fit components that are inherently less susceptible to the effect of stray radiated fields.

19.3 High-intensity/energy radiated fields

High-intensity/energy radiated fields (HIRFs/HERFs) are generated by certain radio-frequency (RF) sources that are **external** to the aircraft. These fields disrupt electronic components and systems within the aircraft via currents that are induced from these fields into the aircraft's wiring. (The terms HIRF and HERF mean the same thing; HIRF will be used for the remainder of this chapter.) Radio, radar and television transmitters (particularly highly directional broadcasts) have the capability of adversely affecting the operation of aircraft electrical and electronic systems. The HIRF environment has become a significant threat to aircraft using the electronic systems described in this book, together with the communications, navigation and flight guidance systems. These systems are potentially very susceptible to the HIRF environment. Accidents and incidents on aircraft with such systems have led to the need for a thorough understanding of, and increased protection from, high-intensity radiated fields.

Practical experience with the effects of HIRF often result in unexplained and unrelated disruption to aircraft systems e.g. simultaneous navigation errors and erroneous engine indications. These systems then start operating normally again without being attributed to any specific equipment failures. HIRF has been cited as the cause of misleading roll and pitch information on electronic displays and the total loss of engine power due to interference with electronic control systems. The need for protection of modern electrical and electronic systems from HIRF is required because of the:

- dependence on these systems used for the continued safe flight and landing of the aircraft
- increased use of composite materials (reducing the natural Faraday cage protection of metallic structures)
- Increased complexity of digital systems (faster operating speeds, higher-density integrated circuits)
- expanded frequency usage of microwave energy
- increased quantity and power of transmitters
- proliferation of RF transmitters.

19.3.1 HIRF environment

The **HIRF environment** is created by the transmission of high-power radio-frequency (RF) energy into free space. These transmissions can be from military systems, television, radio, radars and satellites communicating with ground-based equipment, ships or other aircraft. When an aircraft operates in a HIRF environment, this can have an adverse effect on the aircraft systems and equipment that could result in system failure, malfunction, or misleading information. The process whereby electromagnetic energy from an RF source is induced in a system by radiation is termed **coupling**. It is entirely possible that HIRF will affect individual components through to system level via individual wires (or wire bundles) that connect the items of equipment. In the event of the HIRF environment having an adverse effect on systems, it is essential that the aircraft has the capability for continued controlled flight and landing to a suitable location, albeit under emergency conditions.

19.3.2 HIRF characteristics

The analysis of HIRF is centred on the **frequency** of transmission and **field strength**. The practical considerations of RF transmissions are from approximately 10 kilohertz (10 kHz) through to 100 gigahertz (100 GHz). Field strength is defined as the magnitude of the electromagnetic energy propagating in free space expressed in volts per meter (V/m). Aircraft systems need to be tested and/or analysed across a range of frequencies and field strengths to determine their susceptibility characteristics. Certain systems (or individual items of equipment) are immune to HIRF and have the ability to continue to perform their intended function. This could occur as an inherent or system design feature of the equipment, e.g. if it is located behind a material (**reflection plane**) that reflects RF signals.

Alternatively, a decrease in electromagnetic field strength in transmission from one point to another can occur by **attenuation**, expressed in decibels (dB). Attenuation is the scalar ratio of the RF energy input magnitude and output magnitude. (Further details of attenuation are given in the appendices.). In the HIRF environment, the ratio of current induced in a wire bundle form the external HIRF field strength (as a function of frequency) is termed the **transfer function**. An item of equipment or system that is susceptible to the HIRF environment means that it is unable to perform its intended function; this will be defined by a susceptibility level, where the effects of interference from HIRF become apparent.

19.4 Lightning

Lightning together with precipitation is associated with electrical activity within **cumulonimbus** clouds, see Fig. 19.9. Lightning results from the build up of huge amounts of static charge in the atmosphere. Precipitation may be defined as the result of water vapour **condensing** in the atmosphere that subsequently falls to the earth's surface.

Figure 19.9   Electrical storm

Electrical activity can originate at the top of thunderclouds or the outside edges of the precipitation area. If an aircraft is subjected to a **lightning strike**, or discharge, the structure and bonding are designed to dissipate this energy, see Fig. 19.10.

Key maintenance point

The probability of an aircraft experiencing a lightning strike on any given flight is one. High standards of design and maintenance are required to ensure that serious damage from a lightning strike is a rare occurrence.

19.4.1 Faraday cage

The metallic aircraft structure acts as a **Faraday cage (or Faraday shield)**, named after the British physicist Michael Faraday. This is an enclosure formed by a conducting material that shields the inside of the structure from electromagnetic effects. An external electrical field will cause the electrical charges within the structure to redistribute themselves so as to cancel the field's effects inside the fuselage. This effect is used very effectively in aircraft to protect electronic equipment from lightning strikes and other electrostatic discharges. Lightning normally enters the aircraft at an extremity, e.g. the nose cone or wing tip; currents flow through the conductive structure, and then exits the aircraft at another extremity. This can cause short-term interference with systems, but there should be no permanent damage. Some currents could enter the structure due to the high voltages, the energy will normally be conducted through bonding leads and back into the fuselage skin. The aircraft is categorized into specific areas when planning system and equipment locations, see Fig. 19.11:

- zone one is where lightning can be expected to enter and exit the aircraft
- zones two and three provide the conductive paths through the aircraft.

Key maintenance point

The majority of physical damage on the aircraft occurs at the exit point of the lightning strike.

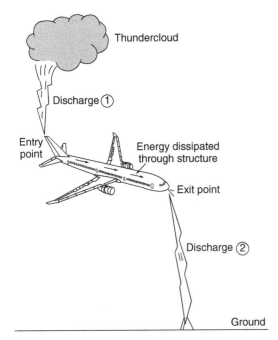

Figure 19.10 Lightning strike; dissipation of energy through the aircraft

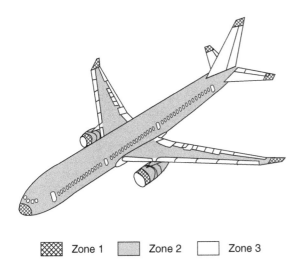

Zone 1   Zone 2   Zone 3

Figure 19.11 Lightning zones

19.4.2 Aircraft construction

The metal structure dissipates electric currents generated from the external electromagnetic fields, thereby reducing or even eliminating electromagnetic interference. More aircraft are now being built using (non-conducting) **composite materials**, ranging from individual items of structure through to the complete exterior. In this case, the Faraday cage effect has to be designed into the composite structure as a mesh of conducting material. Additional design effort, testing and on-going maintenance are required as a consequence. One particular part of the aircraft often affected by lighting is the nose cone; this is made from a composite material and therefore has no inherent conductive paths. Lightning strips can be fitted to protect this area, see Fig. 19.12.

Key maintenance point

The aircraft is designed to provide conductive paths to dissipate the energy from lightning strikes. Over periods of time, certain parts of the aircraft will corrode and be subjected to vibration; it is essential that these conducive paths are maintained, e.g. by inspection of bonding leads.

Even when the lightning discharge is dissipated safely through the aircraft, structural damage can occur at the exit points of the discharge. Furthermore, the high circulating currents can cause EMI as previously discussed. Lightning discharges via the atmosphere from one cloud to another, or directly to the earth's surface. The atmosphere has a very high dielectric breakdown (depending on moisture content), typically three million volts per meter. The voltage generated by lightning depends on the length travelled by the lightning discharge; this will be approximately one gigavolt (10^9 volts) for a 300 m (1000 ft)

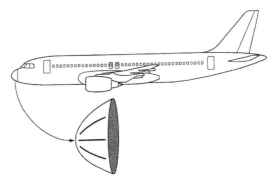

Figure 19.12 Radome protection

lightning bolt. With a typical current of 100 kA, this gives rise to a power dissipation of 100 terawatts (100×10^{12} watts). The high energy radiated by a lightning discharge carries sufficient electromagnetic field strength to couple into the aircraft's wiring. This causes high currents of relatively short duration to be **injected** into electronic equipment. Adequate protection must be designed into the equipment to prevent disruption of signal processing and/or damage to components.

19.4.3 Certification of aircraft for HIRF and lightning protection

Although the sources of HIRF and lightning (HIRF/L) threats are very different, the strategies used in design, certification and maintenance are very similar, if not the same. The external HIRF and lightning (HIRF/L) environment has evolved over time and varies between aircraft types. Various specifications defining HIRF/L threat levels have emerged in response to the increasing understanding of the problem. In Europe, aircraft have been required to comply with HIRF requirements since early 1992. In-service aircraft are therefore specified with varying HIRF standards ranging from no requirement through to the current standards. HIRF threat levels can be broadly thought of in terms of field strength and frequency of the RF signal. Compliance with HIRF requirements typically involves a range of factors including:

- testing the entire aircraft
- equipment tests
- analogy/similarity with other installations.

Equipment and systems being certified for HIRF/L immunity and susceptibility are in fact achieved by any combination of the above. Comprehensive test programmes are required on new aircraft types; this is carried out on entire aircraft, with engines and systems operating. Various techniques are used for HIRF/L testing:

- **bulk current injection** (BCI) via a probe that injects RF signals directly into wire bundles when clamped around them
- **direct drive testing** (DDT) involves connecting an RF signal source directly to the unit being tested.

Testing is an expensive and time-consuming activity, one that has to be planned and agreed with the certifying authorities well in advance of the aircraft going into service. An alternative strategy is to use existing HIRF compliance documentation and data from a similar system (or aircraft) to demonstrate HIRF compliance. Failure conditions resulting from HIRF/L are assessed against certification requirements and are given classifications, i.e. catastrophic, hazardous or major. The classifications for HIRF/L related system failures are assigned certification levels of:

A. **Catastrophic**: where system failure(s) would prevent the continued safe flight and landing of the aircraft
B. **Hazardous**: where system failure would significantly reduce the capability of the aircraft and/or the ability of the flight crew to respond to the failure(s)
C. **Major**: where system failure reduces the capability of the aircraft or the ability of the flight crew to respond to an adverse operating condition(s).

A safety assessment is prepared to address all potential adverse effects. The safety assessment could establish that some systems have different failure conditions in different phases of flight, e.g. a navigation system may have a catastrophic failure condition during the landing phase, but in cruise the classification could be reduced to a major failure condition. To cater for these different classifications on the same system, varying HIRF environments would be applied, see Table 19.1.

The field strength values shown in Table 19.1 are the maximum for any given frequency within the range. Field strength values are expressed in root mean square (RMS) units, measured during the peak of the modulation cycle.

Table 19.1 External HIRF environments

Environment	Frequency range	Field strength (V/m) (peak)	Field strength (V/m) (average)
I	10 kHz–40 GHz	3000	200
II	10 kHz–40 GHz	3000	230
III	10 kHz–40 GHz	7200	490

19.4.4 Maintenance for HIRF/L protection

The maintenance engineer has a number of essential tasks to ensure that HIRF/L protection features and devices are serviceable for continued airworthiness of the aircraft. These tasks include, but are not limited to:

- bonding resistance measurements
- wire/cable resistance or impedance measurements
- inspection and disassembly of connectors to detect corrosion or termination failure.

19.4.5 Aircraft wiring and cabling

When many potential sources of EMI are present in a confined space, aircraft wiring and cabling has a crucial role to play in maintaining electromagnetic compatibility. Adequate wire separation should be maintained between noise source wiring and susceptible equipment. For example, radio-frequency (RF) navigation systems' wiring should be strategically routed in the aircraft to ensure a high level of EMC. Any changes to the routing of RF systems' wiring could have an adverse affect on EMC. Separation requirements for all wire categories must be maintained; wire lengths should be kept as short as possible to maintain coupling at a minimum. Where wire shielding is incorporated to protect against EMI caused by lightning, it is important that the shield grounds (pigtails) be kept to their designed length. An inch or two added to the length will result in degraded lightning protection. Equipment grounds must not be lengthened beyond design specification. A circuit ground with too much impedance may no longer be a true ground.

Key maintenance point

With the aid of the aircraft technical manuals, grounding and bonding integrity must be maintained. This includes proper preparation of the surfaces where electrical bonding is made.

19.5 Grounding and bonding

The aircraft structure provides an extremely important means of reducing EMI and also protecting the aircraft, its passengers, crew and systems, from the effects of lightning strikes and static discharge. Grounding and bonding are specific techniques that are used to achieve electrical integrity, see Fig. 19.13.

Grounding and bonding can also be instrumental in minimizing the effects of high-intensity radiated fields (HIRFs) emanating from high-power radio transmitters and radar equipment. Grounding and bonding resistances of less than 0.001 ohms to 0.003 ohms are usually required.

Grounding can be defined as the technique of electrically connecting components to either a conductive structure or another return path for the purpose of completing a circuit. **Bonding** refers to the electrical connecting of two or more conducting objects that are not otherwise adequately connected. Bonding and grounding connections are made in an aircraft in order to accomplish the following:

- dissipate energy from high-intensity radiated fields (HIRFs) and lightning strikes
- dissipate static electricity
- limit the potential difference between equipment
- provide a low-resistance path for earth return systems.

The following general procedures and precautions apply when making bonding or grounding connections:

- bond or ground parts to the primary aircraft structure where possible
- make bonding or grounding connections so that no part of the aircraft structure is weakened
- bond parts individually if feasible
- install bonding or grounding connections against smooth, clean surfaces
- install bonding or grounding connections so that vibration, expansion or contraction, or relative movement in normal service, will not break or loosen the connection
- check the integrity and effectiveness of a bonded or grounded connection using an approved bonding tester.

The main types of bonding are:

Equipment bonding. Low impedance paths to the aircraft structure are generally required for electronic equipment to provide radio-frequency return circuits and to facilitate reduction in EMI.

(a)

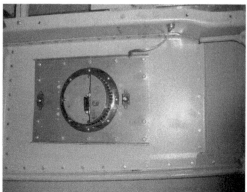

(b)

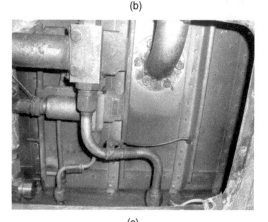

(c)

Figure 19.13   Examples of bonding

Metallic surface bonding. All conducting objects located on the exterior of the airframe should be electrically connected to the airframe through mechanical joints, conductive hinges, or bond straps, which are capable of conducting static charges and currents induced by lightning strikes.

Static bonds. All isolated conducting paths inside and outside the aircraft with an area greater than 3 in.² and a linear dimension over 3 inches that are subjected to electrostatic charging should have a mechanically secure electrical connection to the aircraft structure of adequate conductivity to dissipate possible static charges.

19.6 Multiple choice questions

1. The ability of an item of equipment to operate alongside other items of equipment without causing EMI is called:
   (a) EMC
   (b) ESSD
   (c) HIRF.

2. Higher-frequency signals can lead to cross-talk between wires as a result of the:
   (a) dielectric effect of the wire insulation causing increased capacitance
   (b) dielectric effect of the wire insulation causing reduced capacitance
   (c) dielectric effect of the wire conductors causing reduced capacitance.

3. Radio and radar transmitters in the external environment are sources of:
   (a) EMC
   (b) ESSD
   (c) HIRF.

4. Of the materials used for protecting static sensitive devices:
   (a) conductive and anti-static materials offer the least protection
   (b) conductive materials offer the least protection whilst anti-static materials offer the most protection
   (c) conductive materials offer the greatest protection whilst anti-static materials offer the least protection.

5. Shielding of conductors:
   (a) reduces radiation and minimizes susceptibility
   (b) increases radiation and minimizes susceptibility
   (c) reduces radiation and increases susceptibility.

6. The presence of unwanted voltages or currents in systems is caused by:
   (a) EMC
   (b) ESSD
   (c) EMI.

7. When the lightning discharge is dissipated safely through the aircraft, structural damage is most likely to occur at the:
   (a) bonding between structures
   (b) exit points of the discharge
   (c) entry points of the discharge.

8. In the context of EMI, high-frequency signals should be:
   (a) filtered out of a circuit wherever possible
   (b) introduced into a circuit wherever possible
   (c) used only in unshielded cables.

9. The twisted pair is designed to:
   (a) maximize cross-talk between adjacent pairs of wires
   (b) protect static sensitive devices
   (c) minimize cross-talk between adjacent pairs of wires.

10. The effect of lower humidity on ESSD components in typical working environments will be to:
    (a) create lower voltages and pose more threat to the components
    (b) create higher voltages and pose more threat to the components
    (c) create higher voltages and pose less threat to the components.

Continuing airworthiness

Many processes are required throughout the aircraft's operating life to ensure that it complies with the applicable airworthiness requirements and can be safely operated. The generic term for this range of processes is continuing airworthiness. The term 'maintenance' is used for any combination of overhaul, repair, inspection, replacement, modification or defect rectification of an aircraft or component, with the exception of the pre-flight inspection. Persons responsible for the release of an aircraft or a component after maintenance are the certifying staff. Maintenance of an aircraft and its associated systems requires a variety of test equipment and documentation; these are required by certifying staff to fulfil their obligations in ensuring continued airworthiness. Aircraft wiring cannot be considered as 'fit and forget'. Legislation is being proposed to introduce a new term: **electrical wire interconnection system** (EWIS); this will acknowledge the fact that wiring is just one of many components installed on the aircraft. EWIS relates to any wire, wiring device, or combination of these, including termination devices, installed in the aircraft for transmitting electrical energy between two or more termination points. This chapter reviews some practical installation requirements, documentation and test equipment required by the avionics engineer to ensure the continued airworthiness of aircraft electrical and electronic systems.

20.1 Wire and cable installations

The importance of aircraft wire and cable selection, installation and maintenance cannot be overstated. Modern aircraft (or upgraded older aircraft) are installed with more wiring, carrying more current than earlier generations of aircraft; much of this wiring carries digital signals. Wiring and cables must be treated as integral components of the aircraft; they are not to be treated as 'fit and forget'. Wires are formed from a single solid conductor or stranded conductors,

contained within insulation and protective sheath materials. Cables can be defined as:

- two or more separate wires within the same insulation and protective sheath
- two or more wires twisted together
- any number of wires covered by a metallic braid, or sheath
- a single insulated conductor covered by a metallic outer conductor (co-axial cable).

The terms wires and cables are often interchanged. In this chapter, reference will be made to 'wiring' in the all-embracing generic sense; cables will be referred to in specific terms as and when required. Surveys and inspections of aircraft have revealed a number of issues and problems that require the close attention of system design and maintenance to ensure continuing airworthiness. Wire insulation can deteriorate over time (typically over ten years); exposed conductors create the environment for potential faults, spurious signals and **arcing**.

Wires are vulnerable to their installed environment, e.g. changes in temperature, exposure to moisture and vibration that can lead to **open circuits** and/or **chafing**. In certain areas of the aircraft, e.g. the leading/trailing edges of the wing and wheel wells, the physical environment is harsher than protected areas, e.g. the flight compartment or passenger cabin. The installation of wire and cable is very important; the following must be avoided:

- sharp bend radii
- unsupported wires
- routing high and low power circuits in the same bundle, or loom.

All the above can intensify the ageing and environmental effects on wiring. Certain older standards of wire insulation, notably single-walled aromatic polyamide, have a known vulnerability to high temperature that can lead to the insulation forming small cracks; this can lead to moisture ingress, and overheating. Under these conditions, the carbonized insulation becomes a conductor and the situation propagates leading to the possibility of fire. It is essential that older standards of wiring are inspected in accordance with maintenance schedule; the wiring should be replaced if there are any signs of deterioration. Care must be taken not to disturb or damage wiring during maintenance or inspection of nearby equipment.

Key maintenance point

Faulty wiring is often overlooked as the reason for systems becoming unserviceable. Line replaceable units (LRUs) are often changed first, with the inevitable 'no fault found'.

20.1.1 Cable and wire looms

Grouping individual wires and cables into bundles forms a loom (or harness); these bundles are tied, or strapped together to form a secure assembly. Wire looms must be installed and maintained to ensure maximum integrity. The loom should be formed without twisting or overlapping the cables/wires. They can be tied together with waxed string, lacing cord or nylon straps. These ties are made at regular intervals, equally spaced along the loom; the loom is then secured to the airframe with clamps. The current-carrying capacity of wires reduces in looms since the inner wires are not able to radiate heat efficiently. To illustrate this point, a single 20-gauge wire in free circulating air is rated at 14A; this reduces depending on the number of wires in the loom:

- single wire, 14 A
- three wires, 9 A
- seven wires, 7 A
- twelve wires, 5 A.

Figure 20.1 provides some examples of installed wire looms. The loom will follow a path through the aircraft dictated by existing structure and obstacles; additional support and protection is provided where required. Any bends or branches in the loom must not be so sharp that the loom becomes kinked (never bend or form wires or cables with pliers or any sharp edge). Looms must be installed such that they are supported with clamps and protected from chafing through contact with sharp edges, pipes and other wiring.

Cable clamps and grommets are used throughout the aircraft to provide this protection. General guidelines for the installation of wire looms are as follows:

- Install with a downward slope away from equipment to prevent moisture running into the equipment

- Do not install below fuel pipes
- Avoid areas of high temperature
- Install to minimize EMI
- Make provision for at least one remake of the wire
- Take care not to crush coaxial cables
- Areas of high vibration require additional loom support
- Breakout should not cross over the main loom.

Open looms are formed with bound wires in a bundle; these are supported by p-clips and protected by grommets. The use of open looms is dictated by temperature and length of loom, together with any EMI considerations. **Conduits** (Fig. 20.2) are used in specific areas, e.g. wing leading edges to protect wiring loom from rain and other fluids; these conduits are made from plastic or metal.

When looms can be passed through bulkheads, they must be prevented from chafing; this can be achieved by clamping or potting, see Fig. 20.3. **Ducted looms** are formed in channels made from a suitable material, e.g. aluminium alloy or composite. They provide more support and are used to guide the looms through and around specific areas of the aircraft. When the loom passes through bulkheads, they are often sealed with a rubber bung, see Fig. 20.3.

Certain areas of the aircraft will experience high vibration (Fig. 20.4); wiring in these areas will be subjected to a harsher environment. These areas include:

- wheel wells
- empennage
- wing roots
- wing trailing edges
- wing leading edges
- engine pylons and nacelles.

The wiring in these areas will be exposed to severe wind and moisture problems (**SWAMP**); wiring specifications and inspection requirements must be adhered to.

Key maintenance point

Cracking and breakdown of the insulation material through exposure to moisture, the speed of breakdown depending on both temperature and stress; this phenomenon is known as **hydrolysis**.

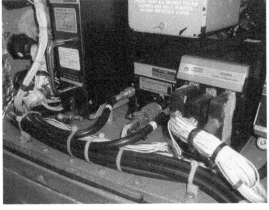
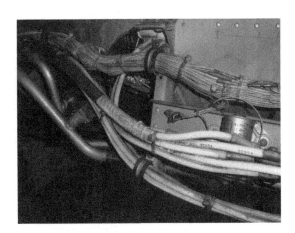

Figure 20.1 Examples of installed wire looms

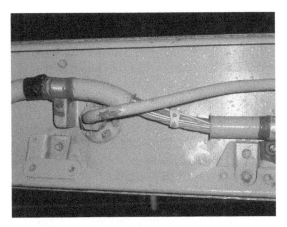
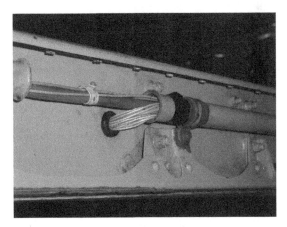

Figure 20.2 Examples of cable-loom conduit

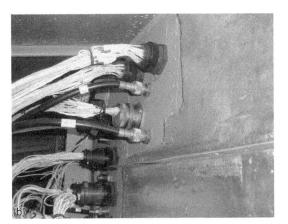

Figure 20.3 Bulkhead wire looms: (a) sealant/potting, (b) connectors

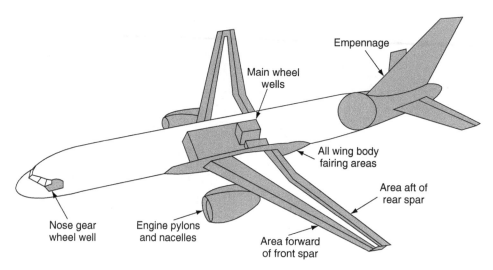

Figure 20.4 High vibration areas

Key maintenance point

Carbon arc tracking occurs when contaminating moisture (including aircraft fluids) creates a short circuit between an exposed conductor and the aircraft structure or an adjacent exposed conductor at a different potential; this phenomenon is known as **wet arc tracking**.

Key maintenance point

Carbon arc tracking occurs in dry conditions when one or more conductors are shorted as a result of abrasion from: the aircraft structure, wire to wire, or installation error. This phenomenon is known **dry arc tracking**.

20.1.2 Wire terminations and connections

It is highly unlikely that a single wire or cable will be routed from the power source directly to the load – it will invariably pass through bulkheads via connectors. Certain areas in the aircraft are not suitable for the routing of wires (e.g. zones exposed to high temperature or EMI). Wiring is invariably installed in sections and joined at intervals. Terminations and connection types used will depend on a number of factors driven by cost and continued airworthiness requirements. The size and configuration of the aircraft will determine where connectors are needed and located. Other considerations are when aircraft sections (wings, fuselage, etc.) are manufactured at various geographical locations and come together at a different place for final assembly. Finally, the need for inspection, removal and installation of equipment needs to be considered; quick-release connectors are used for most **line-replaceable units (LRUs)**, see Fig. 20.5.

Wires can be joined by soldering, although this is normally only used within equipment. (Soldering aircraft wires reduces their flexibility and can lead to premature failure.) The majority of cable and wire terminations in the airframe are made by attaching crimp tags for use with terminal blocks or pin and sockets within connectors, see Fig. 20.6. There are two types of crimp: dimple and confined; the latter is the type normally used on aircraft.

The confined crimp is formed by compressing the crimp's shank onto the conductor. This results in the **cold flow** of metal between shank and conductor forming a homogeneous mass. The crimping operation also causes the crimp to form over the insulating and sheath materials. The crimping operation is performed with crimping pliers that contain two dies to form the crimp in a controlled and preset way. For larger-diameter cables and wires, power tools are required. Connections can be made to terminate a cable with a ring-tag, or join two cables together with a splice. These can be used for repairs or as permanent installations. They can also be used to form junctions, e.g. when modifying

Figure 20.5 Quick-release connectors (LRUs)

Figure 20.6 Terminal blocks with connections

an aircraft to 'tap' into a signal line. General rules for the use of splices will be given in the relevant aircraft documents; these will address the:

- spacing of splices in the same wire
- maximum number of splices in a given length of wire
- support of the splices (they should not be located in curved sections).

Both tags and splices have a plastic or nylon insulating sleeve covering the shank. This insulation is coloured red, blue or yellow depending on size; this colour coding relates to the specific crimping pliers required for the operation. To perform the crimping operation on a tag or splice, the wire must have its insulation/sheath stripped from the end. Stripping is carried out taking the following requirements into account:

- an approved stripping tool must be used with the correct size of cutting blades
- knives or side cutters must not be used since they can sever or damage the conductor(s)
- all insulating and sheath material must be removed to suit the crimp or splice.

If the above is not adhered to, problems will be introduced, e.g. reduced current-carrying capacity and reduced mechanical strength; this could lead to premature failure and/or overheating of the wire. Once stripped, the strands of a wire should be formed by hand so that they lie neatly together. They must be 100% inserted into the barrel of the tag or splice; if any strands are disturbed by the insertion, the wire

must be withdrawn and the strands straightened. Excessive twisting will increase the diameter of the conductor, making it difficult to insert into the barrel. Once formed, the tag can be attached to terminal blocks with connections made to respective circuits. Some terminations benefit from a heat-shrink sleeve; this provides extra mechanical protection and support. Heat-shrink material is polythene based and reduces to a pre-determined diameter (but not length) when heated. The advantages of crimping are:

- good conductivity
- consistent operation
- good strength
- resistance to corrosion.

Key maintenance point

Always use approved crimping tools; crushing a termination does not constitute a good connection!

20.1.3 Connectors

Wires and cables also have to be routed through bulkheads or production breaks; this is normally achieved by using quick-release connectors. General aviation connectors (D-shaped) are the same type as used on personal computers, see Fig. 20.7.

Larger aircraft use circular quick-release connectors as shown in Fig. 20.8. Pins and sockets are crimped onto their respective ends of stripped wires; the connector is assembled usually with a supporting

Figure 20.7 General aviation connectors (D-shaped)

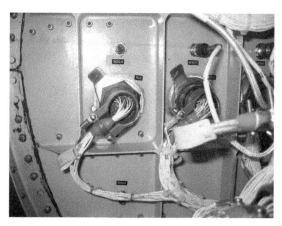

Figure 20.8 Circular quick-release connectors

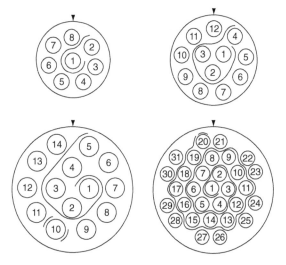

Figure 20.9 Pin/socket identification

The bodies and shells of quick-release connectors are made from light alloy, stainless steel or cadmium-plated alloys. Keyways ensure the correct alignment and prevent any twisting movement. The pins slide-fit into the respective socket; the pins are gripped by springs inside the socket to provide the electrical contact surface. A number identifies each pin/socket; see Fig. 20.9 for examples.

Key maintenance point

Pin/socket numbers of a connector are referenced to a physical point, or marking on the connector.

Key maintenance point

Connector socket spring-grips wear over time and can cause loss of continuity. Inserting anything but the correct pin can cause permanent damage.

bracket. Connectors are released by screws or by twisting the quick-release mechanism. Pins and sockets are removed and inserted using special tools; they are crimped onto the wires in a similar way to crimping of tags and splices.

Key maintenance point

If the connector is to be left undone for a period of time, always protect the pins/sockets with a cap or plastic bag.

Key maintenance point

When re-making connectors, always check the alignment of keyways of mating connectors before making the connection.

20.1.4 Aluminium wires or cables

Some applications use **aluminium** wires or cables to save weight. It is essential that the correct procedures, tools and materials are used when working with aluminium wires or cables. Aluminium oxidizes immediately when exposed to air, leading to high-resistance joints; compounds must be used to prevent this. Petrolatum (semi-solid mixture of hydrocarbons) and granular zinc compounds can either be applied during the crimping process, or be pre-filled within the crimp. When the crimp joint is made, the compound penetrates around each of the strands; the abrasive action of the compound scours the surfaces and removes the oxidation. When the crimping process is completed, the compound remains in the crimp to form a sealed termination. Aluminium wiring crimps joined into the main aircraft distribution system, will be joined with relatively hard materials, e.g. copper or steel. Fixing hardware, e.g. nuts, bolts and washers must be made of materials such as cadmium-plated aluminium that will not cause electrolytic corrosion. It is essential that washers are used to protect the softer aluminium material. The surfaces must first be cleaned and treated with compound; the terminations are secured with nuts that must be torque-loaded.

Key maintenance point

Flexing of aluminium wires during installation together with aircraft vibration causes work-hardening leading to brittle joints.

20.2 Bonding

It is a mandatory requirement that aircraft structure and equipment are electrically bonded. Specific bonding and grounding connections are made in an aircraft to accomplish the following:

- dissipate energy from high-intensity radiated fields (HIRFs) and lightning strikes
- dissipate static electricity
- limit the potential difference between equipment
- provide a low-resistance path for earth return systems.

Bonding connections are made between components and structure using purpose-made straps, see Fig. 20.10

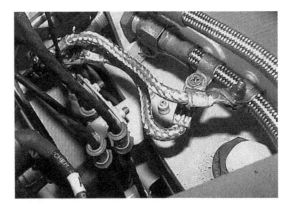

Figure 20.10 Bonding

(further examples are given in Chapter 19). Bonding is categorized as either primary or secondary; this is determined by the magnitude of current being conducted. **Primary** bonding is designed for carrying lightning discharges and to provide electrical return paths. **Secondary** bonding is used to dissipate static electricity and keep all structure at the same potential. Bonding straps (or leads) are pre-fabricated from braided copper or aluminium terminated with crimps.

20.2.1 Composite materials

There is an increasing use of **composite materials** being used in the construction of aircraft because of their good strength-to-weight ratio (compared with aluminium). Composite material has a high electrical resistance and is intrinsically unsuitable for bonding, earth returns and lightning strike dissipation. A ground plane has to be integrated into the airframe; this is normally achieved by bonding an aluminium wire mesh into the composite structure during manufacture. This mesh is accessed at key points around the aircraft to gain access to the ground plane.

Direct bonding (Fig. 20.11) is achieved by exposing the mesh (ground plane) and mounting the equipment directly onto the conductive path. **Indirect bonding** (Fig. 20.12) is achieved by exposing the mesh and installing a bonding wire and connector. The mesh must always be coated after making a connection since the aluminium will oxidize when exposed to air, leading to high resistance and unreliable joints.

Lightning protection in composite aircraft is achieved via aluminium wire integrated into the outer

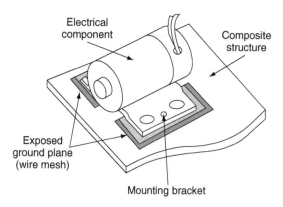

Figure 20.11 Direct bonding on composite structure

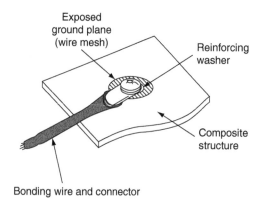

Figure 20.12 Indirect bonding on composite structure

layers of the composite construction. The lightning strike will enter and leave the aircraft at its extremities; the integrated wires are installed in anticipation of this and the energy dissipated through the aircraft along pre-determined routes to the exit point(s).

20.2.2 Maintenance requirements

Maintenance requirements for bonding include inspection of the two crimped tags to ensure that they are secure and not corroded. The braided conductor should be inspected for any visual signs of mechanical damage; it must not interfere with any moving items e.g. flying controls or actuators. The bonding lead must have sufficient slack to allow for aircraft flexing. All contacting surfaces must be bare metal; the crimp tags are then treated and sealed in accordance with the

maintenance manual. Electrolytic corrosion can occur at the bonding lead connection; it is essential that the correct bonding lead materials and treatments are used.

Key maintenance point

If a bonding lead has to be changed, refer to the aircraft documentation to obtain the correct type and part number.

20.3 Static charges

Static charges build up on the airframe during flight and these must be discharged into the atmosphere to avoid interference with avionic equipment. Electric charge is dissipated from the surface of an aircraft during flight by an effect called **corona discharge**. Coronas can generate both audible and radio-frequency (RF) noise, thereby disrupting the avionic systems. (Power lines produce an audible sound because they are producing a corona discharge that is interacting with the surrounding air.) Static electricity is discharged from the aircraft to atmosphere through **static wicks** (Fig. 20.13), sometimes called **static dischargers**. These ensure that the discharge of static electricity is from pre-determined points.

Static dischargers are fabricated with a wick of conductive element that provides a continuous low-resistance discharge path between the aircraft and atmosphere. The corona discharge can cause a faint glow adjacent to the prominent sections of the aircraft, e.g. wing tips; this results from ionization of oxygen and the formation of ozone in the surrounding air. The electric field created by the flow of electrons into the conductive section of the wick contains enough energy to ionize the oxygen and nitrogen in the air. This can produce low-energy **plasma**, the corona discharge. Plasma is a fourth state of matter along with solids, liquids, and gases. It is similar in appearance to a gas or liquid, but the molecules are separated into atoms from which the electrons in the outer shell are released into the plasma.

Key maintenance point

Helicopters (or rotorcraft) build up additional static charges through their rotor blades.

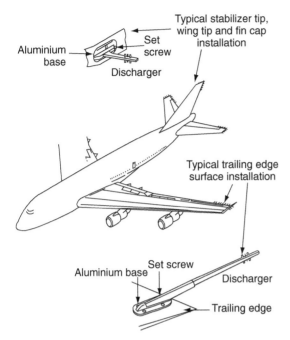

Figure 20.13   Static discharger locations

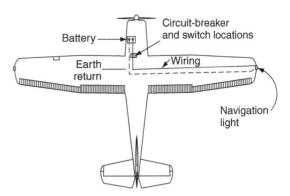

Figure 20.14   Airframe 'earth return' circuit

Despite the precautions taken to dissipate static charge during flight, it is highly likely that some charge will remain in the aircraft after landing. This presents a personal shock hazard for crew, passengers and ground staff. Static electricity is dissipated through tyres when the aircraft lands; the tyres are impregnated with a conductive compound. The retained static electricity also presents a major hazard during refuelling since the aircraft and fuel bowser will almost certainly be at different potentials, thereby causing a spark to jump across the fuel connections. The aircraft and fuel bowser are therefore connected by a length of cable onto specified points before refuelling starts.

Key maintenance point

A loose bonding connection causes more problems than having no bonding at all.

20.4 Earth returns

Aircraft with metallic structure use the airframe as the means of completing the electrical circuit, thereby reducing the cost, weight and installation time of installing a return wire, see Fig. 20.14. In this example, the positive side of the battery power supply is connected to the navigation light via the circuit protection and control switch.

The earth (or ground) return is connected to the negative side of the power supply via the aircraft structure. In some installations, grounding of the load, e.g. the navigation light is via the body of the component's housing. In other installations, all the negative connections are collected at earth stations. (Wires terminating at an earth station are usually identified with a suffix letter N.)

The location of the earth stations depends on a number of factors including the:

* mechanical strength of the structure
* current through the connections
* corrosive effects (dissimilar materials)
* ease of accessing/making the connections.

It is essential that the ground (or earth points) are not mixed between different types of circuit e.g. AC and digital signals. Using a common return path can lead to corrupted signals as illustrated in Fig. 20.15(a). Separate return paths (Fig. 20.15(b)) should be considered on all new system designs; this separation must be maintained whenever specified in the aircraft manuals.

Taking these factors into account, installations will vary considerably; see Fig. 20.16 for some examples. In order to make good electrical contact, there must be minimal resistance between the conductor and structure. Earth stations incorporate an anti-corrosion tag or plate to prevent electrolytic reaction between the base of the assembly and aircraft structure. The stations are usually identified on the adjacent structure by numerals preceded by an asterisk.

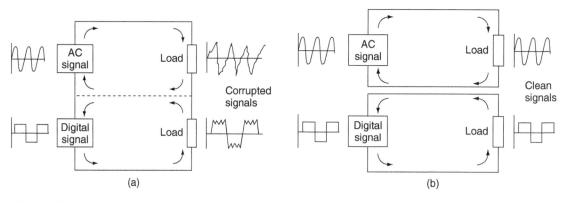

Figure 20.15   Earth/ground loops: (a) common return paths, (b) separate return paths

Figure 20.16   Earth station installations

Certain circuits must be isolated from each other, e.g. AC neutral and DC earth-returns must not be on the same termination; this could lead to current flow from the AC neutral through the DC system. Relay and lamp returns should not be on the same termination; if the relay earth connection has high resistance, currents could find a path through the low resistance of the filament lamp when cold.

For aircraft that have non-metallic **composite** structure, an alternative means of providing the return path must be made. This can be in the form of copper strips running the length of the fuselage or a wire mesh formed into the composite material. The principles of earth return remains the same as with bonding; the method of achieving it will vary.

20.5 Aircraft manuals

There is a wide range of technical manuals required by certifying staff, including, but not limited to:

- maintenance manuals
- illustrated parts catalogues
- wiring diagram manuals
- schematic diagrams.

These documents contain information required to explain how equipment operates, how it is maintained, and the correct part numbers that can be used when replacing components. For electrical and electronic systems, additional information is required to show how the systems are supplied with power, interconnections with other systems, etc. For larger aircraft, a standardized manual format was developed by the Airline Transport Association (ATA) of America called **ATA Spec. 100**. This specification contains format and content guidelines for technical manuals written by aircraft manufacturers and suppliers. The format has widespread international acceptance by aircraft manufacturers, airlines and equipment suppliers for the maintenance of their respective products. The specification is organized into **ATA chapters** that are specific to an aircraft system, e.g. ATA Chapter 24 contains all subjects within the electrical power systems, Chapter 33 contains details of the aircraft lights. (A complete list of ATA chapters is given in Appendix 8.) The ATA Spec. 100 is not widely adopted by the general aviation (GA) industry; the information required by certifying staff on GA aircraft is presented in a number of formats. In 2000, ATA Spec 100 (documentation) and ATA Spec 2100 (for inventory management) were incorporated into **ATA iSpec 2200**.

20.5.1 Maintenance manual

This is the primary source of information and data for certifying staff. When based on ATA Spec. 100, the maintenance manual (MM) is organized into specific sections for complete systems and/or individual components:

- detailed description (from p. 1)
- component locations (from p. 101)
- maintenance practices (from p. 201)
- servicing (from p. 301)
- removal and installation (from p. 401)
- adjustment/test (from p. 501)
- inspection/checking (from p. 601).

Not all of these sections apply to every system and/or individual component.

20.5.2 Wiring diagram manual

Aside from the aircraft maintenance manual, the primary manuals used by avionic engineers are the wiring diagram manuals (WDMs), schematic diagrams and/or circuit diagrams. Wires and cables should be identified to facilitate installation, trouble-shooting and potential modifications. There are various specifications that provide details of how wiring identification is implemented, including those adopted by the aircraft manufacturers. The simplest form of identifying cables is to mark the ends of the wires with the **source** and **destination** of the individual wire or cable. The wire/cable insulation is marked with indelible ink or laser printing. Wires are sometimes identified by the system type, wire number and gauge. Some manufacturers mark the cables at intervals along the cable. On more complex installations, where cables are bundled into looms, the entire loom may be given an identification code; this would be marked on a sleeve or band.

When created under ATA Spec. 100, the WDM provides details of equipment locations, wiring between equipment and connectors (including wire and cable identification). Schematic diagrams provide an overview of the system interconnections and equipment locations; individual wire numbers and connections are omitted for clarity. In some GA applications, basic electrical and electronic system information is contained within the maintenance manual. In smaller GA aircraft, much of the avionic equipment is installed as a customer option; in this case, the equipment installer provides the necessary details as supplements. This also applies in the case where new equipment is installed as part of an equipment upgrade. For the majority of cases, symbols used in wiring diagrams and schematics are standardized; however, some variations do exist among manufacturers. Symbols used for electrical and electronic equipment are provided in Appendix 9.

20.6 Circuit testing

The typical sequence in which electrical and electronic systems are checked and tested is as follows:

- **visual inspection**; this enables obvious faults to be identified and the appropriate action(s) taken
- **bonding test**; this checks the continuity of the earth return path back to the power supply(s)
- **continuity test**; this checks the interconnections between all circuits
- **insulation resistance (IR) test**; this checks for adequate insulation resistance between conductors and between conductors and the airframe (earth)
- **functional test**; there will be varying levels of test depending on complexity of circuit. These range from simple 'self tests' through to deeper tests that can only be carried out in the workshop.

Key maintenance point

For personal safety and to reduce the risk of damage to wiring and/or components, isolate power supplies when changing equipment or inspecting interconnecting wiring.

There are a number of techniques used to identify and locate faults in aircraft wiring and equipment, e.g. continuity and resistance measurements. There is a variety of equipment available to the avionics engineer to check and test aircraft electrical and electronic systems, including multimeters and oscilloscopes.

20.6.1 Multimeters

The most popular item of test equipment used for simple troubleshooting is the **multimeter**; these can be either analogue or digital instruments, see Fig. 20.17. Multimeters combine several functions into a single item of test equipment; the basic functions on simple multimeters include the measurement of current, voltage and resistance.

20.6.1.1 Analogue instruments

A popular analogue instrument used in the aircraft industry is the **Avometer**® (registered trade mark of Megger Group Limited). This instrument is often referred to simply as an **AVO**, deriving its name from the words amperes, volts, ohms. It has been in widespread use in the UK from the 1930s and can still be found in many hangars, workshops and repair stations. Referring to Fig. 20.18, features include the measurement of:

- alternating currents up to 10 A
- voltages up to 1000 V
- resistance from 0.1 Ω up to 200 kΩ.

The instrument is very accurate, typically ±1% of full-scale deflection (FSD) on DC ranges and ±2% on AC ranges. Maximum current consumption for

Figure 20.17   Multimeter; analogue or digital

Figure 20.18 Avometer® (registered trademark of Megger Group Limited)

voltage measurement is 50 μA (corresponding to a sensitivity of 20,000 Ω per volt); thereby minimizing voltage measurement error. Two rotary switches are used to select the function and range to be measured; if the wrong combination of function or range is selected, an overload cut-out switch (similar to a circuit-breaker) disconnects the test circuit. Some multimeters can measure the voltage drop across semiconductor junctions and measure additional circuit/component quantities such as:

- capacitance
- frequency
- temperature
- conductance
- inductance.

20.6.1.2 Digital multimeters

Digital multimeters are generally smaller than the analogue type of multimeter making them a useful hand-held device for basic fault finding on the aircraft. Higher specification devices can measure to a very high degree of accuracy and will commonly be found in repair shops and calibration laboratories. Digital multimeters

offer more features than basic analogue instruments; commonly available measurement features include:

- autoranging
- sample and hold
- graphical representation
- data acquisition
- personal computer interface.

Autoranging selects the appropriate range for the quantity under test so that meaningful digits are shown. For example, if a battery cell terminal voltage of 1.954 V DC was being measured, autoranging by a four-digit multimeter would automatically display this voltage instead of 0.019 (range set too high), or 0.999 (range set too low) via manual range selection. **Sample and hold** retains the most recent display for evaluation after the instrument is disconnected from the circuit being tested. **Graphical representation** of the quantity under test can be displayed in a number of ways, e.g. as a bar graph. This facilities observations of trends. Simple **data acquisition** features are used in some multimeters to record maximum and minimum readings over a given period of time, or to take a number of sample measurements at fixed time intervals. Higher-specification multimeters feature a **personal computer** interface, typically achieved by infrared (IR) links, or datalink connections, e.g. RS-232 or universal serial bus (USB). This interface allows the multimeter to upload measured data into the computer for storage and/or analysis.

20.6.2 Bonding meters

Aircraft bonding is tested with a bonding meter; the maintenance manual will define specific points on the airframe where measurements are made. The maximum resistance between specific points will be defined in the maintenance manual. The bonding meter consists of a low reading ohmmeter and two leads. One lead is 60 feet long with a single prodded end-piece; the other lead is 6 feet long with a double prodded end-piece. (Note that these leads must never be shortened.) The bond meter is first checked by:

- shorting all three prods together to check for zero-reading
- momentarily short the two prods of the 6 feet lead to check for full-scale deflection.

The maximum resistance for between extremities of fixed portions of a metallic aircraft and between bonded components/earth stations will be stated in the maintenance manual; the typical maximum value is 0.05 ohms for an aircraft with metal structure. The maximum value for composite aircraft will vary from type to type, but it will by necessity be very low.

20.6.3 Oscilloscopes

The **oscilloscope** (often referred to as a 'scope') is an item of electronic test equipment used to measure and view signal voltages as a two-dimensional graph (usually signal voltage on the vertical axis versus time on the horizontal axis). The oscilloscope will be found in most workshops, repair stations and calibration laboratories, see Fig. 20.19.

Original equipment was based on the cathode ray tube (CRT) display, making them heavy and bulky. Use of liquid crystal display (LCD) technology makes the oscilloscope useful as a portable device. (More information is provided on CRTs and LCDs in a related title in the series: *Aircraft Digital and Electronic Computer Systems.*) The portable oscilloscope has many uses during troubleshooting, e.g. checking for electrical noise and measurement of digital signals. PC-based oscilloscopes can be configured in an existing laptop via a specialized signal acquisition board and suitable hardware interfaces; the PC-based oscilloscope has a number of features:

- lower cost compared to a stand-alone oscilloscope
- efficient exporting of data into standard PC software, e.g. spreadsheets
- control of the instrument via custom programmes on the PC
- utilization of the PC's networking and disc storage functions

Figure 20.19   Oscilloscope

- larger-/higher-resolution colour displays
- colours can differentiate between waveforms
- portability.

20.7 Automatic test equipment

Automatic test equipment (ATE) is dedicated ground test equipment that provides a variety of different functional checks on line replaceable units (LRUs) or printed circuit boards (PCBs). The equipment being tested is connected to a variety of external circuits that represent the aircraft interfaces; additional connections are often made for diagnostic purposes. ATE is able to gather and analyse a large amount of data very quickly, thus avoiding the need to make a very large number of manual measurements in order to assess the functional status of an item of equipment.

ATE usually incorporates computerized control with displays and printouts that indicate what further action (repair or adjustment) is necessary in order maintain the equipment. Equipment may then require further detailed tests and measurements following initial diagnosis. ATE tends to be dedicated to a particular type of avionic system; it is therefore expensive to develop, manufacture and maintain. Because of this, ATE tends to be only used by original equipment manufacturers (OEMs) and licensed repairers.

20.8 On-board diagnostic equipment

Systems have been developed to match the complexity of electrical and electronic systems to assist the avionics engineer in fault finding on aircraft systems. On-board diagnostics use a range of techniques that are built into and integrated with the aircraft systems.

20.8.1 Built-in test equipment

As the name implies, **built-in test equipment (BITE)** is primarily a self-test feature built into aircraft electrical and electronic equipment as a means of:

- detecting and indicating specific equipment faults
- monitoring equipment performance
- detecting problems
- storing fault data
- isolate faulty sensors/components.

The origins of BITE started with simple on/off displays on the front of **line replaceable units** (LRUs) to assist the avionics engineer with troubleshooting; this display (typically a light-emitting diode, LED) would indicate the **go/no-go status** for a particular unit, either as a result of system test or in-flight fault. If the LED indicates an LRU fault, the engineer either changes the LRU or checks the interfaces with the unit. This simple technique is applied to individual LRUs, and will only indicate **real-time faults**. BITE techniques were developed alongside the increasing complexity of electrical and electronic equipment. Features such as fault storage provide an indication of faults over many flights; this information is often provided in coded form and needs to be interpreted by the engineer.

BITE is usually designed as a signal flow type test. If the signal flow is interrupted or deviates outside accepted levels, warning alerts indicate a fault has occurred. The functions or capabilities of BITE include the following:

- real-time monitoring of systems
- continuous display presentation
- sampled recorder readouts
- module and/or subassembly failure isolation
- verification of systems status
- go/no-go indications
- quantitative displays
- degraded operation status
- percentage of functional deterioration.

20.8.2 Centralized maintenance systems (CMSs)

BITE technology has now been developed into centralized maintenance systems on modern aircraft. The electronic centralized aircraft monitoring system (ECAM) developed for Airbus aircraft oversees a variety of aircraft systems and also collects data on a continuous basis. While ECAM automatically warns of malfunctions, the flight crew can also manually select and monitor individual systems. Failure messages recorded by the flight crew can be followed up by maintenance personnel by using the system test facilities on the maintenance panel in the flight compartment. Printouts can be produced as permanent records for further analysis, see Fig. 20.20.

Dedicated CMS control display units (CDUs) are installed on the centre pedestal. The CDU allows systems to be investigated via a menu selection; this provides a description of the **flight deck effect** (FDE) and likely reasons for the fault. Typical centralized

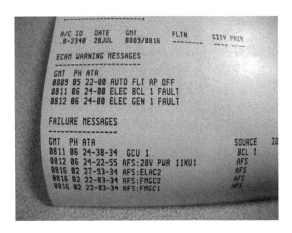

Figure 20.20 ECAM printout

maintenance systems have a dedicated **central maintenance computer** (CMC) to collect, analyse and store fault information. The CMC has up to 50 data bus inputs, 10 data bus outputs and can store up to 500 fault messages into non-volatile memory (NVM).

20.8.3 Aircraft communication addressing and reporting system

Some aircraft are installed with a system called ACARS (aircraft communication addressing and reporting system). This is a digital data link system transmitted in the VHF range (118 MHz to 136 MHz). ACARS provides a means by which aircraft operators can exchange maintenance and operational data directly without human intervention. This makes it possible for airline staff to communicate directly with the aircraft in their fleet in much the same way as it is possible to exchange data using a land-based digital network. ACARS uses an aircraft's unique identifier and the system has some features that are similar to those currently used for electronic mail. Typical ACARS messages are used to convey routine information such as:

- fuel data
- engine performance data
- aircraft fault data
- passenger loads
- departure reports
- arrival reports.

This information can be requested by the company and retrieved from the aircraft at periodic intervals or on demand. Prior to ACARS this type of information would have been transferred via VHF voice communications. (Additional information on ACARS is provided in a related book in the series: *Aircraft Communications and Navigation Systems.*)

20.9 Electrostatic sensitive devices (ESSDs)

Advances in electronic technology bring many new features and benefits, e.g. faster processors, higher-density memory and highly efficient displays. These advances are primarily due to the reduction in the physical size of semiconductor junctions; this leads to higher-density components in given size of integrated circuit. One significant problem associated with the handling of semiconductor devices is that the smaller junctions are susceptible to damage from **electrostatic voltages**. An example of static electricity that the reader might have encountered is the electric shock received when stepping out of a car. The synthetic materials used for clothing as well as the vehicle's interior are capable of producing large amounts of static charge which is only released when the driver or passenger sets foot on the ground!

20.9.1 Triboelectric effect

When two dissimilar non-conducting materials are rubbed together, the friction transfers electrical charge from one material to the other. This raises the electrical potential between the materials, and is known as the **triboelectric effect**. The build-up of charge, and subsequent attraction of materials, can be observed when clingfilm is separated from its roll. The polarity and strength of the electrical charges depend primarily on the materials, surface finish, ambient temperature and humidity. The triboelectric series classifies different materials according to how readily they create static electricity when rubbed with another material. The series is arranged on a **triboelectric scale** of increasingly positive and increasingly negative materials.

Materials that give up electrons and become positive when charged (thereby appearing as positive on the triboelectric scale) include glass, air and dry human skin. Materials that attract electrons become negatively charged (appearing as negative on the triboelectric scale) include:

- polyester
- polystyrene
- polyethylene
- poly(vinyl chloride) (PVC).

Certain materials that do not readily attract or give up electrons when brought in contact or rubbed with other materials are neutral on the triboelectric scale; examples include cardboard, cotton and steel.

The largest amounts of static charge resulting from materials being rubbed together (or separated from each other in the case of cling film) are at the extreme ends of the triboelectric scale. For example, PVC rubbed against glass or polyester rubbed against dry human skin both produce an accumulation of charge. A common occurrence when working in a dry atmosphere is that people rapidly discharge current (sometimes producing a spark) when touching metal objects. This is because they have relatively dry skin (which can become highly positive in charge), and is accentuated when the clothes they wear are made of man-made material (such as polyester), which will acquire a negative charge. The effect is much less pronounced in a humid atmosphere where the stray charge can 'leak away' harmlessly into the atmosphere. (Note that moist skin tends to dissipate charge more readily.) People that build up static charges due to dry skin are advised to wear all-cotton clothes since cotton is neutral on the triboelectric scale.

20.9.2 Working environment

The problem for the electronics industry is that high voltages can be accumulated in people and materials and then discharged into equipment. This discharge can weaken and/or damage electronic components. Representative values of electrostatic voltages generated in some typical working situations are shown in Table 20.1. (Note the significant difference in voltage generated at different values of relative humidity.)

Static voltage susceptibility varies for different types of semiconductor device or components, for example static discharge voltages as low as 20 V to 100 V can affect complex devices such as microprocessors. Single components such as silicon-controlled rectifiers are not affected until levels of between 4 kV to 15 kV are reached.

Table 20.1 Representative values of electrostatic voltages generated in typical work situations

Situation	20% relative humidity	80% relative humidity
Walking over a wool/ nylon carpet	35 kV	1.5 kV
Sliding a plastic box across a carpet	18 kV	1.2 kV
Removing parts from a polystyrene bag	15 kV	1 kV
Walking over vinyl flooring	11 kV	350 V
Removing shrink wrap packaging	10 kV	250 V
Working at a bench wearing overalls	8 kV	150 V

Electrostatic sensitive devices (including printed circuit boards, circuit modules, and plug-in devices) are invariably marked with warning notices. These are usually printed with black text on yellow backgrounds, as shown in Fig. 20.21.

Special precautions must be taken when handling or removing ESSDs. These precautions include use of the following:

- wrist/heel straps
- static dissipative floor and bench mats
- ground jacks
- grounded test equipment
- low-voltage soldering equipment and anti-static soldering stations (low-voltage soldering irons with grounded bits)
- anti-static insertion and removal for integrated circuits
- avoidance of nearby high-voltage sources (e.g. fluorescent light units)
- anti-static packaging (static-sensitive devices and printed circuit boards should be stored in their anti-static packaging until such time as they are required for use)
- Protective materials.

Wrist straps (and also heel straps) are conductive bands that are connected to an effective ground point

Figure 20.21 ESD warning notice

by means of a short wire lead. The lead is usually fitted with an integral 1 MΩ resistor which minimizes currents arising between the person wearing the strap and the equipment being worked on. Wrist straps are usually stored at strategic points on the aircraft, or they may be carried by maintenance technicians.

Figure 20.22 shows a typical wrist strap being used for a bench operation. The bench itself must be grounded; this is achieved by connecting the conductive mat to an external earthing point outside of the building (Fig. 20.23).

There are three main classes of materials used for protecting static sensitive devices. These are:

- conductive materials, e.g. metal foils
- static dissipative materials (a cheaper form of conductive material)
- anti-static materials, e.g. cardboard and cotton.

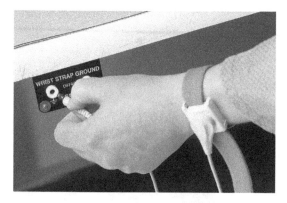

Figure 20.22   ESD Typical wrist strap grounding point

Of these, conductive materials offer the greatest protection whilst anti-static materials offer the least protection. The working environment plays an important part in the safe handling of ESSDs; extra vigilance is required when the relative humidity is low, e.g. in air conditioned workshops. (Further reading on the subject of EMI and ESD can be found in a related book in the series, *Aircraft Digital and Electronic Computer Systems*.)

Key maintenance point

The effect of lower humidity on ESSD components in typical working environments will be to create higher voltages and pose more threat to the components.

20.10 Multiple choice questions

1. ACARS is a digital data link system transmitted in the:
    (a) VHF range
    (b) LF range
    (c) UHF range.

2. Secondary bonding is designed for:
    (a) carrying lightning discharges through the aircraft
    (b) keeping all the structure at the same potential
    (c) discharging static electricity from the aircraft to atmosphere.

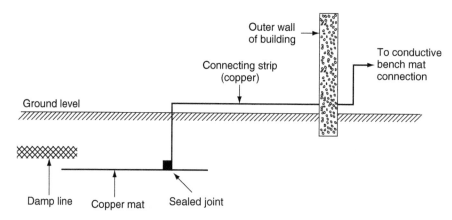

Figure 20.23   ESD work area protection

3. Some electrical installations use aluminium wires or cables to:
   (a) reduce EMI
   (b) save cost
   (c) save weight.

4. Visual inspection of wiring installations:
   (a) enables obvious faults to be identified and the appropriate action(s) taken
   (b) checks for adequate insulation resistance between conductors
   (c) checks the continuity of the earth return path back to the power supply(s).

5. In terms of bonding, composite material has a:
   (a) high resistance and is unsuitable for bonding
   (b) low resistance and is unsuitable for bonding
   (c) high resistance and is suitable for bonding.

6. For a given wire size, when installed in a loom it will be able to safely conduct:
   (a) more current compared to a wire in free circulating air
   (b) less current compared to a wire in free circulating air
   (c) the same current compared to a wire in free circulating air.

7. Bonding is categorized as primary or secondary determined by the:
   (a) use of composite or metal structure
   (b) locations of static wicks
   (c) magnitude of current being conducted.

8. The use of composite materials for aircraft structures results in:
   (a) less natural paths for bonding
   (b) more natural paths for bonding
   (c) higher probability of a lightning strike.

9. Bonding is made between components and structure using:
   (a) coaxial cable
   (b) purpose-made straps
   (c) general-purpose wiring.

10. Static electricity is discharged from the aircraft to atmosphere through:
    (a) composite structure
    (b) earth stations
    (c) static wicks.

Appendix 1 Abbreviations and acronyms

Abbreviations and acronyms are extensively used to describe electrical and electronic systems, as well as the parameters with which they operate. The abbreviations and acronyms used in this book, together with some commonly used terms, are listed below.

ABS	Anti-lock braking system
A/C	Air conditioning
AC	Alternating current
ACARS	Aircraft communication addressing and reporting system
ACAS	Airborne collision avoidance system
ACM	Air cycle machine
ACMS	Aircraft condition monitoring system
ADC	Analogue to digital converter
ADC	Air data computer
ADF	Automatic direction finder
ADI	Attitude director indicator
ADIRS	Air data/inertial reference system
ADS	Air data system
AEEC	Airlines electronic engineering committee
AES	Aircraft earth stations
AFC	Automatic frequency control
AFCS	Automatic flight control system (autopilot)
AFS	Automatic flight system (autopilot)
AGL	Above ground level
AGM	Absorbent glass mat
AHRS	Attitude/heading reference system
AHS	Attitude heading system
AI	Airbus industries
AIAA	American Institute of Aeronautics and Astronautics
AIDS	Aircraft integrated data system
AIMS	Airplane information management system
AIS	Aeronautical information system
AIV	Anti-icing valve
ALT	Altitude
AMLCD	Active matrix liquid crystal display
AMM	Aircraft maintenance manual

AoA	Angle of attack
AP	Autopilot
APP	Approach
APU	Auxiliary power unit
ARINC	Aeronautical radio incorporated
ARMS	Aircraft recording and monitoring system
ASCB	Aircraft system common data bus
ASI	Air speed indicator
ASM	Air separation module
ASR	Aerodrome surveillance radar
ATA	Actual time of arrival or Airline Transport Association (ATA) of America
ATC	Air traffic control
ATE	Automatic test equipment
ATFM	Air traffic flow management
AVLAN	Avionics local area network
AVOD	Audio-video on demand
AWG	American wire gauge (US spelling gage)
BC	Bus controller
BCD	Binary coded decimal
BCF	Bromochlorodifluoromethane
BCI	Bulk current injection
BGA	Ball grid array
BIC	Backplane interface controller
BIOS	Basic input/output system
BIST	Built-in self-test
BIT(E)	Built-in test (equipment)
BIU	Bus interface unit
BJT	Bipolar junction transistor
BNC	Bayonet nut coupling
BOT	Beginning of tape (FDR)
BPCU	Bus power control unit
BPRZ	Bipolar return to zero
BPS	Bits per second
BPV	Bypass valve
BSU	Bypass switch unit
BTB	Bus tie breaker
BTM	Bromotrifluoromethane
CAA	Civil Aviation Authority
CABLAN	Cabin local area network
CAGE	Commercial and government entity
CAI	Computer-aided instruction

CAN	Controller area network		EADI	Electronic attitude director indicator
CAS	Collision avoidance system		EATMS	Enhanced air traffic management system
CAS	Crew alerting system			
CAT	Clear-air turbulence		ECAM	Electronic centralized aircraft monitoring
CBM	Condition-based maintenance		ECM	Electronic countermeasures
CDI	Course deviation indicator		ECS	Environmental control system
CDU	Control display unit		ECU	Electronic control unit
CFDS	Central fault display system		EDP	Engine-driven pump
CFIT	Controlled flight into terrain		EEC	Electronic engine control
CG	Centre of gravity		EEPROM	Electrically erasable programmable read-only memory
CI	Condition indicator			
CIDS	Cabin intercommunication data system		EFIS	Electronic flight instrument system
CMC	Central maintenance computer		EGPWS	Enhanced ground proximity warning system
CMOS	Complementary metal oxide semiconductor			
			EGT	Exhaust gas temperature
CMS	Centralized maintenance system		EHSI	Electronic horizontal situation indicator
CNS/ATM	Communication, navigation, surveillance and air traffic management		EICAS	Engine indication and crew alerting system
CO_2	Carbon dioxide		EMC	Electromagnetic compatibility
COTS	Commercial off-the-shelf		EMF	Electromotive force
CPA	Collision prediction and alerting		EMI	Electromagnetic interference
CPU	Central processing unit		EOT	End of tape (FDR)
CRT	Cathode ray tube		EPR	Engine pressure ratio
CSD	Constant speed drive		EPROM	Erasable programmable read-only memory
CT	Current transformer			
CVR	Cockpit voice recorder		EROPS	Extended range operations
CWT	Centre wing tank		ESD	Electrostatic discharge
			ESSD	Electrostatic sensitive device
DA	Drift angle		ETA	Estimated time of arrival
DAC	Digital to analogue converter		ETFE	Ethylene tetrafluoroethylene
DC	Direct current		ETOPS	Extended-range twin-engine operations
DCU	Data concentrator unit		EVM	Engine vibration monitoring
DDR	Digital data recorder		EWIS	Electrical wire interconnection system
DDT	Direct drive testing			
DECU	Digital engine control unit		FAA	Federal Aviation Administration (USA)
DEOS	Digital engine operating system		FAC	Flight augmentation computer
DFDAU	Digital flight data acquisition unit		FBW	Fly-by-wire
DFDR	Digital flight data recorder		FCC	Flight control computer
DFLD	Database field loadable data		FCGC	Flight control and guidance computer
DFT	Discrete Fourier transform		FCU	Flight control unit
DH	Decision height		FDAU	Flight data acquisition unit
DIL	Dual in-line		FDC	Fight director computer
DIMM	Dual in-line memory module		FDD	Floppy disk drive
DIP	Dual in-line plastic		FDDI	Fibre distributed data interface
DLS	DME-based landing system		FDE	Flight deck effect
DMA	Direct memory access		FDR	Flight data recorder
DMC	Display management computer		FDS	Flight director system
DME	Distance measuring equipment		FEP	Fluorinated ethylene-propylene
DO	Design organization		FET	Field effect transistor
DPM	Data position module		FFT	Fast Fourier transform
DPU	Display processor unit		FG	Flight guidance
DRAM	Dynamic random access memory		FGC	Flight guidance computer
DSC	Digital source collector		FIFO	First-in first-out
DSP	Display select panel		FIR	Flight information region
DVD	Digital versatile disc		FIS	Flight information system

FL	Flight level		ILS	Instrument landing system
FLS	Field loadable software		IMA	Integrated modular avionics
FLIR	Forward-looking infrared		INU	Inertial navigation unit
FLTA	Forward-looking terrain avoidance		INS	Inertial navigation system
FMC	Flight management computer		IOAPIC	Input/output advanced programmable
FMS	Flight management system			input controller
FOG	Fibre optic gyros		IOM	Input/output module
FPM	Feet per minute		IP	Internet protocol
FSD	Full scale deflection		IPC	Instructions per cycle
FSHR	Fuel solenoid holding relay		IPR	Intellectual property rights
FQIS	Fuel quantity indicating system		IPS	Inches per second
FWC	Flight warning computer		IR	Infrared
			IRS	Inertial reference system
GA	General aviation		ISA	International standard atmosphere
GCAM	Ground collision avoidance module		ISAS	Integrated situational awareness system
GCB	Generator circuit-breaker		ISU	Iridium subscriber unit
GCCU	Generator conversion control unit			
GCR	Generator control relay		LAAS	Local area augmentation system
GCU	Generator control unit		LAN	Local area network
GEO	Geostationary earth orbit		LCC	Leadless chip carrier
GND	Ground		LCD	Liquid crystal display
GNSS	Global navigation satellite system		LE	Leading edge
GPCU	Ground power control unit		LED	Light-emitting diode
GPM	Ground position module		LEO	Low earth orbit
GPS	Global positioning system		LFC	Laminar flow control
GPWS	Ground proximity warning system		LGC	Landing gear control
GS	Ground speed or glide slope		LIDAR	Light radar
			LP	Low pressure
HDG	Heading		LRRA	Low-range radio altimeter
HEIU	High-energy ignition unit		LRU	Line replaceable unit
HERF	High-energy radiated field		LSAP	Loadable aircraft software part
HFDS	Head-up flight display system		LSB	Least significant bit
HGS	Head-up guidance system		LSD	Least significant digit
HIRF	High-intensity radiated field		LSI	Large scale integration
HIRF/L	High-intensity radiated field/lightning		LVDT	Linear variable differential transformer
HIRL	High-intensity runway lights			
HM	Health monitoring		MARMS	Modular aircraft recording and
HP	High pressure			monitoring system
HSI	Horizontal situation indicator		MAU	Modular avionics unit
HUD	Head-up display		MCDU	Microprocessor controlled display units
HUGS	Head-up guidance system		MEL	Minimum equipment list
HUMS	Health and usage monitoring systems		MEMS	Micro-electromechanical systems
			MFD	Multi-function display
I/O	Input/output		MFDS	Multi-function display system
IAPS	Integrated avionics processing system		MGB	Main gearbox
IAS	Indicated airspeed		MSL	Mean sea level
IC	Integrated circuit		MLS	Microwave landing system
ICP	Integrated circuit preamplifier		MLW	Maximum landing weight
IDE	Integrated drive electronics		MM	Maintenance manual
IDG	Integrated drive generator		MMI	Man machine interface
IFE	In-flight entertainment		MMM	Mass memory module
IFR	Instrument flight rules		MMS	Mission management system
IGB	Intermediate gearbox		MNPS	Minimum navigation performance
IHUMS	Integrated health and usage monitoring			specification
	system		MOS	Metal oxide semiconductor

MOSFET	Metal oxide semiconductor field effect transistor		PVDF	Polyvinylidene fluoride
MPU	Multifunction processor unit		PVS	Personal video screen
MRO	Maintenance/repair/overhaul		PWM	Pulse-width-modulated
MSB	Most significant bit			
MSD	Most significant digit		QAR	Quick access recorder
MSI	Medium scale integration		QFP	Quad flat package
MTBF	Mean time between failure			
MTBO	Mean time between overhaul		RAM	Random access memory
MTOW	Maximum takeoff weight		RAT	Ram air turbine
MWS	Master warning system		RF	Radio-frequency
			RFI	Radio-frequency interference
NC	Normally closed		ROM	Read-only memory
ND	Navigation display		RMI	Radio magnetic indicator
NEA	Nitrogen-enriched air		RMS	Root mean square
NGS	Nitrogen generating system		RNAV	Area navigation
NO	Normally open		RTB	Rotor track and balance
NOTAR	No tail rotor		RTD	Resistance temperature device
NRV	Non-return valve		RZ	Return to zero
NRZ	Non-return to zero			
NTSB	National transport safety board		SCR	Silicon controlled rectifier
NVM	Non-volatile memory		SDAC	System data acquisition concentrator
			SDD	System definition document
OBI	Omni bearing indicator		SDI	Source/destination identifier
OBIGGS	On-board inert gas generation system		SHP	Shaft horsepower
OBS	Omni bearing selector		SIL	Single in-line
OEA	Oxygen enriched air		SIP	Single in-line package
OEM	Original equipment manufacturer		SLA	Sealed lead-acid
OFD	Optical fire detector		SG	Signal generator
OFV	Outflow valve		SNOC	Satellite network operation center
OS	Operating system		SO	(n) Shaft order (n-1 harmonic)
			SRAM	Synchronous random access memory
PA	Passenger address		SRD	System requirement document
PAPI	Precision approach path indicator		SROM	Serial read only memory
PBGA	Plastic ball grid array		SPDT	Single pole double throw
PCMCIA	Personal computer memory card international association		SPST	Single pole single throw
			SSI	Small scale integration
PC	Personal computer		SSM	Sign/status matrix
PCB	Printed circuit board		STP	Shielded twisted pair
PDL	Portable data loader		SWAMP	Severe wind and moisture problems
PFD	Primary flight display			
PGA	Pin grid array		TACAN	Tactical air navigation
PLCC	Plastic leadless chip carrier		TAS	True air speed
PMG	Permanent magnet generator		TAWS	Terrain awareness warning system
POST	Power-on self-test		TBM	Time-based maintenance
PP	Pre-processor		TBO	Time between overhaul
PQFP	Plastic quad flat package		TCAS	Traffic alert and collision avoidance system
PROM	Programmable read-only memory		TDS	Tail drive shaft
PRSOV	Pressure-regulating shutoff valve		TE	Trailing edge
PSEU	Proximity switch electronic unit		TGB	Tail gearbox
PSM	Power supply module		TNC	Thread-nut coupling
PSU	Passenger service unit		TRU	Transformer rectifier unit
PTFE	Polytetrafluoroethylene		TTAC	Telemetry tracking and command/control
PTT	Press to test		TTL	Transistor-transistor logic

ULB	Underwater locator beacon
ULSI	Ultra-large-scale integration
UMS	User modifiable software
USB	Universal serial bus
UTP	Unshielded twisted pair
UV	Ultraviolet
VAS	Virtual address space
VFR	Visual flight rules
VGA	Video graphics adapter
VHM	Vibration health monitoring
VLSI	Very large scale integration

VMS	Vibration monitoring system
VRLA	Valve-regulated lead-acid
VSCF	Variable speed constant frequency
WAAS	Wide area augmentation system
WDM	Wiring diagram manual
WORM	Write-once read-many
WoW	Weight on wheels
WX	Weather
WXP	Weather radar panel
WXR	Weather radar

These papers are intended to provide you with practice in preparation for examinations. Each paper contains 20 questions; these should be attempted without the use of a calculator or notes. The time allocation for each paper is 25 minutes.

Paper 1

1. Which one of the following gives the symbol and abbreviated units for resistance?
   (a) symbol, R; unit, Ω
   (b) symbol, V; unit, V
   (c) symbol, R; unit, A.

2. When testing a diode with an ohmmeter, what resistance will be measured between the anode and the cathode?
   (a) high resistance one way
   (b) high resistance both ways
   (c) low resistance both ways.

3. A DC generator produces an output of 28 V at 20 A. The power supplied by the generator will be:
   (a) 14 W
   (b) 560 W
   (c) 1.4 kW.

4. A silicon transistor has a base-emitter voltage of 0 V. In this condition the transistor will be:
   (a) conducting heavily
   (b) conducting slightly
   (c) turned off.

5. Battery capacity is measured in:
   (a) volts
   (b) amperes
   (c) ampere-hours.

6. TRUs are used to convert what forms of electrical power:
   (a) AC into DC
   (b) AC into AC
   (c) DC into AC.

7. For a given AWG, a wire will have a specified:
   (a) diameter and hence a known conductance
   (b) length
   (c) screening.

8. The main distribution bus is sometimes called the:
   (a) essential bus
   (b) emergency bus
   (c) non-essential bus.

9. Metal elements used in RTDs have a temperature coefficient that is:
   (a) negative, temperature increases cause an increase in resistance
   (b) positive, temperature decreases cause an increase in resistance
   (c) positive, temperature increases cause an increase in resistance.

10. Engine pressure ratio (EPR) is used to measure a gas turbine engine's:
    (a) torque
    (b) thrust
    (c) temperature.

11. When fuel level decreases, the capacitance of the fuel quantity sensor:
    (a) decreases and the reactance increases
    (b) Increases and the reactance increases
    (c) Increases and the reactance decreases.

12. The starboard wing tip navigation light has the following colour and divergence:
    (a) red and 110 degrees
    (b) green and 110 degrees
    (c) white and ±70 degrees.

13. Satellite communication systems use a low earth orbit to:
    (a) provide greater coverage
    (b) maintain a geostationary position
    (c) minimize voice delays.

14. Indications of landing gear fully down and locked are:
    (a) red lights on, green lights off
    (b) red lights on, green lights on
    (c) red lights off, green lights on.

15. Stall warning systems provide the crew with a clear and distinctive warning:
    (a) before the stall is reached
    (b) after the stall is reached
    (c) at all angles of attack.

16. The action of twisting a fire handle closes micro-switches that:
    (a) activate the engine fire extinguisher
    (b) cancels the alarm
    (c) shuts off the fuel.

17. Red areas are used on TAWS displays to indicate terrain that is:
    (a) above the aircraft's current altitude
    (b) level with the aircraft's current altitude
    (c) safe in terms of required terrain clearance.

18. The flight data recorder must start to record data automatically:
    (a) after the aircraft is capable of moving under its own power
    (b) after take off
    (c) prior to the aircraft being capable of moving under its own power.

19. Shielding of conductors:
    (a) reduces radiation and minimizes susceptibility
    (b) increases radiation and minimizes susceptibility
    (c) reduces radiation and increases susceptibility.

20. For a given wire size, when installed in a loom it will be able to safely conduct:
    (a) more current compared to a wire in free circulating air
    (b) less current compared to a wire in free circulating air
    (c) the same current compared to a wire in free circulating air.

Paper 2

1. The relationship between power, P, current, I, and resistance, R, is:
   (a) $P = IR$
   (b) $P = \dfrac{R}{I}$
   (c) $P = I^2R$.

2. In order to make a silicon-controlled rectifier (thyristor) conduct a:
   (a) small current pulse must be applied to its gate
   (b) high voltage must be applied to its anode
   (c) load must be connected between its anode and cathode.

3. A two-input NAND gate with both inputs inverted will become which of these:
   (a) NOR gate
   (b) AND gate
   (c) OR gate.

4. Decreasing the current in the field coil of a generator will:
   (a) decrease the output current
   (b) increase the output voltage
   (c) decrease the output frequency.

5. Static electricity is discharged from the aircraft to atmosphere through:
   (a) composite structure
   (b) earth stations
   (c) static wicks.

6. When the lightning discharge is dissipated safely through the aircraft, structural damage is most likely to occur at the:
   (a) bonding between structures
   (b) exit points of the discharge
   (c) entry points of the discharge.

7. The mandatory parameters required for an aircraft DFDR depend on the:
   (a) speed and weight of the aircraft
   (b) maximum weight of the aircraft
   (c) size of the aircraft and the prevailing regulatory rules applied to that aircraft.

8. Mode 1 terrain awareness cautions are given for:
   (a) negative climb rate or altitude loss after take-off
   (b) excessive descent rate
   (c) altitude callout at 500 feet.

9. Open area smoke detectors rely on the transfer of particulate matter from the source of fire to the detector by:
   (a) convection
   (b) radiation
   (c) conduction.

10. In the automatic braking system, when a wheel is approaching a skid condition, this is detected when the speed is:
    (a) decreasing at a given rate
    (b) increasing at a given rate
    (c) constant.

11. Engine fire or loss of cabin pressure would be displayed on ECAM as:
    (a) level 3 failures
    (b) level 2 failures
    (c) level 1 failures.

12. Audio-video on demand (AVOD) entertainment enables passengers to:
    (a) pause, rewind, fast-forward or stop a programme
    (b) make phone calls via satellite communication
    (c) ignore PA system voice announcements and chime signals.

13. Incandescence is the radiation of light from:
    (a) a gas-discharge device
    (b) an electrical filament due to an increase in its temperature
    (c) a combined optical and electrical phenomenon.

14. The volume of fuel in a tank varies with temperature; as the temperature changes:
    (a) the mass and volume of fuel remains the same
    (b) the mass of fuel changes but the volume remains the same
    (c) the mass of fuel remains the same, but the volume changes.

15. Ground idle speed occurs when the engine has:
    (a) stabilized (slightly above self-sustaining speed)
    (b) stabilized (slightly below self-sustaining speed)
    (c) Just been started.

16. The thermocouple principle is based on the Seebeck effect, when heat is applied:
    (a) a change of resistance is measured
    (b) this causes the element to bend
    (c) an electromotive force (EMF) is generated.

17. The reverse current relay is needed on any DC generation system to prevent the:
    (a) battery from feeding excess current back through to the generator's armature
    (b) generator from feeding excess current back through to the battery
    (c) battery from feeding excess current to the starter motor.

18. Two or more separate wires within the same insulation and protective sheath is referred to as a:
    (a) screened wire
    (b) coaxial cable
    (c) cable.

19. The desired output frequency of a rotary inverter is determined by the:
    (a) AC input voltage
    (b) input frequency
    (c) DC input voltage.

20. The energy storage capacity of a cell is determined by the:
    (a) terminal voltage
    (b) electrolyte specific gravity
    (c) amount of material available for chemical reaction.

Paper 3

1. A cabin reading lamp consumes 10 W from a 24 V DC supply. The current supplied will be:
   (a) 0.42 A
   (b) 0.65 A
   (c) 2.4 A.

2. The output frequency of an AC generator driven at a constant speed:
   (a) increases with the number of pole pairs
   (b) decreases with the number of pole pairs
   (c) is independent of the number of pole pairs.

3. A compound-wound generator has:
   (a) only a series field winding
   (b) only a shunt field winding
   (c) both a series and a shunt field winding.

4. The decimal equivalent of the natural binary code 10110 is:
   (a) 12
   (b) 22
   (c) 38.

5. Lead acid batteries are recharged by constant:
   (a) voltage
   (b) current
   (c) ampere-hours.

6. Inverters are used to convert what forms of electrical power?
   (a) DC into DC
   (b) AC into AC
   (c) DC into AC.

7. Trip-free circuit-breaker contacts:
   (a) can always be closed whilst a fault exists
   (b) cannot be closed whilst a fault exists
   (c) are only used during maintenance.

8. Different size pins are used on external DC power connectors to:
   (a) prevent a reverse polarity voltage being applied
   (b) prevent excessive power being applied
   (c) prevent power being applied when the battery is discharged.

9. The rotors of a torque synchro transmitter and receiver are supplied from:
   (a) the same power supply (normally 26 V AC)
   (b) Different power supplies (normally 26 V AC)
   (c) the same power supply (normally 26 V DC).

10. HEIUs can remain charged for several:
    (a) seconds
    (b) minutes
    (c) hours.

11. Air and fuel have dielectrics of approximately:
    (a) unity and zero respectively
    (b) unity and two respectively
    (c) two and unity respectively.

12. Green or indicator blue lights in the instrument panel are used inform the crew that:
    (a) a safe condition exists
    (b) an unsafe condition exists
    (c) an abnormal condition exists.

13. The passenger address (PA) system is primarily a safety system that provides passengers with:
    (a) in-flight entertainment
    (b) reduced amount of IFE wiring to a seat position
    (c) voice announcements and chime signals.

14. An electrical flap drive system uses a:
    (a) reversible DC motor
    (b) variable-speed DC motor
    (c) unidirectional DC motor.

15. The stall identification system contains an actuator that:
    (a) maintains the angle of attack
    (b) pulls the control column rearward
    (c) pushes the control column forward.

16. When a pneumatic fire detector is rapidly heated; hydrogen is liberated causing:
    (a) sufficient gas pressure to close the alarm switch
    (b) the integrity switch to open
    (c) the alarm switch to reset.

17. Forward-looking terrain avoidance (FLTA) looks:
    (a) ahead of, and below the aircraft's lateral and vertical flight path
    (b) ahead of, and above the aircraft's lateral and vertical flight path
    (c) either side of, and below the aircraft's lateral and vertical flight path.

18. The DFDR on large aircraft has to be able to retain the recorded data for a minimum of the last:
    (a) 30 minutes of its operation
    (b) 25 hours of its operation
    (c) 25 flights.

19. The ability of an item of equipment to operate alongside other items of equipment without causing EMI is called:
    (a) EMC
    (b) ESSD
    (c) HIRF.

20. Visual inspection of wiring installations:
    (a) enables obvious faults to be identified and the appropriate action(s) taken
    (b) checks for adequate insulation resistance between conductors
    (c) checks the continuity of the earth return path back to the power supply(s).

Paper 4

1. A generator delivers 250 W of power to a 50 Ω load. The current flowing in the load will be:
   (a) 0.2 A
   (b) 5 A
   (c) 10 A.

2. The power factor in an AC circuit is defined as the ratio of:
   (a) true power to apparent power
   (b) apparent power to true power
   (c) reactive power to true power.

3. In a PNP transistor, conventional current:
   (a) flows into the collector
   (b) flows out of the collector
   (c) there is no current flow in the collector.

4. Gillham code from an altitude encoder is based on:
   (a) natural binary code
   (b) a form of Gray code
   (c) self-correcting hexadecimal code.

5. In terms of bonding, composite material has a:
   (a) high resistance and is unsuitable for bonding
   (b) low resistance and is unsuitable for bonding
   (c) high resistance and is suitable for bonding.

6. Higher frequency signals can lead to cross-talk between wires as a result of the:
   (a) dielectric effect of the wire insulation causing increased capacitance
   (b) dielectric effect of the wire insulation causing reduced capacitance
   (c) dielectric effect of the wire conductors causing reduced capacitance.

7. Lateral acceleration and radio altitude are typical parameters recorded on the:
   (a) FDR
   (b) CVR
   (c) ULB.

8. Premature descent alert compares the aircraft's:
   (a) ground speed with the proximity of the nearest airport
   (b) lateral and vertical position with the proximity of the nearest airport
   (c) lateral and vertical position with the proximity of high terrain.

9. To operate an engine fire extinguisher, the fire handle is:
   (a) twisted
   (b) twisted and then pulled
   (c) pulled and then twisted.

10. When the AoA reaches a certain angle, the airflow over the wing:
    (a) becomes turbulent and the lift is dramatically decreased
    (b) becomes streamlined and the lift is dramatically decreased
    (c) becomes turbulent and the lift is dramatically increased.

11. When an electrically operated landing gear is fully retracted, the up-lock switch contacts:
    (a) open thereby removing power from the motor
    (b) close thereby removing power from the motor
    (c) open thereby applying power to the motor.

12. Fibre optic cable bends need to have a sufficiently large radius to:
    (a) minimize losses and damage
    (b) maximize immunity to electromagnet interference (EMI)
    (c) accurately align the connector optical components.

13. The starboard wing tip navigation light has the following colour and divergence:
    (a) red and 110 degrees
    (b) green and 110 degrees
    (c) white and ±70 degrees.

14. Intrinsic safety is a technique used for:
    (a) safe operation of electrical/electronic equipment in explosive atmospheres
    (b) ensuring fuel temperature does not become too low/high
    (c) reducing fuel quantity.

15. The starting sequence for a gas turbine engine is to:
    (a) turn on the ignition, develop sufficient airflow to compress the air, and then open the fuel valves
    (b) develop sufficient airflow to compress the air, open the fuel valves and then turn on the ignition
    (c) develop sufficient airflow to compress the air, turn on the ignition and then open the fuel valves.

16. When a foil strain gauge is deformed, this causes:
    (a) an electromotive force (EMF) to be generated
    (b) its electrical resistance to change
    (c) different coefficients of thermal expansion.

17. The split bus system is sometimes called a:
    (a) non-parallel system
    (b) parallel system
    (c) standby and essential power system.

18. The white collar just below a circuit-breaker button provides visual indication of the:
    (a) circuit-breaker trip current
    (b) system being protected
    (c) circuit-breaker being closed or tripped.

19. The voltage coil of a carbon pile regulator contains a:
    (a) large number of copper wire turns connected across the generator output
    (b) low number of copper wire turns connected across the generator output
    (c) large number of copper wire turns connected in series with the generator output.

20. The only accurate and practical way to determine the condition of the nickel-cadmium battery is with a:
    (a) specific gravity check of the electrolyte
    (b) measured discharge in the workshop
    (c) check of the terminal voltage.

Appendix 3 Answers to multiple choice questions

Chapter 1 (Electrical fundamentals)

1. b
2. c
3. a
4. a
5. b
6. b
7. b
8. a
9. b
10. c
11. b
12. c
13. c
14. a
15. c

Chapter 2 (Electronic fundamentals)

1. b
2. b
3. c
4. c
5. a
6. a
7. b
8. a
9. c
10. b
11. a
12. a
13. a
14. b
15. c

Chapter 3 (Digital fundamentals)

1. b
2. c
3. b
4. a
5. a
6. a
7. a
8. b
9. a
10. c
11. b
12. c
13. b
14. c
15. a

Chapter 4 (Electrical machines)

1. a
2. a
3. a
4. a
5. a
6. a
7. b
8. b
9. b
10. b
11. a
12. b
13. b
14. c
15. c

Chapter 5 (Batteries)

1. a
2. c
3. a
4. b
5. b
6. a
7. b
8. c
9. a
10. a
11. b
12. c

Chapter 6 (Power supplies)

1. a
2. c
3. c
4. a
5. b
6. c
7. c
8. b
9. a
10. a

Chapter 7 (Wiring and circuit protection)

1. a
2. c
3. a
4. c
5. b
6. a
7. c
8. b
9. a
10. c

Chapter 8 (Distribution of power supplies)

1. a
2. b
3. c
4. a
5. c
6. a
7. c
8. b
9. a
10. c
11. c
12. b

Chapter 9 (Controls and transducers)

1. b
2. a
3. b
4. c
5. a
6. c
7. b
8. a
9. a
10. c

11. a
12. b
13. a
14. b
15. b

Chapter 10 (Engine systems)

1. b
2. a
3. a
4. a
5. c
6. a
7. b
8. a
9. a
10. c
11. b
12. c

Chapter 11 (Fuel management)

1. c
2. a
3. b
4. a
5. a
6. b
7. c
8. a
9. b
10. b

Chapter 12 (Lights)

1. a
2. b
3. b
4. a
5. a
6. b
7. c
8. b
9. c
10. a
11. a
12. b
13. c
14. c
15. b
16. a

Chapter 13 (Cabin systems)

1. b
2. a
3. b
4. a
5. c
6. a
7. c
8. a
9. b
10. a
11. a
12. c
13. a
14. c
15. a
16. c

Chapter 14 (Airframe monitoring, control and indicating systems)

1. c
2. b
3. a
4. a
5. c
6. a
7. a
8. c
9. b
10. a

Chapter 15 (Warning and protection systems)

1. a
2. b
3. c
4. b
5. a
6. a
7. c
8. a
9. b
10. b

Chapter 16 (Fire and overheat protection)

1. a
2. c
3. a
4. a
5. b

6. c
7. a
8. a
9. a
10. c

Chapter 17 (Terrain awareness warning systems)

1. a
2. b
3. c
4. c
5. a
6. b
7. a
8. a
9. b
10. c
11. c
12. a
13. b
14. b
15. c

Chapter 18 (Flight data and cockpit voice recording)

1. a
2. c
3. b
4. b
5. c
6. a
7. b
8. a
9. a
10. c
11. b
12. a
13. c

Chapter 19 (Electrical and magnetic fields)

1. a
2. b
3. c
4. c
5. a
6. c
7. b

8. a
9. c
10. b

Chapter 20 (Continuing airworthiness)

1. a
2. b
3. c
4. a
5. a
6. b
7. c
8. a
9. b
10. c

Revision paper 1

1. a
2. a
3. b
4. c
5. c
6. c
7. a
8. c
9. c
10. b
11. a
12. b
13. c
14. c
15. a
16. a
17. a
18. c
19. a
20. b

Revision paper 2

1. c
2. a
3. c
4. a
5. c
6. b
7. c
8. b
9. a
10. a
11. a
12. a

13. b
14. c
15. a
16. c
17. a
18. c
19. c
20. c

Revision paper 3

1. a
2. a
3. c
4. b
5. a
6. c
7. b
8. a
9. a
10. b
11. b
12. a
13. c
14. b
15. c
16. a
17. a
18. b
19. a
20. a

Revision paper 4

1. b
2. a
3. b
4. b
5. a
6. b
7. a
8. b
9. c
10. a
11. a
12. a
13. b
14. a
15. c
16. b
17. a
18. c
19. a
20. b

Appendix 4
Electrical quantities, symbols and units

Quantity	Symbol	Unit	Abbreviated units
Angle	ϕ	radian or degree	Rad or °
Capacitance	C	farad	F
Charge	Q	coulomb	C
Conductance	G	siemen	S
Current	I	ampere	A
Energy	W	joule	J
Flux	Φ	weber	Wb
Flux density	B	tesla	T
Frequency	f	hertz	Hz
Impedance	Z	ohm	Ω
Inductance	L	henry	H
Power	P	watt	W
Reactance	X	ohm	Ω
Resistance	R	ohm	Ω
Time	t	second	s
Voltage	V	volt	V

Electrical formulae

Charge, current and voltage

$$Q = I \times t$$

Ohm's law

$$V = I \times R \text{ and } I = V/R \text{ and } R = V/I$$

Similarly if *resistance* is replaced by *reactance* or *impedance*:

$$V = I \times X \text{ and } I = V/X \text{ and } X = V/I$$
$$V = I \times Z \text{ and } I = V/Z \text{ and } Z = V/I$$

Power and energy

$$P = I \times V \text{ and } P = V^2/R \text{ and } P = I^2R$$
$$J = P \times t \text{ and since } P = I \times V \text{ so } J = IVt$$

Resistors in series

$$R_T = R_1 + R_2 + R_3$$

Resistors in parallel

$$\frac{1}{R_T} = \frac{1}{R_1} + \frac{1}{R_2} + \frac{1}{R_3}$$

but where there are *only two* resistors,

$$R_T = \frac{R_1 \times R_2}{R_1 + R_2}$$

Capacitance

$$C = \frac{\varepsilon A}{d}$$

where ε is the *permittivity* of the dielectric and $\varepsilon = \varepsilon_0 \varepsilon_r$

Capacitance, charge and voltage

$$Q = CV$$

Inductance

$$L = n^2 \frac{\mu A}{l}$$

where μ is the *permeability* of the magnetic medium and $\mu = \mu_0 \mu_r$.

Energy stored in a capacitor

$$W = \tfrac{1}{2}CV^2$$

Energy stored in an inductor

$$W = \tfrac{1}{2}LI^2$$

Inductors in series

$$L_T = L_1 + L_2 + L_3$$

Inductors in parallel

$$\frac{1}{L_T} = \frac{1}{L_1} + \frac{1}{L_2} + \frac{1}{L_3}$$

but where there are *only* two inductors

$$L_T = \frac{L_1 \times L_2}{L_1 + L_2}$$

Capacitors in series

$$\frac{1}{C_T} = \frac{1}{C_1} + \frac{1}{C_2} + \frac{1}{C_3}$$

but where there are *only* two capacitors

$$C_T = \frac{C_1 \times C_2}{C_1 + C_2}$$

Capacitors in parallel

$$C_T = C_1 + C_2 + C_3$$

Induced e.m.f. in an inductor

$$e = -L\frac{di}{dt}$$

where $\dfrac{di}{dt}$ is the rate of change of current with time.

Current in a capacitor

$$i = C\frac{dv}{dt}$$

where $\dfrac{dv}{dt}$ is the rate of change of voltage with time.

Sine wave voltage

$$v = V_{max}\sin(\omega t) \text{ or } v = V_{max}\sin(2\pi ft)$$

because $\quad \omega = 2\pi f$

$$f = 1/T$$

where T is the periodic time.

For a *sine wave*, to convert:

RMS to peak multiply by **1.414**
Peak to RMS multiply by **0.707**
Peak to average multiply by **0.636**
Peak to peak-peak multiply by **2**

Capacitive reactance

$$X_C = \frac{V_C}{I_C} = \frac{1}{2\pi fC}$$

Inductive reactance

$$X_L = \frac{V_L}{I_L} = 2\pi fL$$

Resistance and reactance in series

$$Z = \sqrt{(R^2 + X^2)} \text{ and } \phi = \arctan\left(\frac{X}{R}\right)$$

Resonance

$$X_L = X_C \text{ thus } \omega L = \frac{1}{\omega C} \text{ or } 2\pi f_o L = \frac{1}{2\pi f_o C}$$

$$f_o = \frac{1}{2\pi\sqrt{LC}}$$

Power factor

Power factor = true power/apparent power
$\qquad\qquad$ = watts/volt-amperes = W/VA
$\quad$ True power = $V \times (I \times \cos\phi) = VI\cos\phi$
$\quad$ Power factor = $\cos\phi = R/Z$
Reactive power = $V \times (I \times \sin\phi) = VI\sin\phi$

Motors and generators

$$F = BIl$$

$$f = pn/60$$

where p is the number of pole *pairs* and n is the speed in r.p.m.

Three phase
Star connection

$$V_L = 1.732 \times V_P \text{ and } I_L = I_P$$

Note that $1.732 = \sqrt{3}$

Delta connection

$$V_L = V_P \text{ and } I_L = 1.732 \times I_P$$

Power in a three-phase load

$$P = 3 \times V_P I_P \cos\phi = 1.732 \times V_L I_L \cos\phi$$

Appendix 6 Decibels

Decibels (dB) are a convenient means of comparing power, voltage and current in electrical and electronic circuits. They are also commonly used for expressing gain (amplification) and loss (attenuation) in electronic circuits. They are used as a *relative* measure (i.e. comparing one voltage with another, one current with another, or one power with another). In conjunction with other units, decibels are sometimes also used as an *absolute* measure. Hence dB V are decibels relative to 1 V, dB m are decibels relative to 1 mW, etc. The decibel is one-tenth of a bel which, in turn, is defined as the logarithm (base 10) of the ratio of power, P, to reference power, P_R. Provided that the resistances/impedances are identical, voltage and current ratios may be similarly defined, i.e. one-twentieth of the logarithm (base 10) of the ratio of voltage, V, to the reference voltage, V_R, or one-twentieth of the logarithm (base 10) of the ratio of current, I, to the reference current, I_R.

Decibels (dB)	Power ratio	Voltage ratio	Current ratio
0	1	1	1
1	1.26	1.12	1.12
2	1.58	1.26	1.26
3	2	1.41	1.41
4	2.51	1.58	1.58
5	3.16	1.78	1.78
6	3.98	2	2
7	5.01	2.24	2.24
8	6.31	2.51	2.51
9	7.94	2.82	2.82
10	10	3.16	3.16
13	19.95	3.98	3.98
16	39.81	6.31	6.31
20	100	10	10
30	1,000	31.62	31.62
40	10,000	100	100
50	100,000	316.23	316.23
60	1,000,000	1,000	1,000
70	10,000,000	3,162.3	3,162.3

Appendix 7

Wire and cable sizes

The sizing of most, if not all, aircraft wiring installations is based on the **American wire gauge (AWG)**. The size of wire (or gauge) is a function of its diameter and cross-sectional area. The cross-sectional area of each gauge is an important factor for determining its current-carrying capacity. Increasing gauge numbers give decreasing wire diameters.

Measurement of wire gauge is based on a linear measurement of one thousandth of an inch (**one mil**). A **square mil** is a unit of area equal to the area of a square with sides of one thousandth of an inch. This unit of area is usually used in specifying the area of the cross-section of rectangular conductors, e.g. bus bars or terminal strips.

The **circular mil** is a unit of area, equal to the area of a circle with a diameter of one mil. Note that a circular mil is not the true cross sectional area of a wire (given by πr^2). One circular mil is equal to the cross sectional area of a 0.001 in. diameter wire. (Although the term 'mil' is used, there is no relationship to any metric units.)

Circular mils are a convenient unit for comparing wires and cables, without the need to reference π. Circular mils were introduced into the aircraft industry from the USA and should be used with caution since potential error will result from the lack of π in its definition! This is particularly relevant when working with the resistivity of metals; these figures are based on the true cross-sectional area.

When the wire is contained within a bundle (or loom) the current-carrying capacity reduces since heat dissipation is reduced. The voltage drop along a given length of wire will also be a limiting factor for the current-carrying capacity; tables are published that allow the circuit designer to calculate the wire gauge for a given power supply voltage, wire length and whether the current is intermittent or continuous.

Table A7.1 Copper wire, American Wire Gauge (AWG)

Wire size (AWG)	Diameter (mils)	Cross-section (Circular mils)	Maximum current (A) in free air
20	32.0	1,024	11
18	40.3	1,624	16
16	50.8	2,581	22
14	64.1	4,109	32
12	81.0	6,561	41
10	102.0	10,404	55
8	128.5	16,512	73
6	162.0	26,244	101
4	204.0	41,616	135
2	258.0	66,564	181

Key maintenance point

The maximum current (A) in Table A7.1 is for training purposes and guidance only; always consult the approved data.

ATA chapter/subsystem list

For larger aircraft, a standardized manual format was developed by the Airline Transport Association (ATA) of America called ATA Spec. 100. This specification contains format and content guidelines for technical manuals written by aircraft manufacturers and suppliers. Note that in 2000, ATA Spec 100 (documentation) and ATA Spec 2100 (for inventory management) were incorporated into ATA iSpec 2200. Many aircraft documents that were written via the ATA Spec. 100 will be in service for many years to come.

The following gives a list of ATA chapters and subsections within the chapter.

Chapter 21 Air Conditioning

21-00-00 General
21-10-00 Compression
21-20-00 Distribution
21-30-00 Pressurization control
21-40-00 Heating
21-50-00 Cooling
21-60-00 Temperature control
21-70-00 Moisture/air contamination

Chapter 22 Autoflight

22-00-00 General
22-10-00 Autopilot
22-20-00 Speed-attitude correction
22-30-00 Autothrottle
22-40-00 System monitor
22-50-00 Aerodynamic load alleviating

Chapter 23 Communications

23-00-00 General
23-10-00 Speech communication
23-20-00 Data transmission, auto. calling
23-30-00 Passenger address and entertainment
23-40-00 Interphone
23-50-00 Audio integrating
23-60-00 Static discharging
23-70-00 Audio & video monitoring
23-80-00 Integrated automatic tuning

Chapter 24 Electrical Power

24-00-00 General
24-10-00 Generator drive
24-20-00 AC generation
24-30-00 DC generation
24-40-00 External power
24-50-00 AC electrical load distribution
24-60-00 DC electrical load distribution

Chapter 25 Equipment & Furnishings

25-00-00 General
25-10-00 Flight compartment
25-20-00 Passenger compartment
25-30-00 Buffet/galley
25-40-00 Lavatories
25-50-00 Cargo compartments
25-60-00 Emergency
25-70-00 Accessory compartments
25-80-00 Insulation

Chapter 26 Fire Protection

26-00-00 General
26-10-00 Detection
26-20-00 Extinguishing
26-30-00 Explosion suppression

Chapter 27 Flight Controls

27-00-00 General
27-10-00 Aileron & tab
27-20-00 Rudder & tab
27-30-00 Elevator & tab

27-40-00 Horizontal stabilizer
27-50-00 Flaps
27-60-00 Spoiler, drag devices, fairings
27-70-00 Gust lock & damper
27-80-00 Lift augmenting

Chapter 28 Fuel

28-00-00 General
28-10-00 Storage
28-20-00 Distribution
28-30-00 Dump
28-40-00 Indicating

Chapter 29 Hydraulic Power

29-00-00 General
29-10-00 Main
29-20-00 Auxiliary
29-30-00 Indicating

Chapter 30 Ice & Rain Protection

30-00-00 General
30-10-00 Airfoil
30-20-00 Air intakes
30-30-00 Pitot and static
30-40-00 Windows, windshields & doors
30-50-00 Antennas & radomes
30-60-00 Propellers & rotors
30-70-00 Water lines
30-80-00 Detection

Chapter 31 Indicating & Recording Systems

31-00-00 General
31-10-00 Instrument & control panels
31-20-00 Independent instruments
31-30-00 Recorders
31-40-00 Central computers
31-50-00 Central warning systems
31-60-00 Central display systems
31-70-00 Automatic data reporting systems

Chapter 32 Landing Gear

32-00-00 General
32-10-00 Main gear & doors
32-20-00 Nose gear & doors

32-21-01 Actuators
32-30-00 Extension & retraction
32-40-00 Wheels & brakes
32-50-00 Steering
32-60-00 Position and warning
32-70-00 Supplementary gear

Chapter 33 Lights

33-00-00 General
33-10-00 General compartment
33-20-00 Passenger compartments
33-30-00 Cargo and service compartments
33-40-00 Exterior
33-50-00 Emergency lighting

Chapter 34 Navigation

34-00-00 General
34-10-00 Flight environment data
34-20-00 Attitude & direction
34-30-00 Landing & taxiing aids
34-40-00 Independent position determining
34-50-00 Dependent position determining
34-60-00 Flight management computing
34-70-00 ATC Transponder

Chapter 35 Oxygen

35-00-00 General
35-10-00 Crew
35-20-00 Passenger
35-30-00 Portable

Chapter 38 Water/waste

38-00-00 General
38-10-00 Potable
38-20-00 Wash
38-30-00 Waste disposal
38-40-00 Air supply

Chapter 51 Structure

51-00-00 General

Chapter 52 Doors

52-00-00 General
52-10-00 Passenger/crew

52-20-00 Emergency exit
52-30-00 Cargo
52-40-00 Service
52-50-00 Fixed interior
52-60-00 Entrance stairs
52-70-00 Door warning
52-80-00 Landing gear

Chapter 53 Fuselage

53-00-00 General

Chapter 54 Nacelles/pylons

54-00-00 General
54-10-00 Nacelle section
54-50-00 Pylon section

Chapter 55 Horizontal & vertical stabilizers

55-00-00 General
55-10-00 Horizontal stabilizer or canard
55-20-00 Elevator
55-30-00 Vertical stabilizer
55-40-00 Rudder

Chapter 56 Windows

56-00-00 General
56-10-00 Flight compartment
56-20-00 Passenger compartment
56-30-00 Door
56-40-00 Inspection & observation

Chapter 57 Wings

57-00-00 General
57-10-00 Centre wing
57-20-00 Outer wing
57-30-00 Wing tip
57-40-00 Leading edge and leading edge
57-50-00 Trailing edge and trailing edge
57-60-00 Ailerons and elevons
57-70-00 Spoilers

Chapter 61 Propellers

61-00-00 General
61-10-00 Propeller assembly
61-20-00 Controlling

61-25-01 Governor, propeller
61-30-00 Braking
61-40-00 Indicating
61-50-00 Propulsor duct

Chapter 71 Power plant

71-00-00 General
71-10-00 Cowling
71-20-00 Mounts
71-30-00 Fireseals
71-40-00 Attach fittings
71-50-00 Electrical harness
71-60-00 Air intakes
71-70-00 Engine drains

Chapter 72 Engine – turbine

72-00-00 General

Chapter 73 Engine fuel and control

73-00-00 General
73-10-00 Distribution
73-15-00 Divider flow
73-20-00 Controlling
73-25-00 Unit fuel control
73-30-00 Indicating

Chapter 74 Ignition

74-00-00 General
74-10-00 Electrical power supply
74-15-01 Box, ignition exciter
74-20-00 Distribution

Chapter 75 Air

75-00-00 General
75-10-00 Engine anti-icing
75-20-00 Cooling
75-30-00 Compressor control
75-35-01 Valve HP & LP bleed
75-40-00 Indicating

Chapter 76 Engine controls

76-00-00 General
76-10-00 Power control
76-20-00 Emergency shutdown

Chapter 77 Engine indicating

77-00-00 General
77-10-00 Power
77-20-00 Temperature
77-30-00 Analysers
77-40-00 Integrated engine instrument systems

Chapter 78 Exhaust

78-00-00 General
78-10-00 Collector/nozzle
78-20-00 Noise suppressor
78-30-00 Thrust reverser
78-40-00 Supplementary air

Chapter 79 Oil

79-00-00 General
79-10-00 Storage
79-20-00 Distribution
79-30-00 Indicating

Chapter 80 Starting

80-00-00 General
80-10-00 Cranking

Appendix 9 — Electrical and electronic symbols

Batteries

+| |— Single-cell

+| | |— Multi-cell (battery)

Audible devices

Bell

Horn

Busbar

Crossing
conductor

Test jack

Slip ring

Connector test point

Single pin connector

Resistors

Fixed

Tapped

Variable

Heater

Ballast

Fuse

Gearbox

Grounds

Internal

External

Chassis

Warning lights

R

R — with press
to test

Meters

A — Amps

V — Volts

F — Frequency

Shunt

A — B

C D

Complete connector

Receptacle plug

Wires

Co-axial

Single

Shielded

Grounded
shield

Twisted

Semiconductors

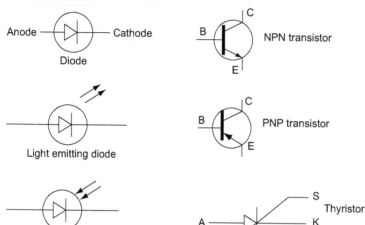

Anode — Diode — Cathode

NPN transistor

Light emitting diode

PNP transistor

Photo sensitive diode

Thyristor

Transformers

Basic   Step-down   • No phase shift   Step-up   • Phase shift 180°

Auto

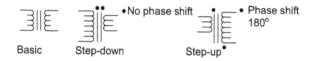

Fixed   Variable   Current

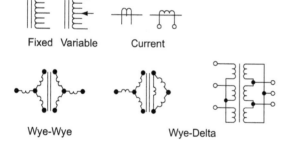

Wye-Wye   Wye-Delta

Thermal devices

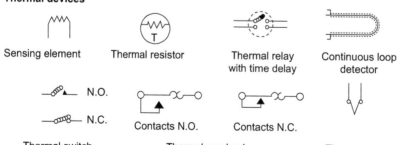

Sensing element   Thermal resistor   Thermal relay with time delay   Continuous loop detector

—— N.O.
—— N.C.

Contacts N.O.   Contacts N.C.

Thermal switch   Thermal overlaod   Thermocouple

Switches

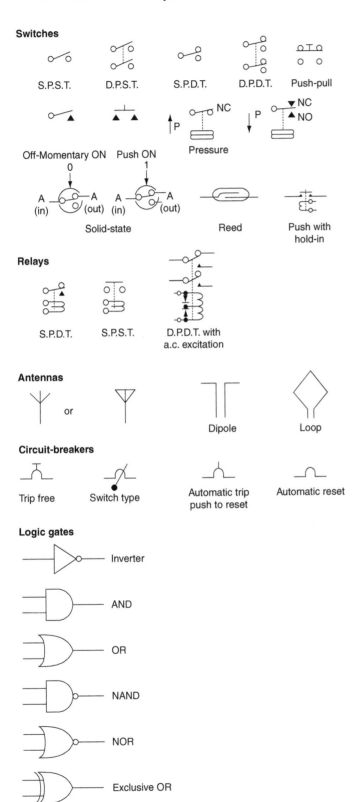

S.P.S.T.   D.P.S.T.   S.P.D.T.   D.P.D.T.   Push-pull

Off-Momentary ON   Push ON

Pressure

A (in)   A (out)   A (in)   A (out)

Solid-state   Reed   Push with hold-in

Relays

S.P.D.T.   S.P.S.T.   D.P.D.T. with a.c. excitation

Antennas

or

Dipole   Loop

Circuit-breakers

Trip free   Switch type   Automatic trip push to reset   Automatic reset

Logic gates

Inverter

AND

OR

NAND

NOR

Exclusive OR

Wire numbering/coding

Wires and cables should be identified to facilitate installation, trouble-shooting and potential modifications. There are various specifications that provide details of how wiring identification is implemented, including those adopted by the aircraft manufacturers. The simplest form of identifying cables is to mark the ends of the wires with the **source** and **destination** of the individual wire or cable. The wire/cable insulation is marked with indelible ink or laser printing. Wires are sometimes identified by the system type, wire number and gauge. Some manufacturers mark the cables at intervals along the cable. On more complex installations, where cables are bundled into looms, the entire loom may be given an identification code; this would be marked on a sleeve or band. The ATA specification is illustrated here to describe the principles of wire coding.

The code consists of a six-position combination of letters and numbers that is marked on the wire or cable, and is documented in the wiring diagram manuals. If the number cannot be marked on the wire or cable, e.g. due to its small diameter, it is printed onto sleeves that are fitted to the ends of the wire or cable. The following examples illustrate the principles of coding; the wire code would be marked thus:

$$2 \; P \; 4 \; B \; 20 \; N$$

Each of the six positions has a specific purpose:

$$2_1 \; P_2 \; 4_3 \; B_4 \; 20_5 \; N_6$$

Position 1 indicates the system number, e.g. in a twin-engine aircraft; this would be the wiring associated with the number two engine.

Position 2 indicates the function of the circuit or system, e.g. P = electrical power. Other examples of circuit functions/system codes are given below.

Position 3 indicates the sequential number of wires in the same circuit; three wires used by the stator windings of synchro system would be numbered 1, 2 and 3.

Position 4 indicates the segment of wiring in a circuit, i.e. that portion of wire between two connections. This normally starts with A and builds up sequentially through the circuit.

Position 5 indicates the AWG size of the wire or cable; on certain cable types, e.g. coaxial and thermocouples, this number is not used, and is replaced by a (-).

Position 6 indicates whether the wire is being used as either: a connection to a ground/neutral point, an AC supply or a thermocouple. Codes used for position 6 are as follows:

N	ground/neutral
V	single-phase AC supply
A, B, C	three-phase AC supply
AL	alumel thermocouple
CH	chromel thermocouple
CU	aluminium thermocouple
CN	constantan thermocouple

Examples of circuit functions/system codes:

C	Flight control
D	De-icing/anti-icing
E	Engine instruments
F	Flight instruments
H	Heating/ventilation
J	Ignition
K	Engine control
L	Lighting
M	Miscellaneous
P	Power
Q	Fuel/oil
R	Radio navigation/communication
V	Inverter
W	Warning devices
X	AC power

Index

A/C pack 258
AC 19–28
ACARS 375, 376
ACM 258
ADC 59
AES 253
AGM 123
AHRS 314
AIDS 336
AND logic 65
APU 111, 140, 200; starter logic 70; starting 199
ARINC-429 80, 81, 146, 350
ATA 371
ATE 374
AVO 372
AVOD 252
AWG 145
Absorbent glass mat 123
Accelerometer 188, 189
Active-low 69
Address 82; bus 82
Advisory messages 217
Aerofoil 273, 277
Air: conditioning 257–260; cyclic machine 258; distribution system 201
Aircraft communication and reporting system 375; earth station 253; integrated data system 336
Aircraft manual 370
Airflow 274
Airstairs 262
Alarm switch 297
Alternating current 20
Alternator 95, 111; busbar 155
Altitude 313–314
Aluminium 144, 367
American Wire Gauge 145
Amplifier 56
Amplitude 21
Analogue: to digital converter 59; signal 59
Angle of attack 273, 275
Anode 42, 117, 118
Anti: -collision lights 244, 246; -lock braking 285; -skid system 285
AoA 273–277; sensor 278
Apparent power 27, 168
Arcing 361

Argand diagram 29
Armature 91, 174
Assisted recovery 326
Asynchronous data 80
Audio: entertainment system 251; -video on demand 252
Aural warnings 289
Automatic test equipment 374
Autoranging 373
Autosyn 187
Autotransformer 139, 140
Auxiliary power unit 111, 140
Avionic bus 154

BCD 72, 74, 338, 340
BCF 291
BITE 374
BJT 53, 54, 57
BNC 147
BPRZ 350
BPS 351
BPV 228
BTM 291
Balanced load 99
Bandwidth 351
Battery 6, 8, 117–119; bus 158; busbar 153, 161; charging 156, 161; connections 130; location 128–129; maintenance 121, 124; master switch 153; venting 128, 130; -backed memory 83
Beryllium oxide 42
Bi: -metallic strip 181, 182; -phase rectifier 49, 50
Bias 43, 53; current 58
Bidirectional bus 78
Binary 74, 82; coded decimal 72, 338, 340
Bipolar junction transistor 53
Bistable 71, 72
Bits per second 351
Bladder tanks 221
Bleed air 283; leaks 291
Bohr model 1
Bonding 357, 367, 368; lead 229; meter 373
Boost pump 226
Boundary layer separation 275
Bourdon tube 180, 214
Breakers 165, 166

Bridge rectifier 51, 52
Brushes 88, 97
Brushless AC generator 112, 114
Buffer 65, 257
Built-in test equipment 374
Bus: cable 80; coupler 78, 80; interface 81; speed 81; system 78, 79; terminator 80
Bypass valve 228

CAS 216
CFIT 307
CHT 209
CMOS RAM 83
CMS 375
CPU 82, 83
CRT 215, 374
CSD 113
CT 164
CVR 229, 340
CWT 229
Cabin: pressure 261; systems 249
Cable 143–147; installation 361; loom 362
Capacitance fringing 225
Capacitive fuel quantity system 223, 225; reactance 23, 24
Capacitor 4; run 109; start 108, 109
Capacity check 121, 122
Capsule 180
Carbon-pile regulator 133–136
Cargo bay fire extinguisher 304
Cargo smoke detector 301
Cathode 42, 117, 118; ray tube 215, 374
Caution: Alerting System 216; light 236, 238; messages 217; panel 239
Cell 6, 117–119
Centralized: Aircraft Monitoring 215; maintenance computer 375; maintenance system 375
Centre wing tank 229
Chafing 361
Changeover relay 174, 175
Charge 2, 6; carrier 2, 39
Charging 120, 123
Circuit: breaker 150–152; protection 148; testing 371
Cladding 256
Class: A amplifier 58; of fire 292

Coaxial cable 147; connector 148
Cockpit voice recorder 229, 330, 340
Cold: flow 364; junction 184
Combinational logic 66
Common: base 56, 58; collector 56, 58; emitter 54, 56, 58
Commutator 89, 93
Comparator 61
Complex notation 28
Compound-wound DC motor 94
Computer system 82
Conductor 2
Conduits 362
Connectors 365, 366
Constant speed drive 113
Contactors 165, 174
Control: bus 82; surfaces 269; system 61
Controlled flight into terrain 307
Conventional current 6
Cooling 260
Copper loss 34, 35
Corona discharge 368
Coulomb 2
Couple 91
Coupler 80
Coupling 353
Crash-proof recorder 334, 335
Crimping 147
Cross talk 350, 351
Current 2, 6–8; gain 56, 58; law 10; limiter 149, 154, 156; regulation 133, 134; transformer 112, 115, 163, 164
Cylinder head temperature 209

D-type bistable 71
DAC 59
DC 6–10
DFDR 331, 352, 335–337
Data 82; acquisition 373; bus 82; communication 252; frame 338; selector 76, 77, 83
De-icing: boots 282; fluid 281, 282
Decoder 72, 74
Deep stall 279
Defuelling 229
Delta connection 99
Depletion region 42
Destination 371
Desyn 186
Detection loop 293
Diagnostic equipment 374
Dielectric 3
Differential: bus 80; circuit protection 163, 164
Digital: data bus 336; flight data recorder 331, 352; multimeter 373; signal 59; to analogue converter 59

Diode 42–45; characteristic 43
Direct: bonding 367; cranking 197; current 6, 20
Discharge 6, 120, 121, 124, 127
Displacement transducer 177
Distribution of power 153
Distributor 193
Dome light 234
Doppler radar 320, 323
Drag cup 204, 205
Drop-out relay 175
Ducted looms 362

E-field 349
E.m.f. 7, 87
ECAM 214–218, 270, 271, 375
ECS 257
EDP 226
EI sensor 178, 179
EICAS 214–217, 270
EMC 347, 352
EMI 347–352; filter 115; reduction 352; susceptibility 352
EPR 209, 210
ESD 377, 378
ESSD 376–378
ETFE 145
EWIS 143, 361
Earth: loop 370; return 369
Eddy current 204; loss 34, 35
Effective value 21
Efficiency 35
Electric: field 3, 4; shock 36
Electrical: noise 352; power control panel 161; wire interconnection system 143, 361
Electroluminescence 233
Electrolysis 123
Electrolyte 120, 122, 125, 145; spill 122, 125
Electromagnetic: circuit-breaker 150; compatibility 347, 352; interference 347; filter 115; wave 349
Electromotive force 7
Electron 1, 2, 39
Electronic: Centralized Aircraft Monitoring 215, 270; voltage regulator 136
Electrostatic 1, 3; sensitive devices 376; voltages 376, 377
Elevation 313
Emergency: exits 238; lighting 234; power 142
Emitter follower 56
Energy 12
Engine 193; Indicating and Crew Alerting System 214, 270; driven fuel pump 226; fire extinguisher 303, 304; instruments 203; pressure ratio 209; speed indicator 203, 205, 206; temperature

206; temperature measurement 207; temperature system 207, 208
Environmental control system 257
Equipment: bonding 357; cooling 260
Erase head 336
Essential: loads 117; power 161; services bus 158
Exclusive: NOR 66; OR 66
Exhaust gas temperature 206
Exit light 238
Exterior: lights 240, 242, 243; power 137, 156, 157

FDDI 81
FDR 229, 331, 334
FEP 145
FLTA 307–310, 321, 324
FQIS 223
FSHR 200
FWC 270
Failure mode 145
Fan 260
Faraday: cage 354; shield 354
Faraday's law 17
Ferromagnetic 14
Fibre optic 256
Field: strength 353; winding 92, 93
Fire 291; bottle 303; classification 292; detection loop 293; detector 293, 294; extinguishing 302
Flameout 200
Fleming's left-hand rule 92, 93
Flight: control surfaces 269; data recorder 229, 334; idle speed 201; warning computer 270
Float: gauge 222; stick 223
Flood lighting 234
Flooded cell battery 119
Floor path lighting 238
Fluid pressure transducer 180
Fluorescent lamp 233
Flux 14–17, 33, 87; density 15
Forward: bias 43; conduction voltage 43; looking terrain: avoidance 307; looking awareness 321
Fourier analysis 344
Free electron 2
Frequency 20
Fuel: density compensation 225; distribution 226–228; dumping 229; flow 209, 212; flow-metering 211; jettison 229, 230; management 221; quantity indicating system; quantity measurement 221; quantity sensor 224; tanks 224; temperature 213, 228; transfer system 226, 227
Full: -load 34; -wave rectifier 49, 52
Fuse 148, 149; holder 149; location 150
Fusible link 148, 149

GCCU 115
GCR 115
GCU 112–115, 164
GEO 254
GPS 254, 308
GPWS 307, 309, 314–321
Galley equipment 251
Gas turbine engine 197
Gasper fan 260
Generator 87, 88, 95, 98, 110–113; control logic 71; control relay 115; control unit 112–115, 164; conversion control unit 115
Geostationary earth orbit 254
Germanium 40, 44
Gillham 75
Global positioning system 254
Gray code 72, 74
Ground: idle 200; loop 350, 370; power connector 157, 158; power relay 156; proximity modes 315; proximity warning system 307
Grounding 357

H-field 349
HEIU 198, 200
HIRF 353–357
HUMS 342
Half-wave rectifier 47
Halogenated hydrocarbons 291
Halon 291
Hand-held fire extinguisher 303
Harvard bi-phase format 337
Hazards 36, 37
Health and usage monitoring system 342
Helicopter: lights 245; torque indicator 213
Hertz 20
Heterodyning 214
High 82; energy radiated fields 353; intensity radiated fields 353; energy ignition unit 198
Holdover time 281, 282
Holes 40
Hot: battery busbar 161; junction 184
Hydrolysis 362
Hysteresis loss 34, 35

I/O 82, 83
IFE 253
ILS 313
IRS 314
ISU 253
Ice: accretion 279; detection 279, 281; inspection light 241; protection 279–283
Icing 279
Igniter 199
Ignition: cable 195; system 193, 202; unit 198

Impedance 25; bridge 224; triangle 25
Impurity atom 39
In: -flight entertainment 251; -line splice 147; start 198
Incandescence 233
Indicating: cap 148; systems 203
Indirect bonding 367
Inductance 18; loop 173
Induction 16; motor 104, 107
Inductive reactance 23, 24
Inductor 19
Inert 231
Inertial reference system 314
Inlet temperature 206
Input characteristic 54, 55
Instrument: landing system 313; lighting 235
Instrumentation system 61
Insulator 2, 40
Integral fuel tanks 221
Integrated circuit 59, 60
Integrity switch 297
Inter-turbine temperature 206
Internal resistance 118, 123
Internet 252
Interphone system 249
Intrinsic safety 225
Inverter 65, 106, 137
Ion 1
Ionization smoke detector 299
Ionized gas 233
Iridium 252–254; ground network 255; subscriber unit 253
Iron loss 34, 35

J-K bistable 72, 73
J-operator 29
Jet pipe temperature 206
Jettison pump 229
Joule 12
Junction diode 43, 43

Kilobyte 83
Kilowatt-hour 12
Kirchhoff's laws 10
Knife edge cutter 279, 280

LAN 256
LCD 252, 374
LED 47, 233; lighting 238
LEO 254
LRRA 312
LRU 78, 79, 364, 365, 374
LVDT 177, 178, 209
Landing: gear 265; gear control system 266; gear logic 66, 68; gear position 265, 267; light 241, 242
Lattice structure 39
Lead-acid battery 117, 120

Lenz's law 17, 105
Lift 274, 275
Light: control 235, 236; -emitting diode 47, 233
Lightning 354, 356; strike 355; zones 355
Lights 233–240
Limiting resistor 152
Line replaceable unit 78, 79, 364, 374
Linear: fire detector 294, 295; variable differential transducer 177, 178, 209
Liquid: crystal display 252, 374; nitrogen 231
Lithium: battery 117, 126; -ion 126
Load 8; sharing 168; shedding 158, 165
Local area network 256
Logic: 0 82; 1 82; gates 65
Logo light 240
Loom 362, 363
Low 82; earth orbit 254; -pressure cock 226; -range radio altimeter 312

MEMS 189, 314
MFD navigation display 311
MIL-W-M22759E 145, 146
MSL 313
MW 236
Magnesyn 188
Magnetic: field 14, 350; flux 14, 15, 87; induction 16; recording 330
Magnetism 14
Magneto 193; ignition 193–196
Main: bus 158; engine starting 200
Maintenance manual 371
Majority vote logic 68
Manuals 370, 371
Mass flow 210
Master: switch 153; warning 236, 238, 293
Maximum: reverse voltage 43; value 21
Mean sea level 313
Measured discharge 124
Memory effect 118
Merz Price circuit 164, 165
Metallic surface bonding 357
Micro: -electrical-mechanical sensor 189; -switch 172, 173, 265; -burst 322
Molecular sieve 231
Molecule 1
Moment 91
Monostable 69
Motor 87–94, 101; -driven impeller 210
Moving-map system 252
Multi-stranding 144
Multimeter 372
Multiphase supply 22
Multiplexer 76–78, 255
Mutual inductance 18

N1 speed indication 205
N2 speed indication 206
NAND logic 66
NOR logic 66
NPN 53, 54
NRZ format 337
NTC 41
NVM 216, 340
Natural binary 72, 74
Navigation lights 243, 244
Negative 118; ion 1; temperature coefficient 41
Neutron 1
Ni-MH battery 126
Nickel: -cadmium battery 123, 125; -metal hydride 117; -metal hydride battery 126
No load 34
Node 10
Noise 352
Nominal voltage 121
Non-parallel system 159
Non-volatile: memory 216, 340; storage 83
North pole 14

OFV 261
OR logic 65
Octal 72, 74; to binary encoder 76
Ohm 9
Ohm's law 8, 9, 144
Oil temperature 213
Oleo leg 265
On-board diagnostic equipment 374
One-shot 69
Open: area smoke detector 299, 300; circuit 361; looms 362
Operational amplifier 59
Optical: fibre 256, 257; fire detection 292–300
Oscilloscope 374
Out-of-balance condition 112
Outflow valve 261
Outlet temperature 206
Output characteristic 55
Overhead switch 293

P-N junction 43
P.d. 7
PA 249, 250
PFVD 145
PIV 43
PMG 112
PNP 53, 54
PRSOV 257
PSEU 173, 174
PTC 41
PTFE 145
PVS 252

PWM 106
Parallel: bus 78; bus system 161; circuit 10, 11; data 79; load distribution 161
Passenger: address system 249; cabin lights 238; telephone system 252
Peak: inverse voltage 43; value 21
Pentavalent impurity 40
Per-unit: regulation 34; slip 106
Period 20
Periodic time 20, 21
Permanent magnet generator 112
Personal video screen 252
Phase: angle 25; protection 164, 165
Phasor 28, 29; diagram 99
Phonic wheel 212
Phosphorous lighting 237
Photoelectric smoke detector 300
Piezo: -resistive strain gauge 186; -electric crystal 214
Plaques 123
Plasma 233, 368
Plate group 119, 120
Pneumatic fire detector 297
Pneumatic ice protection 282
Polar inductor 194
Polarized relay 175
Poles 15, 171, 175
Positive 118; ion 1; temperature coefficient 41, 182
Potential difference 4, 7, 9
Potentiometer 177
Power 12, 13; conversion 137; distribution 153, 167; factor 27, 168; supplies 133
Pressure: capsule 180; transducer 181; -regulating shutoff valve 257
Pressurization 260, 261, 262
Primary 33; bonding 367; cell 6; coil 194; power 159
Priority encoder 75
Propeller: de-icing 283; synchronization 214
Protection 148
Proton 1, 39
Proximity switch 172–174, 265
Public address system 250
Pulse width modulation 106
Pyro-electric cell 298

QAR 336
Quadrature 25
Quick: access recorder 336; -release connector 129, 366

R-S bistable 71
R.m.s. 21
RAM 82, 83
RAT 141, 142
ROM 82, 83, 84

RTD 183
Radiated EMI 352
Radio altimeter 312
Rain: protection 283; repellent 285–287
Ram air turbine 141, 142
Ratiometer 180–183, 222
Re-circulation fan 260
Reactance 23
Reactive: metals 126; power 168
Read-only memory 83
Real power 168
Real-time faults 375
Receiver 349
Recording tape 335
Rectifier 47
Reed: relay 175; sensor 275; switch 173
Reflected code 72
Refuelling 229
Regulation 34
Regulator 133
Relays 174, 176
Reservoir capacitor 48, 49
Reset 71
Resistance 8, 9
Resistivity 144
Resonance 26
Responder 297
Reverse: bias 43; current relay 154, 155
Rheostat 177, 235
Rigid tanks 221
Ring tongue terminal 147
Ripple 49
Rocker switch 172
Root mean square 21
Rotary: inverter 137; position transducer 186
Rotating beacon light 246
Rotor 92, 97, 105, 187

SDAC 270
SPST 175
STP 80
SWAMP 362
Safety 36
Sample and hold 373
Satellite communication 253
Scratched foil 330
Screened cable 146
Screening 144, 146
Sealed battery 119, 122
Secondary 33; bonding 367; cell 6; power 159
Seebeck effect 184
Segments 90
Self: -clocking 81; -clocking bus 81; -inductance 18; -oxidizing 144; -sustaining speed 197
Selsyn 186
Semiconductor 39

Sensor 60, 62; vane 277
Separation point 274
Serial: bus 78; data 79; interface module 78
Series: circuit 10, 11; wound DC motor 93, 94
Service light 241
Set 71
Severe wind and moisture problems 362
Shaded pole 109
Shading coil 109
Shaft horsepower 212
Shannon-Hartley theorem 351
Shell 1, 39
Shielded twisted pair 80
Shielding 144, 146, 348
Shock 36
Shunt-wound DC motor 94
Side tone 249
Sight glass 222
Signal 59
Silicon 40, 44; -controlled rectifier 46
Sine wave 22
Single-phase: AC generator 97; DC generator 95; induction motor 107; supply 22
Single walled cable 146
Situational awareness 311
Slip 105, 106; -rings 88, 97, 187
Slow-blow fuse 149
Slugged relay 175
Smoke detector 299, 301, 302
Soldering 364
Solenoid 15, 16, 92, 177
Solid-state data recorder 331
Source 8, 371
South pole 15
Spark plugs 195, 196
Specific gravity 121, 122
Speed 105
Splicing 147
Split: avionic bus 154; bus 159, 160; parallel bus system 161
Spongy lead 120
Squat switch 265
Squirrel cage rotor 104
Stagnation point 274
Stall 273, 274, 275, 279; identification 278; warning 273, 276–278
Standby power 161
Star connection 99
Starter-generator 95, 96, 197
Starting system 193, 194, 202
Static: bonds 357; charge 368; discharger 3, 368, 369; inverter 138; wick 3, 368
Stator 97, 105, 187
Step: -down transformer 33; -up transformer 33
Stick: push 279; shaker 275, 277

Storage cell 118
Strain 185; gauge 185, 186, 212; transducer 185
Streamline airflow 273
Stress 185
Strobe 234, 235; light 233, 248
Stub cable 80
Sulphation 122
Surge tank 229
Switches 171, 172
Synchro 186, 187, 188
Synchronous: data 80; motor 104; speed 106
System data acquisition concentrator 270

TAWS 307–310, 324, 325
TCAS 75
TNC 147
TRU 139
TTL 69
Tachometer system 203, 204
Take-off warning system 287, 288
Target wheel 205
Taxi light 240, 242
Tee-tailed aircraft 278
Telephone gateway 255
Temperature: bulb 182; control zone 258; sensor 183; transducer 181
Terminal blocks 365
Termination 147
Terminator 80
Terrain awareness warning system 307
Thermal: anti-icing 284; circuit-breaker 150; fire detection 293; fire detector 292; ice protection 282; runaway 41, 124
Thermistor 41, 42, 182, 295
Thermocouple 62, 63, 184, 185, 206
Three-phase: AC generator 98, 99, 111; AC motor 102; autotransformer 140; induction motor 104; supply 22, 23; systems 99–101
Throw 171, 175
Thyristor 46
Timer 59
Toggle switch 171, 172
Torque 91, 105, 210; synchro 186, 188; transducer 213; speed characteristic 94
Trailing edge flaps 268
Transducer 62, 171, 177, 185
Transfer characteristic 56
Transformer 18, 33, 139; efficiency 35; rectifier unit 139
Transistor 57, 236: characteristic 54; lighting control 237; transistor logic 69
Transition point 274
Transmitter 349
Triboelectric: effect 376; scale 376
Trivalent impurity 40
True power 27

Truth table 65
Turbine: engine starter 197, 198; gas temperature 206
Turns: ratio 33; per volt 33
Twin walled cable 146
Twisted pair 350, 351
Two: -phase AC generator 98; -phase induction motor 107; -shot fire extinguisher 303, 305

ULB 334, 340, 342
Ullage 229
Ultrasonic ice detector 279
Underwater location: transmitter 334; beacon 334, 340
Unidirectional bus 78
Unit: fire detector 293; of electricity 12
Unusable fuel 231

VRLA 122
VSCF 113–115
Valence shell 39
Valve-regulated lead-acid battery 122
Vane sensor 276
Variable: reluctance transducer 204; resistors 176; speed/constant frequency 113
Ventilation 260
Venting 128, 130
Vibrating contact regulator 133, 134
Vibration 214, 364; sensor system 215
Video projector 251
Volt-ampere 27
Voltage 7; divider 177; law 10; regulation 34, 133; regulator 45, 134, 136
Volume flow 210

Warning: messages 217; panel 239
Watt-second 12
Waveform 19–23
Weight on wheels 265
Wet: arc tracking 145, 364; start 201; cell 119
Wheatstone bridge 184, 214
Windscreen 285; ice protection 283; rain protection 283; wipers 284, 286
Wire 143, 146; gauge 145; installation 361; loom 362; size 145; termination 147, 364
Wiring 361; diagram 371
WoW 265
Wound rotor 104
Wrist strap 378

Xenon 233–235

Zener diode 44–46